GEOLOGICAL OBSERVATIONS ON SOUTH AMERICA

Charles Darwin

Copyright © 2017

ISBN-13: 978-1543232684

ISBN-10: 154323268X

GEOLOGICAL OBSERVATIONS ON SOUTH AMERICA
by CHARLES DARWIN
EDITORIAL NOTE.

Although in some respects more technical in their subjects and style than Darwin's "Journal," the books here reprinted will never lose their value and interest for the originality of the observations they contain. Many parts of them are admirably adapted for giving an insight into problems regarding the structure and changes of the earth's surface, and in fact they form a charming introduction to physical geology and physiography in their application to special domains. The books themselves cannot be obtained for many times the price of the present volume, and both the general reader, who desires to know more of Darwin's work, and the student of geology, who naturally wishes to know how a master mind reasoned on most important geological subjects, will be glad of the opportunity of possessing them in a convenient and cheap form.

The three introductions, which my friend Professor Judd has kindly furnished, give critical and historical information which makes this edition of special value.

G.T.B.
PLATE I. GEOLOGICAL SECTIONS THROUGH THE CORDILLERAS.
SECTION 1/1. SECTION OF THE PEUQUENES OR PORTILLO PASS OF THE CORDILLERA.

SECTION 1/2. SECTION OF THE CUMBRE OR USPALLATA PASS.
SECTION 1/3. SECTION OF THE VALLEY OF COPIAPO TO THE BASE OF THE MAIN CORDILLERA.

PLATE II. MAP OF SOUTHERN PORTION OF SOUTH AMERICA.
TABLE OF CONTENTS.

CRITICAL INTRODUCTION.

CHAPTER I.—ON THE ELEVATION OF THE EASTERN COAST OF SOUTH AMERICA.
Upraised shells of La Plata.—Bahia Blanca, Sand-dunes and Pumice-pebbles.- -Step-formed plains of Patagonia, with upraised shells.—Terrace-bounded valley of Santa Cruz, formerly a sea-strait.—Upraised shells of Tierra del Fuego.—Length and breadth of the elevated area.—Equability of the movements, as shown by the similar heights of the plains.—Slowness of the elevatory process.—Mode of formation of the step-formed plains.—Summary.- -Great shingle formation of Patagonia; its extent, origin, and distribution.—Formation of sea-cliffs.

CHAPTER II.—ON THE ELEVATION OF THE WESTERN COAST OF SOUTH AMERICA.
Chonos Archipelago.—Chiloe, recent and gradual elevation of, traditions of the inhabitants on this subject.—Concepcion, earthquake and elevation of.- -VALPARAISO, great elevation of, upraised shells, earth or marine origin, gradual rise of the land within the historical period.—COQUIMBO, elevation of, in recent times; terraces of marine origin, their inclination, their escarpments not horizontal.—Guasco, gravel terraces of.—Copiapo.—PERU.— Upraised shells of Cobija, Iquique, and Arica.—Lima, shell-beds and sea- beach on San Lorenzo.—Human remains, fossil earthenware, earthquake debacle, recent subsidence.—On the decay of upraised shells.—General summary.

CHAPTER III.—ON THE PLAINS AND VALLEYS OF CHILE:—SALIFEROUS

SUPERFICIAL DEPOSITS.
Basin-like plains of Chile; their drainage, their marine origin.—Marks of sea-action on the eastern flanks of the Cordillera.—Sloping terrace-like fringes of stratified shingle within the valleys of the Cordillera; their marine origin.—Boulders in the valley of Cachapual.—Horizontal elevation of the Cordillera.—Formation of valleys.—Boulders moved by earthquake- waves.—Saline superficial deposits.—Bed of nitrate of soda at Iquique.— Saline incrustations.—Salt-lakes of La Plata and Patagonia; purity of the salt; its origin.

CHAPTER IV.—ON THE FORMATIONS OF THE PAMPAS.
Mineralogical constitution.—Microscopical structure.—Buenos Ayres, shells embedded in tosca-rock.—Buenos Ayres to the Colorado.—S. Ventana.—Bahia Blanca; M. Hermoso, bones and infusoria of; P. Alta, shells, bones, and infusoria of; co-existence of the recent shells and extinct mammifers.— Buenos Ayres to St. Fe.—Skeletons of Mastodon.—Infusoria.—Inferior marine tertiary strata, their age.—Horse's tooth. BANDA ORIENTAL.— Superficial Pampean formation.—Inferior tertiary strata, variation of, connected with volcanic action; Macrauchenia Patachonica at S. Julian in Patagonia, age of, subsequent to living mollusca and to the erratic block period. SUMMARY.—Area of Pampean formation.—Theories of origin.—Source of sediment.—Estuary origin.—Contemporaneous with existing mollusca.— Relations to underlying tertiary strata. Ancient deposit of estuary origin.—Elevation and successive deposition of the Pampean formation.— Number and state of the remains of mammifers; their habitation, food, extinction, and range.—Conclusion.—Supplement on the thickness of the Pampean formation.—Localities in Pampas at which mammiferous remains have been found.

CHAPTER V.—ON THE OLDER TERTIARY FORMATIONS OF PATAGONIA AND CHILE.
Rio Negro.—S. Josef.—Port Desire, white pumiceous mudstone with infusoria.—Port S. Julian.—Santa Cruz, basaltic lava of.—P. Gallegos.— Eastern Tierra del Fuego; leaves of extinct beech-trees.—Summary on the Patagonian tertiary formations.—Tertiary formations of the Western Coast.- -Chonos and Chiloe groups, volcanic rocks of.—Concepcion.—Navidad.— Coquimbo.—Summary.—Age of the tertiary formations.—Lines of elevation.—Silicified wood.—Comparative ranges of the extinct and living mollusca on the West Coast of S. America.—Climate of the tertiary period.—On the causes of the absence of recent conchiferous deposits on the coasts of South America.—On the contemporaneous deposition and preservation of sedimentary formations.
CHAPTER VI.—PLUTONIC AND METAMORPHIC ROCKS:—CLEAVAGE AND FOLIATION. Brazil, Bahia, gneiss with disjointed metamorphosed dikes.—Strike of foliation.—Rio de Janeiro, gneiss-granite, embedded fragment in, decomposition of.—La Plata, metamorphic and old volcanic rocks of.—S. Ventana.—Claystone porphyry formation of Patagonia; singular metamorphic rocks; pseudo-dikes.—Falkland Islands, palaeozoic fossils of.—Tierra del Fuego, clay-slate formation, cretaceous fossils of; cleavage and foliation; form of land.—Chonos Archipelago, mica-schists, foliation disturbed by granitic axis; dikes.—Chiloe.—Concepcion, dikes, successive formation of.—Central and Northern Chile.—Concluding remarks on cleavage and foliation.—Their close analogy and similar origin.—Stratification of metamorphic schists.—Foliation of intrusive rocks.—Relation of cleavage and foliation to the lines of tension during metamorphosis.

CHAPTER VII.—CENTRAL CHILE:—STRUCTURE OF THE CORDILLERA.
Central Chile.—Basal formations of the Cordillera.—Origin of the porphyritic claystone conglomerate.—Andesite.—Volcanic rocks.—Section of the Cordillera by the Peuquenes or Portillo Pass.—Great gypseous formation.—Peuquenes line; thickness of strata, fossils of.—Portillo line.—Conglomerate, orthitic granite, mica-schist, volcanic rocks of.— Concluding remarks on the denudation and elevation of the Portillo line.— Section by

the Cumbre, or Uspallata Pass.—Porphyries.—Gypseous strata.— Section near the Puente del Inca; fossils of.—Great subsidence.—Intrusive porphyries.—Plain of Uspallata.—Section of the Uspallata chain.— Structure and nature of the strata.—Silicified vertical trees.—Great subsidence.—Granitic rocks of axis.—Concluding remarks on the Uspallata range; origin subsequent to that of the main Cordillera; two periods of subsidence; comparison with the Portillo chain.

CHAPTER VIII.—NORTHERN CHILE.—CONCLUSION.
Section from Illapel to Combarbala; gypseous formation with silicified wood.—Panuncillo.—Coquimbo; mines of Arqueros; section up valley; fossils.—Guasco, fossils of.—Copiapo, section up valley; Las Amolanas, silicified wood.—Conglomerates, nature of former land, fossils, thickness of strata, great subsidence.—Valley of Despoblado, fossils, tufaceous deposit, complicated dislocations of.—Relations between ancient orifices of eruption and subsequent axes of injection.—Iquique, Peru, fossils of, salt-deposits.—Metalliferous veins.—Summary on the porphyritic conglomerate and gypseous formations.—Great subsidence with partial elevations during the cretaceo-oolitic period.—On the elevation and structure of the Cordillera.—Recapitulation on the tertiary series.—Relation between movements of subsidence and volcanic action.—Pampean formation.—Recent elevatory movements.—Long-continued volcanic action in the Cordillera.—Conclusion.

INDEX.
GEOLOGICAL OBSERVATIONS ON SOUTH AMERICA
BY
CHARLES DARWIN.

CRITICAL INTRODUCTION.
Of the remarkable "trilogy" constituted by Darwin's writings which deal with the geology of the "Beagle," the member which has perhaps attracted least attention, up to the present time is that which treats of the geology of South America. The actual writing of this book appears to have occupied Darwin a shorter period than either of the other volumes of the series; his diary records that the work was accomplished within ten months, namely, between July 1844 and April 1845; but the book was not actually issued till late in the year following, the preface bearing the date "September 1846." Altogether, as Darwin informs us in his "Autobiography," the geological books "consumed four and a half years' steady work," most of the remainder of the ten years that elapsed between the return of the "Beagle," and the completion of his geological books being, it is sad to relate, "lost through illness!"

Concerning the "Geological Observations on South America," Darwin wrote to his friend Lyell, as follows:—"My volume will be about 240 pages, dreadfully dull, yet much condensed. I think whenever you have time to look through it, you will think the collection of facts on the elevation of the land and on the formation of terraces pretty good."

"Much condensed" is the verdict that everyone must endorse, on rising from the perusal of this remarkable book; but by no means "dull." The three and a half years from April 1832 to September 1835, were spent by Darwin in South America, and were devoted to continuous scientific work; the problems he dealt with were either purely geological or those which constitute the borderland between the geological and biological sciences. It is impossible to read the journal which he kept during this time without being impressed by the conviction that it contains all the germs of thought which afterwards developed into the

"Origin of Species." But it is equally evident that after his return to England, biological speculations gradually began to exercise a more exclusive sway over Darwin's mind, and tended to dispossess geology, which during the actual period of the voyage certainly engrossed most of his time and attention. The wonderful series of observations made during those three and a half years in South America could scarcely be done justice to, in the 240 pages devoted to their exposition. That he executed the work of preparing the book on South America in somewhat the manner of a task, is shown by many references in his letters. Writing to Sir Joseph Hooker in 1845, he says, "I hope this next summer to finish my South American Geology, then to get out a little Zoology, and HURRAH FOR MY SPECIES WORK!"

It would seem that the feeling of disappointment, which Darwin so often experienced in comparing a book when completed, with the observations and speculations which had inspired it, was more keenly felt in the case of his volume on South America than any other. To one friend he writes, "I have of late been slaving extra hard, to the great discomfiture of wretched digestive organs, at South America, and thank all the fates, I have done three-fourths of it. Writing plain English grows with me more and more difficult, and never attainable. As for your pretending that you will read anything so dull as my pure geological descriptions, lay not such a flattering unction on my soul, for it is incredible." To another friend he writes, "You do not know what you threaten when you propose to read it—it is purely geological. I said to my brother, 'You will of course read it,' and his answer was, 'Upon my life, I would sooner even buy it.'"

In spite of these disparaging remarks, however, we are strongly inclined to believe that this book, despised by its author, and neglected by his contemporaries, will in the end be admitted to be one of Darwin's chief titles to fame. It is, perhaps, an unfortunate circumstance that the great success which he attained in biology by the publication of the "Origin of Species" has, to some extent, overshadowed the fact that Darwin's claims as a geologist, are of the very highest order. It is not too much to say that, had Darwin not been a geologist, the "Origin of Species" could never have been written by him. But apart from those geological questions, which have an important bearing on biological thought and speculation, such as the proofs of imperfection in the geological record, the relations of the later tertiary faunas to the recent ones in the same areas, and the apparent intermingling of types belonging to distant geological epochs, when we study the palaeontology of remote districts,—there are other purely geological problems, upon which the contributions made by Darwin are of the very highest value. I believe that the verdict of the historians of science will be that if Darwin had not taken a foremost place among the biologists of this century, his position as a geologist would have been an almost equally commanding one.

But in the case of Darwin's principal geological work—that relating to the origin of the crystalline schists,—geologists were not at the time prepared to receive his revolutionary teachings. The influence of powerful authority was long exercised, indeed, to stifle his teaching, and only now, when this unfortunate opposition has disappeared, is the true nature and importance of Darwin's purely geological work beginning to be recognised.

The two first chapters of the "Geological Observations on South America," deal with the proofs which exist of great, but frequently interrupted, movements of elevation during very recent geological times. In connection with this subject, Darwin's particular attention was directed to the relations between the great earthquakes of South America—of some of which he had impressive experience—and the permanent changes of elevation which were taking place. He was much struck by the rapidity with which the evidence of such great earth movements is frequently obliterated; and especially with the remarkable way in which the action of rain-water, percolating through deposits on the earth's surface, removes all traces of shells and other calcareous organisms. It was these considerations which were the parents of the generalisation that a palaeontological record can only be preserved during those periods in which long-continued slow subsidence is going on. This in turn, led to the still wider and more suggestive conclusion that the geological record as a whole is, and never can be more than, a series of more or less isolated fragments. The recognition of this important fact constitutes the keystone to any theory of evolution which seeks to find a basis in the actual study of the types of life that have formerly inhabited our globe.

In his third chapter, Darwin gives a number of interesting facts, collected during his visits to the plains and valleys of Chili, which bear on the question of the origin of saliferous deposits—the accumulation of salt, gypsum, and nitrate of soda. This is a problem that has excited much discussion among geologists, and which, in spite of many valuable observations, still remains to a great extent very obscure. Among the important considerations insisted upon by Darwin is that relating to the absence of marine shells in beds associated with such deposits. He justly argues that if the strata were formed in shallow waters, and then exposed by upheaval to subaerial action, all shells and other calcareous organisms would be removed by solution.

Following Lyell's method, Darwin proceeds from the study of deposits now being accumulated on the earth's surface, to those which have been formed during the more recent periods of the geological history.

His account of the great Pampean formation, with its wonderful mammalian remains—Mastodon, Toxodon, Scelidotherium, Macrauchenia, Megatherium, Megalonyx, Mylodon, and Glyptodon—this full of interest. His discovery of the remains of a true Equus afforded a remarkable confirmation of the fact- -already made out in North America—that species of horse had existed and become extinct in the New World, before their introduction by the Spaniards in the sixteenth century. Fully perceiving the importance of the microscope in studying the nature and origin of such deposits as those of the Pampas, Darwin submitted many of his specimens both to Dr. Carpenter in this country, and to Professor Ehrenberg in Berlin. Many very important notes on the microscopic organisms contained in the formation will be found scattered through the chapter.

Darwin's study of the older tertiary formations, with their abundant shells, and their relics of vegetable life buried under great sheets of basalt, led him to consider carefully the question of climate during these earlier periods. In opposition to prevalent views on this subject, Darwin points out that his observations are opposed to the conclusion that a higher temperature prevailed universally over the globe during early geological periods. He argues that "the causes which gave to the older tertiary productions of the quite temperate zones of Europe a tropical character, WERE OF A LOCAL CHARACTER AND DID NOT AFFECT THE WHOLE GLOBE." In this, as in many similar instances, we see the beneficial influence of extensive travel in freeing Darwin's mind from prevailing prejudices. It was this widening of experience which rendered him so especially qualified to deal with the great problem of the origin of species, and in doing so to emancipate himself from ideas which were received with unquestioning faith by geologists whose studies had been circumscribed within the limits of Western Europe.

In the Cordilleras of Northern and Central Chili, Darwin, when studying still older formations, clearly recognised that they contain an admixture of the forms of life, which in Europe are distinctive of the Cretaceous and Jurassic periods respectively. He was thus led to conclude that the classification of geological periods, which fairly well expresses the facts that had been discovered in the areas where the science was first studied, is no longer capable of being applied when we come to the study of widely distant regions. This important conclusion led up to the further generalisation that each great geological period has exhibited a geographical distribution of the forms of animal and vegetable life, comparable to that which prevails in the existing fauna and flora. To those who are familiar with the extent to which the doctrine of universal formations has affected geological thought and speculation, both long before and since the time that Darwin wrote, the importance of this new standpoint to which he was able to attain will be sufficiently apparent. Like the idea of the extreme imperfection of the Geological Record, the doctrine of LOCAL geological formations is found permeating and moulding all the palaeontological reasonings of his great work.

In one of Darwin's letters, written while he was in South America, there is a passage we have already quoted, in which he expresses his inability to decide between the rival claims upon his attention of "the old crystalline group of rocks," and "the softer fossiliferous beds" respectively. The sixth chapter of the work before us, entitled "Plutonic and Metamorphic Rocks—Cleavage and Foliation," contains a brief summary of a series of observations and reasonings upon these crystalline rocks, which are, we believe, calculated to effect a revolution in geological science, and— though their value and importance have long been overlooked—are likely to entitle Darwin in the future to a position among geologists, scarcely, if at all, inferior to that which he already occupies among biologists.

Darwin's studies of the great rock-masses of the Andes convinced him of the close relations between the granitic or Plutonic rocks, and those which were undoubtedly poured forth as lavas. Upon his return, he set to work, with the aid of Professor Miller, to make a careful study of the minerals composing the granites and those which occur in the lavas, and he was able to show that in all essential respects they are identical. He was further able to prove that there is a complete gradation between the highly crystalline or granitic rock-masses, and those containing more or less glassy matter between their crystals, which constitute ordinary lavas. The importance of this conclusion will be realised when we remember that it was then the common creed of geologists—and still continues to be so on the Continent—that all highly crystalline rocks are of great geological antiquity, and that the igneous ejections which have taken place since the beginning of the tertiary periods differ essentially, in their composition, their structure, and their mode of occurrence, from those which have made their appearance at earlier periods of the world's history.

Very completely have the conclusions of Darwin upon these subjects been justified by recent researches. In England, the United States, and Italy, examples of the gradual passage of rocks of truly granitic structure into ordinary lavas have been described, and the reality of the transition has been demonstrated by the most careful studies with the microscope. Recent researches carried on in South America by Professor Stelzner, have also shown the existence of a class of highly crystalline rocks—the "Andengranites"—which combine in themselves many of the characteristics which were once thought to be distinctive of the so-called Plutonic and volcanic rocks. No one familiar with recent geological literature—even in Germany and France, where the old views concerning the distinction of igneous products of different ages have been most stoutly maintained—can fail to recognise the fact that the principles contended for by Darwin bid fair at no distant period to win universal acceptance among geologists all over the globe.

Still more important are the conclusions at which Darwin arrived with respect to the origin of the schists and gneisses which cover so large an area in South America.

Carefully noting, by the aid of his compass and clinometer, at every point which he visited, the direction and amount of inclination of the parallel divisions in these rocks, he was led to a very important generalisation— namely, that over very wide areas the direction (strike) of the planes of cleavage in slates, and of foliation in schists and gneisses, remained constant, though the amount of their inclination (dip) often varied within wide limits. Further than this it appeared that there was always a close correspondence between the strike of the cleavage and foliation and the direction of the great axes along which elevation had taken place in the district.

In Tierra del Fuego, Darwin found striking evidence that the cleavage intersecting great masses of slate-rocks was quite independent of their original stratification, and could often, indeed, be seen cutting across it at right angles. He was also able to verify Sedgwick's observation that, in some slates, glossy surfaces on the planes of cleavage arise from the development of new minerals, chlorite, epidote or mica, and that in this way a complete graduation from slates to true schists may be traced.

Darwin further showed that in highly schistose rocks, the folia bend around and encircle any foreign bodies in the mass, and that in some cases they exhibit the most tortuous forms and complicated puckerings. He clearly saw that in all cases the forces by which these striking phenomena must have been produced were persistent over wide areas, and were connected with the great movements by which the rocks had been upheaved and folded.

That the distinct folia of quartz, feldspar, mica, and other minerals composing the metamorphic schists could not have been separately deposited as sediment was strongly insisted upon by Darwin; and in doing so he opposed the view generally prevalent among geologists at that time. He was thus driven to the conclusion that foliation, like cleavage, is not an original, but a superinduced structure in rock-masses, and that it is the result of re-crystallisation, under the controlling influence of great pressure, of the materials of which the rock was composed.

In studying the lavas of Ascension, as we have already seen, Darwin was led to recognise the circumstance that, when igneous rocks are subjected to great differential movements during the period of their consolidation, they acquire a foliated structure, closely analogous to that of the crystalline schists. Like his predecessor in this field of inquiry, Mr. Poulett Scrope, Charles Darwin seems to have been greatly impressed by these facts, and he argued from them that the rocks exhibiting the foliated structure must have been in a state of plasticity, like that of a cooling mass of lava. At that time the suggestive experiments of Tresca, Daubree, and others, showing that solid masses under the influence of enormous pressure become actually plastic, had not been published. Had Darwin been aware of these facts he would have seen that it was not necessary to assume a state of imperfect solidity in rock-masses in order to account for their having yielded to pressure and tension, and, in doing so, acquiring the new characters which distinguish the crystalline schists.

The views put forward by Darwin on the origin of the crystalline schists found an able advocate in Mr. Daniel Sharpe, who in 1852 and 1854 published two papers, dealing with the geology of the Scottish Highlands and of the Alps respectively, in which he showed that the principles arrived at by Darwin when studying the South American rocks afford a complete explanation of the structure of the two districts in question.

But, on the other hand, the conclusions of Darwin and Sharpe were met with the strongest opposition by Sir Roderick Murchison and Dr. A. Geikie, who in 1861 read a paper before the Geological Society "On the Coincidence between Stratification and Foliation in the Crystalline Rocks of the Scottish Highlands," in which they insisted that their observations in Scotland tended to entirely disprove the conclusions of Darwin that foliation in rocks is a secondary structure, and entirely independent of the original stratification of the rock-masses.

Now it is a most significant circumstance that, no sooner did the officers of the Geological Survey commence the careful and detailed study of the Scottish Highlands than they found themselves compelled to make a formal retraction of the views which had been put forward by Murchison and Geikie in opposition to the conclusions of Darwin. The officers of the Geological Survey have completely abandoned the view that the foliation of the Highland rocks has been determined by their original stratification, and admit that the structure is the result of the profound movements to which the rocks have been subjected. The same conclusions have recently been supported by observations made in many different districts—among which we may especially refer to those of Dr. H. Reusch in Norway, and those of Dr. J. Lehmann in Saxony. At the present time the arguments so clearly stated by Darwin in the work before us, have, after enduring opposition or neglect for a whole generation, begun to "triumph all along the line," and we may look forward confidently to the near future, when his claim to be regarded as one of the greatest of geological discoverers shall be fully vindicated.

JOHN W. JUDD.
CHAPTER I. ON THE ELEVATION OF THE EASTERN COAST OF SOUTH AMERICA.
Upraised shells of La Plata.
Bahia Blanca, Sand-dunes and Pumice-pebbles.

Step-formed plains of Patagonia, with upraised Shells.
Terrace-bounded Valley of Santa Cruz, formerly a Sea-strait.
Upraised shells of Tierra del Fuego.
Length and breadth of the elevated area.
Equability of the movements, as shown by the similar heights of the plains.
Slowness of the elevatory process.
Mode of formation of the step-formed plains.
Summary.
Great Shingle Formation of Patagonia; its extent, origin, and distribution.
Formation of sea-cliffs.

In the following Volume, which treats of the geology of South America, and almost exclusively of the parts southward of the Tropic of Capricorn, I have arranged the chapters according to the age of the deposits, occasionally departing from this order, for the sake of geographical simplicity.

The elevation of the land within the recent period, and the modifications of its surface through the action of the sea (to which subjects I paid particular attention) will be first discussed; I will then pass on to the tertiary deposits, and afterwards to the older rocks. Only those districts and sections will be described in detail which appear to me to deserve some particular attention; and I will, at the end of each chapter, give a summary of the results. We will commence with the proofs of the upheaval of the eastern coast of the continent, from the Rio Plata southward; and, in the Second Chapter, follow up the same subject along the shores of Chile and Peru.

On the northern bank of the great estuary of the Rio Plata, near Maldonado, I found at the head of a lake, sometimes brackish but generally containing fresh water, a bed of muddy clay, six feet in thickness, with numerous shells of species still existing in the Plata, namely, the Azara labiata, d'Orbigny, fragments of Mytilus eduliformis, d'Orbigny, Paludestrina Isabellei, d'Orbigny, and the Solen Caribaeus, Lam., which last was embedded vertically in the position in which it had lived. These shells lie at the height of only two feet above the lake, nor would they have been worth mentioning, except in connection with analogous facts.

At Monte Video, I noticed near the town, and along the base of the mount, beds of a living Mytilus, raised some feet above the surface of the Plata: in a similar bed, at a height from thirteen to sixteen feet, M. Isabelle collected eight species, which, according to M. d'Orbigny, now live at the mouth of the estuary. ("Voyage dans l'Amerique Merid.: Part. Geolog.") At Colonia del Sacramiento, further westward, I observed at the height of about fifteen feet above the river, there of quite fresh water, a small bed of the same Mytilus, which lives in brackish water at Monte Video. Near the mouth of Uruguay, and for at least thirty-five miles northward, there are at intervals large sandy tracts, extending several miles from the banks of the river, but not raised much above its level, abounding with small bivalves, which occur in such numbers that at the Agraciado they are sifted and burnt for lime. Those which I examined near the A. S. Juan were much worn: they consisted of Mactra Isabellei, d'Orbigny, mingled with few of Venus sinuosa, Lam., both inhabiting, as I am informed by M. d'Orbigny, brackish water at the mouth of the Plata, nearly or quite as salt as the open sea. The loose sand, in which these shells are packed, is heaped into low, straight, long lines of dunes, like those left by the sea at the head of many bays. M. d'Orbigny has described an analogous phenomenon on a greater scale, near San Pedro on the river Parana, where he found widely extended beds and hillocks of sand, with vast numbers of the Azara labiata, at the height of nearly 100 feet (English) above the surface of that river. (Ibid) The Azara inhabits brackish water, and is not known to be found nearer to San Pedro than Buenos Ayres, distant above a hundred miles in a straight line. Nearer Buenos Ayres, on the road from that place to San Isidro, there are extensive beds, as I am informed by Sir Woodbine Parish, of the Azara labiata, lying at about forty feet above the level of the river, and distant between two and three miles from it. ("Buenos Ayres" etc. by Sir Woodbine Parish) These shells are always found on the highest banks in the district: they are embedded in a stratified earthy mass, precisely like that of the great Pampean deposit hereafter to be described. In one collection of these shells, there were some valves of the Venus sinuosa, Lam., the same species found with the Mactra on the banks of the Uruguay. South of Buenos Ayres, near Ensenada, there are other beds of the Azara, some of which seem to have been embedded in yellowish, calcareous, semi-crystalline matter; and Sir W. Parish has given me from the banks of the Arroyo del Tristan, situated in this same neighbourhood, at the distance of about a league from the Plata, a specimen of a pale- reddish, calcereo-argillaceous stone (precisely like parts of the Pampean deposit the importance of which fact will be referred to in a succeeding chapter), abounding with shells of an Azara, much worn, but which in general form and appearance closely resemble, and are probably identical with, the A. labiata. Besides these shells, cellular, highly crystalline rock, formed of the casts of small bivalves, is found near Ensenada; and likewise beds of sea-shells, which from their appearance appear to have lain on the surface. Sir W. Parish has given me some of these shells, and M. d'Orbigny pronounces them to be:—

1. Buccinanops globulosum, d'Orbigny.
2. Olivancillaria auricularia, d'Orbigny.
3. Venus flexuosa, Lam.
4. Cytheraea (imperfect).
5. Mactra Isabellei, d'Orbigny.
6. Ostrea pulchella, d'Orbigny.

Besides these, Sir W. Parish procured ("Buenos Ayres" etc. by Sir W. Parish) (as named by Mr. G.B. Sowerby) the following shells:—

7. Voluta colocynthis.
8. Voluta angulata.
9. Buccinum (not spec.?).

All these species (with, perhaps, the exception of the last) are recent, and live on the South American coast. These shell-beds extend from one league to six leagues from the Plata, and must lie many feet above its level. I heard, also, of beds of shells on the Somborombon, and on the Rio Salado, at which latter place, as M. d'Orbigny informs me, the Mactra Isabellei and Venus sinuosa are found.

During the elevation of the Provinces of La Plata, the waters of the ancient estuary have but little affected (with the exception of the sand- hills on the banks of the Parana and Uruguay) the outline of the land. M. Parchappe, however, has described groups of sand dunes scattered over the wide extent of the Pampas southward of Buenos Ayres (D'Orbigny "Voyage Geolog."), which M. d'Orbigny attributes with much probability to the action of the sea, before the plains were raised above its level. (Before proceeding to the districts southward of La Plata, it may be worth while just to state, that there is some evidence that the coast of Brazil has participated in a small amount of elevation. Mr. Burchell informs me, that he collected at Santos (latitude 24 degrees S.) oyster-shells, apparently recent, some miles from the shore, and quite above the tidal action. Westward of Rio de Janeiro, Captain Elliot is asserted (see Harlan "Med. and Phys. Res." and Dr. Meigs in "Transactions of the American Philosophical Society"), to have found human bones, encrusted with sea-shells, between fifteen and twenty feet above the level of the sea. Between Rio de Janeiro and Cape Frio I crossed sandy tracts abounding with sea-shells, at a distance of a league from the coast; but whether these tracts have been formed by upheaval, or through the mere accumulation of drift sand, I am not prepared to assert. At Bahia (latitude 13 degrees S.), in some parts near the coast, there are traces of sea-action at the height of about twenty feet above its present level; there are also, in many parts, remnants of beds of sandstone and conglomerate with numerous recent shells, raised a little above the sea-level. I may add, that at the head of Bahia Bay there is a formation, about forty feet in thickness, containing tertiary shells apparently of fresh-water origin, now washed by the sea and encrusted with Balini; this appears to indicate a small amount of subsidence subsequent to its deposition. At

Pernambuco (latitude 8 degrees S.), in the alluvial or tertiary cliffs, surrounding the low land on which the city stands, I looked in vain for organic remains, or other evidence of changes in level.)

SOUTHWARD OF THE PLATA.

The coast as far as Bahia Blanca (in latitude 39 degrees S.) is formed either of a horizontal range of cliffs, or of immense accumulations of sand-dunes. Within Bahia Blanca, a small piece of tableland, about twenty feet above high-water mark, called Punta Alta, is formed of strata of cemented gravel and of red earthy mud, abounding with shells (with others lying loose on the surface), and the bones of extinct mammifers. These shells, twenty in number, together with a Balanus and two corals, are all recent species, still inhabiting the neighbouring seas. They will be enumerated in the Fourth Chapter, when describing the Pampean formation; five of them are identical with the upraised ones from near Buenos Ayres. The northern shore of Bahia Blanca is, in main part, formed of immense sand-dunes, resting on gravel with recent shells, and ranging in lines parallel to the shore. These ranges are separated from each other by flat spaces, composed of stiff impure red clay, in which, at the distance of about two miles from the coast, I found by digging a few minute fragments of sea-shells. The sand-dunes extend several miles inland, and stand on a plain, which slopes up to a height of between one hundred and two hundred feet. Numerous, small, well-rounded pebbles of pumice lie scattered both on the plain and sand-hillocks: at Monte Hermoso, on the flat summit of a cliff, I found many of them at a height of 120 feet (angular measurement) above the level of the sea. These pumice pebbles, no doubt, were originally brought down from the Cordillera by the rivers which cross the continent, in the same way as the river Negro anciently brought down, and still brings down, pumice, and as the river Chupat brings down scoriae: when once delivered at the mouth of a river, they would naturally have travelled along the coasts, and been cast up during the elevation of the land, at different heights. The origin of the argillaceous flats, which separate the parallel ranges of sand-dunes, seems due to the tides here having a tendency (as I believe they have on most shoal, protected coasts) to throw up a bar parallel to the shore, and at some distance from it; this bar gradually becomes larger, affording a base for the accumulation of sand-dunes, and the shallow space within then becomes silted up with mud. The repetition of this process, without any elevation of the land, would form a level plain traversed by parallel lines of sand-hillocks; during a slow elevation of the land, the hillocks would rest on a gently inclined surface, like that on the northern shore of Bahia Blanca. I did not observe any shells in this neighbourhood at a greater height than twenty feet; and therefore the age of the sea-drifted pebbles of pumice, now standing at the height of 120 feet, must remain uncertain.

The main plain surrounding Bahia Blanca I estimated at from two hundred to three hundred feet; it insensibly rises towards the distant Sierra Ventana. There are in this neighbourhood some other and lower plains, but they do not abut one at the foot of the other, in the manner hereafter to be described, so characteristic of Patagonia. The plain on which the settlement stands is crossed by many low sand-dunes, abounding with the minute shells of the Paludestrina australis, d'Orbigny, which now lives in the bay. This low plain is bounded to the south, at the Cabeza del Buey, by the cliff-formed margin of a wide plain of the Pampean formation, which I estimated at sixty feet in height. On the summit of this cliff there is a range of high sand-dunes extending several miles in an east and west line.

Southward of Bahia Blanca, the river Colorado flows between two plains, apparently from thirty to forty feet in height. Of these plains, the southern one slopes up to the foot of the great sandstone plateau of the Rio Negro; and the northern one against an escarpment of the Pampean deposit; so that the Colorado flows in a valley fifty miles in width, between the upper escarpments. I state this, because on the low plain at the foot of the northern escarpment, I crossed an immense accumulation of high sand-dunes, estimated by the Gauchos at no less than eight miles in breadth. These dunes range westward from the coast, which is twenty miles distant, to far inland, in lines parallel to the valley; they are separated from each other by argillaceous flats, precisely like those on the northern shore of Bahia Blanca. At present there is no source whence this immense accumulation of sand could proceed; but if, as I believe, the upper escarpments once formed the shores of an estuary, in that case the sandstone formation of the river Negro would have afforded an inexhaustible supply of sand, which would naturally have accumulated on the northern shore, as on every part of the coast open to the south winds between Bahia Blanca and Buenos Ayres.

At San Blas (40 degrees 40' S.) a little south of the mouth of the Colorado, M. d'Orbigny found fourteen species of existing shells (six of them identical with those from Bahia Blanca), embedded in their natural positions. ("Voyage" etc.) From the zone of depth which these shells are known to inhabit, they must have been uplifted thirty-two feet. He also found, at from fifteen to twenty feet above this bed, the remains of an ancient beach.

Ten miles southward, but 120 miles to the west, at Port S. Antonio, the Officers employed on the Survey assured me that they saw many old sea-shells strewed on the surface of the ground, similar to those found on other parts of the coast of Patagonia. At San Josef, ninety miles south in nearly the same longitude, I found, above the gravel, which caps an old tertiary formation, an irregular bed and hillock of sand, several feet in thickness, abounding with shells of Patella deaurita, Mytilus Magellanicus, the latter retaining much of its colour; Fusus Magellanicus (and a variety of the same), and a large Balanus (probably B. Tulipa), all now found on this coast: I estimated this bed at from eighty to one hundred feet above the level of the sea. To the westward of this bay, there is a plain estimated at between two hundred and three hundred feet in height: this plain seems, from many measurements, to be a continuation of the sandstone platform of the river Negro. The next place southward, where I landed, was at Port Desire, 340 miles distant; but from the intermediate districts I received, through the kindness of the Officers of the Survey, especially from Lieutenant Stokes and Mr. King, many specimens and sketches, quite sufficient to show the general uniformity of the whole line of coast. I may here state, that the whole of Patagonia consists of a tertiary formation, resting on and sometimes surrounding hills of porphyry and quartz: the surface is worn into many wide valleys and into level step-formed plains, rising one above another, all capped by irregular beds of gravel, chiefly composed of porphyritic rocks. This gravel formation will be separately described at the end of the chapter.

My object in giving the following measurements of the plains, as taken by the Officers of the Survey, is, as will hereafter be seen, to show the remarkable equability of the recent elevatory movements. Round the southern parts of Nuevo Gulf, as far as the River Chupat (seventy miles southward of San Josef), there appear to be several plains, of which the best defined are here represented.

(In the following Diagrams: 1. Baseline is Level of sea. 2. Scale is 1/20 of inch to 100 feet vertical. 3. Height is shown in feet thus: An. M. always stands for angular or trigonometrical measurement. Ba. M. always stands for barometrical measurement. Est. always stands for estimation by the Officers of the Survey.

DIAGRAM 1. SECTION OF STEP-FORMED PLAINS SOUTH OF NUEVO GULF.
From East (sea level) to West (high):
Terrace 1. 80 Est.
Terrace 2. 200-220 An. M.
Terrace 3. 350 An. M.)

The upper plain is here well defined (called Table Hills); its edge forms a cliff or line of escarpment many miles in length, projecting over a lower plain. The lowest plain corresponds with that at San Josef with the recent shells on its surface. Between this lowest and the uppermost plain, there is probably more than one step-formed terrace: several measurements show the existence of the intermediate one of the height given in Diagram 1.

(DIAGRAM 2. SECTION OF PLAINS IN THE BAY OF ST. GEORGE.
From East (sea level) to West (high):

Terrace 1. 250 An. M.
Terrace 2. 330 An. M.
Terrace 3. 580 An. M.
Terraces 4, 5 and 6 not measured.
Terrace 7. 1,200 Est.)

Near the north headland of the great Bay of St. George (100 miles south of the Chupat), two well-marked plains of 250 and 330 feet were measured: these are said to sweep round a great part of the Bay. At its south headland, 120 miles distant from the north headland, the 250 feet plain was again measured. In the middle of the bay, a higher plain was found at two neighbouring places (Tilli Roads and C. Marques) to be 580 feet in height. Above this plain, towards the interior, Mr. Stokes informs me that there were several other step-formed plains, the highest of which was estimated at 1,200 feet, and was seen ranging at apparently the same height for 150 miles northward. All these plains have been worn into great valleys and much denuded. The section in Diagram 3 is illustrative of the general structure of the great Bay of St. George. At the south headland of the Bay of St. George (near C. Three Points) the 250 plain is very extensive.

(DIAGRAM 3. SECTION OF PLAINS AT PORT DESIRE.
From East (sea level) to West (high):
Terrace 1. 100 Est.
Terrace 2. 245-255 Ba. M. Shells on surface.
Terrace 3. 330 Ba. M. Shells on surface.
Terrace 4. Not measured.)

At Port Desire (forty miles southward) I made several measurements with the barometer of a plain, which extends along the north side of the port and along the open coast, and which varies from 245 to 255 feet in height: this plain abuts against the foot of a higher plain of 330 feet, which extends also far northward along the coast, and likewise into the interior. In the distance a higher inland platform was seen, of which I do not know the height. In three separate places, I observed the cliff of the 245-255 feet plain, fringed by a terrace or narrow plain estimated at about one hundred feet in height. These plains are represented in the section Diagram 3.

In many places, even at the distance of three and four miles from the coast, I found on the gravel-capped surface of the 245-255 feet, and of the 330 feet plain, shells of Mytilus Magellanicus, M. edulis, Patella deaurita, and another Patella, too much worn to be identified, but apparently similar to one found abundantly adhering to the leaves of the kelp. These species are the commonest now living on this coast. The shells all appeared very old; the blue of the mussels was much faded; and only traces of colour could be perceived in the Patellas, of which the outer surfaces were scaling off. They lay scattered on the smooth surface of the gravel, but abounded most in certain patches, especially at the heads of the smaller valleys: they generally contained sand in their insides; and I presume that they have been washed by alluvial action out of thin sandy layers, traces of which may sometimes be seen covering the gravel. The several plains have very level surfaces; but all are scooped out by numerous broad, winding, flat-bottomed valleys, in which, judging from the bushes, streams never flow. These remarks on the state of the shells, and on the nature of the plains, apply to the following cases, so need not be repeated.

(DIAGRAM 4. SECTION OF PLAINS AT PORT S. JULIAN.
From East (sea level) to West (high):
Terrace 1. Shells on surface. 90 Est.
Terrace 2. 430 An. M.
Terrace 3. 560 An. M.
Terrace 4. 950 An. M.)

Southward of Port Desire, the plains have been greatly denuded, with only small pieces of tableland marking their former extension. But opposite Bird Island, two considerable step-formed plains were measured, and found respectively to be 350 and 590 feet in height.

This latter plain extends along the coast close to Port St. Julian (110 miles south of Port Desire); see Diagram 4.

The lowest plain was estimated at ninety feet: it is remarkable from the usual gravel-bed being deeply worn into hollows, which are filled up with, as well as the general surface covered by, sandy and reddish earthy matter: in one of the hollows thus filled up, the skeleton of the Macrauchenia Patachonica, as will hereafter be described, was embedded. On the surface and in the upper parts of this earthy mass, there were numerous shells of Mytilus Magellanicus and M. edulis, Patella deaurita, and fragments of other species. This plain is tolerably level, but not extensive; it forms a promontory seven or eight miles long, and three or four wide. The upper plains in Diagram 4 were measured by the Officers of the Survey; they were all capped by thick beds of gravel, and were all more or less denuded; the 950 plain consists merely of separate, truncated, gravel-capped hills, two of which, by measurement, were found to differ only three feet. The 430 feet plain extends, apparently with hardly a break, to near the northern entrance of the Rio Santa Cruz (fifty miles to the south); but it was there found to be only 330 feet in height.

(DIAGRAM 5. SECTION OF PLAINS AT THE MOUTH OF THE RIO SANTA CRUZ.

From East (sea level) to West (high):
Terrace 1. (sloping) 355 Ba. M. Shells on surface. 463 Ba. M.
Terrace 2. 710 An. M.
Terrace 3. 840 An. M.)

On the southern side of the mouth of the Santa Cruz we have Diagram 5, which I am able to give with more detail than in the foregoing cases.

The plain marked 355 feet (as ascertained by the barometer and by angular measurement) is a continuation of the above-mentioned 330 feet plain: it extends in a N.W. direction along the southern shores of the estuary. It is capped by gravel, which in most parts is covered by a thin bed of sandy earth, and is scooped out by many flat-bottomed valleys. It appears to the eye quite level, but in proceeding in a S.S.W. course, towards an escarpment distant about six miles, and likewise ranging across the country in a N.W. line, it was found to rise at first insensibly, and then for the last half-mile, sensibly, close up to the base of the escarpment: at this point it was 463 feet in height, showing a rise of 108 feet in the six miles. On this 355-463 feet plain, I found several shells of Mytilus Magellanicus and of a Mytilus, which Mr. Sowerby informs me is yet unnamed, though well-known as recent on this coast; Patella deaurita; Fusus, I believe, Magellanicus, but the specimen has been lost; and at the distance of four miles from the coast, at the height of about four hundred feet, there were fragments of the same Patella and of a Voluta (apparently V. ancilla) partially embedded in the superficial sandy earth. All these shells had the same ancient appearance with those from the foregoing localities. As the tides along this part of the coast rise at the Syzygal period forty feet, and therefore form a well-marked beach-line, I particularly looked out for ridges in crossing this plain, which, as we have seen, rises 108 feet in about six miles, but I could not see any traces of such. The next highest plain is 710 feet above the sea; it is very narrow, but level, and is capped with gravel; it abuts to the foot of the 840 feet plain. This summit-plain extends as far as the eye can range, both inland along the southern side of the valley of the Santa Cruz, and southward along the Atlantic.

THE VALLEY OF THE R. SANTA CRUZ.

This valley runs in an east and west direction to the Cordillera, a distance of about one hundred and sixty miles. It cuts through the great Patagonian tertiary formation, including, in the upper half of the valley, immense streams of basaltic lava, which as well as the softer beds, are capped by gravel; and this gravel, high up the river, is associated with a vast boulder formation. (I have described this formation in a paper in the "Geological Transactions" volume 6) In ascending the valley, the plain which at the mouth on the southern side is 355 feet high, is seen to trend towards the corresponding plain on the northern side, so that their escarpments appear like the shores of a former estuary, larger than the existing one: the escarpments, also, of the 840 feet summit-plain (with a corresponding northern one, which is met with some way up the valley), appear like the shores of a still larger estuary. Farther up the valley, the sides are bounded throughout its entire length by level, gravel-capped terraces, rising above each other in steps. The width between the upper escarpments is on an average between seven and ten miles; in one spot, however, where cutting through the basaltic lava, it was only one mile and a half. Between the escarpments of the second highest terrace the average width is about four or five miles. The bottom of the valley, at the distance of 110 miles from its mouth, begins sensibly to expand, and soon forms a considerable plain, 440 feet above the level of the sea, through which the river flows in a gut from twenty to forty feet in depth. I here found, at a point 140 miles from the Atlantic, and seventy miles from the nearest creek of the Pacific, at the height of 410 feet, a very old and worn shell of Patella deaurita. Lower down the valley, 105 miles from the Atlantic (longitude 71 degrees W.), and at an elevation of about 300 feet, I also found, in the bed of the river, two much worn and broken shells of the Voluta ancilla, still retaining traces of their colours; and one of the Patella deaurita. It appeared that these shells had been washed from the banks into the river; considering the distance from the sea, the desert and absolutely unfrequented character of the country, and the very ancient appearance of the shells (exactly like those found on the plains nearer the coast), there is, I think, no cause to suspect that they could have been brought here by Indians.

The plain at the head of the valley is tolerably level, but water-worn, and with many sand-dunes on it like those on a sea-coast. At the highest point to which we ascended, it was sixteen miles wide in a north and south line; and forty-five miles in length in an east and west line. It is bordered by the escarpments, one above the other, of two plains, which diverge as they approach the Cordillera, and consequently resemble, at two levels, the shores of great bays facing the mountains; and these mountains are breached in front of the lower plain by a remarkable gap. The valley, therefore, of the Santa Cruz consists of a straight broad cut, about ninety miles in length, bordered by gravel-capped terraces and plains, the escarpments of which at both ends diverge or expand, one over the other, after the manner of the shores of great bays. Bearing in mind this peculiar form of the land—the sand-dunes on the plain at the head of the valley—the gap in the Cordillera, in front of it—the presence in two places of very ancient shells of existing species—and lastly, the circumstance of the 355-453 feet plain, with the numerous marine remains on its surface, sweeping from the Atlantic coast, far up the valley, I think we must admit, that within the recent period, the course of the Santa Cruz formed a sea-strait intersecting the continent. At this period, the southern part of South America consisted of an archipelago of islands 360 miles in a north and south line. We shall presently see, that two other straits also, since closed, then cut through Tierra del Fuego; I may add, that one of them must at that time have expanded at the foot of the Cordillera into a great bay (now Otway Water) like that which formerly covered the 440 feet plain at the head of the Santa Cruz.

(DIAGRAM 6. NORTH AND SOUTH SECTION ACROSS THE TERRACES BOUNDING THE VALLEY OF THE RIVER SANTA CRUZ, HIGH UP ITS COURSE.

The height of each terrace, above the level of the river (furthest to nearest to the river) in feet:

A, north and south: 1,122
B, north and south: 869
C, north and south: 639
D, north: not measured. D, north? (suggest south): 185
E: 20
Bed of River.

Vertical scale 1/20 of inch to 100 feet; but terrace E, being only twenty feet above the river, has necessarily been raised. The horizontal distances much contracted; the distance from the edge of A North to A South being on an average from seven to ten miles.) I have said that the valley in its whole course is bordered by gravel-capped plains. The section (Diagram 6), supposed to be drawn in a north and south line across the valley, can scarcely be considered as more than illustrative; for during our hurried ascent it was impossible to measure all the plains at any one place. At a point nearly midway between the Cordillera and the Atlantic, I found the plain (A north) 1,122 feet above the river; all the lower plains on this side were here united into one great broken cliff: at a point sixteen miles lower down the stream, I found by measurement and estimation that B (north) was 869 above the river: very near to where A (north) was measured, C (north) was 639 above the same level: the terrace D (north) was nowhere measured: the lowest E (north) was in many places about twenty feet above the river. These plains or terraces were best developed where the valley was widest; the whole five, like gigantic steps, occurred together only at a few points. The lower terraces are less continuous than the higher ones, and appear to be entirely lost in the upper third of the valley. Terrace C (south), however was traced continuously for a great distance. The terrace B (north), at a point fifty-five miles from the mouth of the river, was four miles in width; higher up the valley this terrace (or at least the second highest one, for I could not always trace it continuously) was about eight miles wide. This second plain was generally wider than the lower ones—as indeed follows from the valley from A (north) to A (south) being generally nearly double the width of from B (north) to B (south). Low down the valley, the summit-plain A (south) is continuous with the 840 feet plain on the coast, but it is soon lost or unites with the escarpment of B (south). The corresponding plain A (north), on the north side of the valley, appears to range continuously from the Cordillera to the head of the present estuary of the Santa Cruz, where it trends northward towards Port St. Julian. Near the Cordillera the summit-plain on both sides of the valley is between 3,200 and 3,300 feet in height; at 100 miles from the Atlantic, it is 1,416 feet, and on the coast 840 feet, all above the sea-beach; so that in a distance of 100 miles the plain rises 576 feet, and much more rapidly near to the Cordillera. The lower terraces B and C also appear to rise as they run up the valley; thus D (north), measured at two points twenty-four miles apart, was found to have risen 185 feet. From several reasons I suspect, that this gradual inclination of the plains up the valley, has been chiefly caused by the elevation of the continent in mass, having been the greater the nearer to the Cordillera.

All the terraces are capped with well-rounded gravel, which rests either on the denuded and sometimes furrowed surface of the soft tertiary deposits, or on the basaltic lava. The difference in height between some of the lower steps or terraces seems to be entirely owing to a difference in the thickness of the capping gravel. Furrows and inequalities in the gravel, where such occur, are filled up and smoothed over with sandy earth. The pebbles, especially on the higher plains, are often whitewashed, and even cemented together by a white aluminous substance, and I occasionally found this to be the case with the gravel on the terrace D. I could not perceive any trace of a similar deposition on the pebbles now thrown up by the river, and therefore I do not think that terrace D was river-formed. As the terrace E generally stands about twenty feet above the bed of the river, my first impression was to doubt whether even this lowest one could have been so formed; but it should always be borne in mind, that the horizontal upheaval of a district, by increasing the total descent of the streams, will always tend to increase, first near the sea-coast and then further and further up the valley, their corroding and deepening powers: so that an alluvial plain, formed almost on a level with a stream, will, after an elevation of this kind, in time be cut through, and left standing at a height never again to be reached by the water. With respect to the three upper terraces of the Santa Cruz, I think there can be no doubt, that they were modelled by the sea, when the valley was occupied by a strait, in the same manner (hereafter to be discussed) as the greater step-formed, shell-strewed plains along the coast of Patagonia.

To return to the shores of the Atlantic: the 840 feet plain, at the mouth of the Santa Cruz, is seen extending horizontally far to the south; and I am informed by the Officers of the Survey, that bending round the head of Coy Inlet (sixty-five miles southward), it trends inland. Outliers of apparently the same height are seen forty miles farther south, inland of the river Gallegos; and a plain comes down to Cape Gregory (thirty-five miles southward), in the Strait of Magellan, which was estimated at between eight hundred and one thousand feet in height, and which, rising towards the interior, is capped by the boulder formation. South of the Strait of Magellan, there are large outlying masses of apparently the same great tableland, extending at intervals along the eastern coast of Tierra del Fuego: at two places here, 110 miles a part, this plain was found to be 950 and 970 feet in height.

From Coy Inlet, where the high summit-plain trends inland, a plain estimated at 350 feet in height, extends for forty miles to the river Gallegos. From this point to the Strait of Magellan, and on each side of that Strait, the country has been much denuded and is less level. It consists chiefly of the boulder formation, which rises to a height of between one hundred and fifty and two hundred and fifty feet, and is often capped by beds of gravel. At N.S. Gracia, on the north side of the Inner Narrows of the Strait of Magellan, I found on the summit of a cliff, 160 feet in height, shells of existing Patellae and Mytili, scattered on the surface and partially embedded in earth. On the eastern coast, also, of Tierra del Fuego, in latitude 53 degrees 20' south, I found many Mytili on some level land, estimated at 200 feet in height. Anterior to the elevation attested by these shells, it is evident by the present form of the land, and by the distribution of the great erratic boulders on the surface, that two sea-channels connected the Strait of Magellan both with Sebastian Bay and with Otway Water. ("Geological Transactions" volume 6

CONCLUDING REMARKS ON THE RECENT ELEVATION OF THE SOUTH-EASTERN COASTS OF AMERICA, AND ON THE ACTION OF THE SEA ON THE LAND.

Upraised shells of species, still existing as the commonest kinds in the adjoining sea, occur, as we have seen, at heights of between a few feet and 410 feet, at intervals from latitude 33 degrees 40' to 53 degrees 20' south. This is a distance of 1,180 geographical miles—about equal from London to the North Cape of Sweden. As the boulder formation extends with nearly the same height 150 miles south of 53 degrees 20', the most southern point where I landed and found upraised shells; and as the level Pampas ranges many hundred miles northward of the point, where M. d'Orbigny found at the height of 100 feet beds of the Azara, the space in a north and south line, which has been uplifted within the recent period, must have been much above the 1,180 miles. By the term "recent," I refer only to that period within which the now living mollusca were called into existence; for it will be seen in the Fourth Chapter, that both at Bahia Blanca and P. S. Julian, the mammiferous quadrupeds which co-existed with these shells belong to extinct species. I have said that the upraised shells were found only at intervals on this line of coast, but this in all probability may be attributed to my not having landed at the intermediate points; for wherever I did land, with the exception of the river Negro, shells were found: moreover, the shells are strewed on plains or terraces, which, as we shall immediately see, extend for great distances with a uniform height. I ascended the higher plains only in a few places, owing to the distance at which their escarpments generally range from the coast, so that I am far from knowing that 410 feet is the maximum of elevation of these upraised remains. The shells are those now most abundant in a living state in the adjoining sea. (Captain King "Voyages of 'Adventure' and 'Beagle'" volume 1 and 133.) All of them have an ancient appearance; but some, especially the mussels, although lying fully exposed to the weather, retain to a considerable extent their colours: this circumstance appears at first surprising, but it is now known that the colouring principle of the Mytilus is so enduring, that it is preserved when the shell itself is completely disintegrated. (See Mr. Lyell "Proofs of a Gradual Rising in Sweden" in the "Philosophical Transactions" 1835. See also Mr. Smith of Jordan Hill in the "Edinburgh New Philosophical Journal" volume 25.) Most of the shells are broken; I nowhere found two valves united; the fragments are not rounded, at least in none of the specimens which I brought home.

With respect to the breadth of the upraised area in an east and west line, we know from the shells found at the Inner Narrows of the Strait of Magellan, that the entire width of the plain, although there very narrow, has been elevated. It is probable that in this southernmost part of the continent, the movement has extended under the sea far eastward; for at the Falkland Islands, though I could not find any shells, the bones of whales have been noticed by several competent observers, lying on the land at a considerable distance from the sea, and at the height of some hundred feet above it. ("Voyages of the 'Adventure' and 'Beagle'" volume 2. And Bougainville's "Voyage" tome 1) Moreover, we know that in Tierra del Fuego the boulder formation has been uplifted within the recent period, and a similar formation occurs on the north-western shores (Byron Sound) of these islands. (I owe this fact to the kindness of Captain Sulivan, R.N., a highly competent observer. I mention it more especially, as in my Paper on the Boulder Formation, I have, after having examined the northern and middle parts of the eastern island, said that the formation was here wholly absent.) The distance from this point to the Cordillera of Tierra del Fuego, is 360 miles, which we may take as the probable width of the recently upraised area. In the latitude of the R. Santa Cruz, we know from the shells found at the mouth and head, and in the middle of the valley, that the entire width (about 160 miles) of the surface eastward of the Cordillera has been upraised. From the slope of the plains, as shown by the course of the rivers, for several degrees northward of the Santa Cruz, it is probable that the elevation attested by the shells on the coast has likewise extended to the Cordillera. When, however, we look as far northward as the provinces of La Plata, this conclusion would be very hazardous; not only is the distance from Maldonado (where I found upraised shells) to the Cordillera great, namely, 760 miles, but at the head of the estuary of the Plata, a N.N.E. and S.S.W. range of tertiary volcanic rocks has been observed (This volcanic formation will be described in Chapter IV. It is not improbable that the height of the upraised shells at the head of the estuary of the Plata, being greater than at Bahia Blanca or at San Blas, may be owing to the upheaval of these latter places having been connected with the distant line of the Cordillera, whilst that of the provinces of La Plata was in connection with the adjoining tertiary volcanic axis.), which may well indicate an axis of elevation quite distinct from that of the Andes. Moreover, in the centre of the Pampas in the chain of Cordova, severe earthquakes have been felt (See Sir W. Parish's work on "La Plata". For a notice of an earthquake which drained a lake near Cordova, see also Temple's "Travels in Peru." Sir W. Parish informs me, that a town between Salta and Tucuman (north of Cordova) was formerly utterly overthrown by an earthquake.); whereas at Mendoza, at the eastern foot of the Cordillera, only gentle oscillations, transmitted from the shores of the Pacific, have ever been experienced. Hence the elevation of the Pampas may be due to several distinct axes of movement; and we cannot judge, from the upraised shells round the estuary of the Plata, of the breadth of the area uplifted within the recent period.

Not only has the above specified long range of coast been elevated within the recent period, but I think it may be safely inferred from the similarity in height of the gravel-capped plains at distant points, that there has been a remarkable degree of equality in the elevatory process. I may premise, that when I measured the plains, it was simply to ascertain the heights at which shells occurred; afterwards, comparing these measurements with some of those made during the Survey, I was struck with their uniformity, and accordingly tabulated all those which represented the summit-edges of plains. The extension of the 330 to 355 feet plain is very striking, being found over a space of 500 geographical miles in a north and south line. A table (Table 1) of the measurements is given below. The angular measurements and all the estimations (in feet) are by the Officers of the Survey; the barometrical ones by myself:—

TABLE 1.
Gallegos River to Coy Inlet (partly angular partly estimation) 350
South Side of Santa Cruz (angular and barometric) 355
North Side of Santa Cruz (angular and barometric) 330

Bird Island, plain opposite to (angular) 350
Port Desire, plain extending far along coast (barometric) 330
St. George's Bay, north promontory (angular) 330
Table Land, south of New Bay (angular) 350

A plain, varying from 245 to 255 feet, seems to extend with much uniformity from Port Desire to the north of St. George's Bay, a distance of 170 miles; and some approximate measurements (in feet), also given in Table 2 below, indicate the much greater extension of 780 miles:—

TABLE 2.
Coy Inlet, south of (partly angular and partly estimation) 200 to 300
Port Desire (barometric) 245 to 255
C. Blanco (angular) 250
North Promontory of St. George's Bay (angular) 250
South of New Bay (angular) 200 to 220
North of S. Josef (estimation) 200 to 300
Plain of Rio Negro (angular) 200 to 220
Bahia Blanca (estimation) 200 to 300

The extension, moreover, of the 560 to 580, and of the 80 to 100 feet, plains is remarkable, though somewhat less obvious than in the former cases. Bearing in mind that I have not picked these measurements out of a series, but have used all those which represented the edges of plains, I think it scarcely possible that these coincidences in height should be accidental. We must therefore conclude that the action, whatever it may have been, by which these plains have been modelled into their present forms, has been singularly uniform.

These plains or great terraces, of which three and four often rise like steps one behind the other, are formed by the denudation of the old Patagonian tertiary beds, and by the deposition on their surfaces of a mass of well-rounded gravel, varying, near the coast, from ten to thirty-five feet in thickness, but increasing in thickness towards the interior. The gravel is often capped by a thin irregular bed of sandy earth. The plains slope up, though seldom sensibly to the eye, from the summit edge of one escarpment to the foot of the next highest one. Within a distance of 150 miles, between Santa Cruz to Port Desire, where the plains are particularly well developed, there are at least seven stages or steps, one above the other. On the three lower ones, namely, those of 100 feet, 250 feet, and 350 feet in height, existing littoral shells are abundantly strewed, either on the surface, or partially embedded in the superficial sandy earth. By whatever action these three lower plains have been modelled, so undoubtedly have all the higher ones, up to a height of 950 feet at S. Julian, and of 1,200 feet (by estimation) along St. George's Bay. I think it will not be disputed, considering the presence of the upraised marine shells, that the sea has been the active power during stages of some kind in the elevatory process.

We will now briefly consider this subject: if we look at the existing coast-line, the evidence of the great denuding power of the sea is very distinct; for, from Cape St. Diego, in latitude 54 degrees 30' to the mouth of the Rio Negro, in latitude 31 degrees (a length of more than eight hundred miles), the shore is formed, with singularly few exceptions, of bold and naked cliffs: in many places the cliffs are high; thus, south of the Santa Cruz, they are between eight and nine hundred feet in height, with their horizontal strata abruptly cut off, showing the immense mass of matter which has been removed. Nearly this whole line of coast consists of a series of greater or lesser curves, the horns of which, and likewise certain straight projecting portions, are formed of hard rocks; hence the concave parts are evidently the effect and the measure of the denuding action on the softer strata. At the foot of all the cliffs, the sea shoals very gradually far outwards; and the bottom, for a space of some miles, everywhere consists of gravel. I carefully examined the bed of the sea off the Santa Cruz, and found that its inclination was exactly the same, both in amount and in its peculiar curvature, with that of the 355 feet plain at this same place. If, therefore, the coast, with the bed of the adjoining sea, were now suddenly elevated one or two hundred feet, an inland line of cliffs, that is an escarpment, would be formed, with a gravel-capped plain at its foot gently sloping to the sea, and having an inclination like that of the existing 355 feet plain. From the denuding tendency of the sea, this newly formed plain would in time be eaten back into a cliff: and repetitions of this elevatory and denuding process would produce a series of gravel-capped sloping terraces, rising one above another, like those fronting the shores of Patagonia.

The chief difficulty (for there are other inconsiderable ones) on this view, is the fact,—as far as I can trust two continuous lines of soundings carefully taken between Santa Cruz and the Falkland Islands, and several scattered observations on this and other coasts,—that the pebbles at the bottom of the sea QUICKLY and REGULARLY decrease in size with the increasing depth and distance from the shore, whereas in the gravel on the sloping plains, no such decrease in size was perceptible.

Table 3 below gives the average result of many soundings off the Santa Cruz:—

TABLE 3.

Under two miles from the shore, many of the pebbles were of large size, mingled with some small ones.

Column 1. Distance in miles from the shore.
Column 2. Depth in fathoms.
Column 3. Size of Pebbles.
1. 2. 3.
3 to 4 11 to 12 As large as walnuts; mingled in every case with some smaller ones.
6 to 7 17 to 19 As large as hazel-nuts.
10 to 11 23 to 25 From three- to four-tenths of an inch in diameter.
12 30 to 40 Two-tenths of an inch.
22 to 150 45 to 65 One-tenth of an inch, to the finest sand.

I particularly attended to the size of the pebbles on the 355 feet Santa Cruz plain, and I noticed that on the summit-edge of the present sea cliffs many were as large as half a man's head; and in crossing from these cliffs to the foot of the next highest escarpment, a distance of six miles, I could not observe any increase in their size. We shall presently see that the theory of a slow and almost insensible rise of the land, will explain all the facts connected with the gravel-capped terraces, better than the theory of sudden elevations of from one to two hundred feet.

M. d'Orbigny has argued, from the upraised shells at San Blas being embedded in the positions in which they lived, and from the valves of the Azara labiata high on the banks of the Parana being united and unrolled, that the elevation of Northern Patagonia and of La Plata must have been sudden; for he thinks, if it had been gradual, these shells would all have been rolled on successive beach-lines. But in PROTECTED bays, such as in that of Bahia Blanca, wherever the sea is accumulating extensive mud-banks, or where the winds quietly heap up sand-dunes, beds of shells might assuredly be preserved buried in the positions in which they had lived, even whilst the land retained the same level; any, the smallest, amount of elevation would directly aid in their preservation. I saw a multitude of spots in Bahia Blanca where this might have been effected; and at Maldonado it almost certainly has been effected. In speaking of the elevation of the land having been slow, I do not wish to exclude the small starts which accompany earthquakes, as on the coast of Chile; and by such movements beds of shells might easily be uplifted, even in positions exposed to a heavy surf, without undergoing any attrition: for instance, in 1835, a rocky flat off the island of Santa Maria was at one blow upheaved above high-water mark, and was left covered with gaping and putrefying mussel-shells, still attached to the bed on which they had lived. If M. d'Orbigny had been aware of the many long parallel lines of sand-hillocks, with infinitely numerous shells of the Mactra and Venus, at a low level near the Uruguay; if he had seen at Bahia Blanca the immense sand-dunes, with water-worn pebbles of pumice, ranging in parallel lines, one behind the other, up a height of at least 120 feet; if he had seen the sand-dunes, with the countless Paludestrinas, on the low plain near the Fort at this place, and that long line on the edge of the cliff, sixty feet higher up; if he had crossed that long and great belt of parallel sand-dunes, eight miles in width, standing at the height of from forty to fifty feet above the Colorado, where sand could not now collect,—I cannot believe he would have thought that the elevation of this great district had been sudden. Certainly the sand-dunes (especially when abounding with shells), which stand in ranges at so many different levels, must all have required long time for their accumulation; and hence I do not doubt that the last 100 feet of elevation of La Plata and Northern Patagonia has been exceedingly slow.

If we extend this conclusion to Central and Southern Patagonia, the inclination of the successively rising gravel-capped plains can be explained quite as well, as by the more obvious view already given of a few comparatively great and sudden elevations; in either case we must admit long periods of rest, during which the sea ate deeply into the land. Let us suppose the present coast to rise at a nearly equable, slow rate, yet sufficiently quick to prevent the waves quite removing each part as soon as brought up; in this case every portion of the present bed of the sea will successively form a beach-line, and from being exposed to a like action will be similarly affected. It cannot matter to what height the tides rise, even if to forty feet as at Santa Cruz, for they will act with equal force and in like manner on each successive line. Hence there is no difficulty in the fact of the 355 feet plain at Santa Cruz sloping up 108 feet to the foot of the next highest escarpment, and yet having no marks of any one particular beach-line on it; for the whole surface on this view has been a beach. I cannot pretend to follow out the precise action of the tidal-waves during a rise of the land, slow, yet sufficiently quick to prevent or check denudation: but if it be analogous to what takes place on protected parts of the present coast, where gravel is now accumulating in large quantities, an inclined surface, thickly capped by well-rounded pebbles of about the same size, would be ultimately left. (On the eastern side of Chiloe, which island we shall see in the next chapter is now rising, I observed that all the beaches and extensive tidal-flats were formed of shingle.) On the gravel now accumulating, the waves, aided by the wind, sometimes throw up a thin covering of sand, together with the common coast-shells. Shells thus cast up by gales, would, during an elevatory period, never again be touched by the sea. Hence, on this view of a slow and gradual rising of the land, interrupted by periods of rest and denudation, we can understand the pebbles being of about the same size over the entire width of the step-like plains,—the occasional thin covering of sandy earth,—and the presence of broken, unrolled fragments of those shells, which now live exclusively near the coast.

SUMMARY OF RESULTS.

It may be concluded that the coast on this side of the continent, for a space of at least 1,180 miles, has been elevated to a height of 100 feet in La Plata, and of 400 feet in Southern Patagonia, within the period of existing shells, but not of existing mammifers. That in La Plata the elevation has been very slowly effected: that in Patagonia the movement may have been by considerable starts, but much more probably slow and quiet. In either case, there have been long intervening periods of comparative rest, during which the sea corroded deeply, as it is still corroding, into the land. (I say COMPARATIVE and not ABSOLUTE rest, because the sea acts, as we have seen, with great denuding power on this whole line of coast; and therefore, during an elevation of the land, if excessively slow (and of course during a subsidence of the land), it is quite possible that lines of cliff might be formed.) That the periods of denudation and elevation were contemporaneous and equable over great spaces of coast, as shown by the equable heights of the plains; that there have been at least eight periods of denudation, and that the land, up to a height of from 950 to 1,200 feet, has been similarly modelled and affected: that the area elevated, in the southernmost part of the continent, extended in breadth to the Cordillera, and probably seaward to the Falkland Islands; that northward, in La Plata, the breadth is unknown, there having been probably more than one axis of elevation; and finally, that, anterior to the elevation attested by these upraised shells, the land was divided by a Strait where the River Santa Cruz now flows, and that further southward there were other sea-straits, since closed. I may add, that at Santa Cruz, in latitude 50 degrees S., the plains have been uplifted at least 1,400 feet, since the period when gigantic boulders were transported between sixty and seventy miles from their parent rock, on floating icebergs.

Lastly, considering the great upward movements which this long line of coast has undergone, and the proximity of its southern half to the volcanic axis of the Cordillera, it is highly remarkable that in the many fine sections exposed in the Pampean, Patagonian tertiary, and Boulder formations, I nowhere observed the smallest fault or abrupt curvature in the strata.

GRAVEL FORMATION OF PATAGONIA.

I will here describe in more detail than has been as yet incidentally done, the nature, origin, and extent of the great shingle covering of Patagonia: but I do not mean to affirm that all of this shingle, especially that on the higher plains, belongs to the recent period. A thin bed of sandy earth, with small pebbles of various porphyries and of quartz, covering a low plain on the north side of the Rio Colorado, is the extreme northern limit of this formation. These little pebbles have probably been derived from the denudation of a more regular bed of gravel, capping the old tertiary sandstone plateau of the Rio Negro. The gravel-bed near the Rio Negro is, on an average, about ten or twelve feet in thickness; and the pebbles are larger than on the northern side of the Colorado, being from one or two inches in diameter, and composed chiefly of rather dark-tinted porphyries. Amongst them I here first noticed a variety often to be referred to, namely, a peculiar gallstone-yellow siliceous porphyry, frequently, but not invariably, containing grains of quartz. The pebbles are embedded in a white, gritty, calcareous matrix, very like mortar, sometimes merely coating with a whitewash the separate stones, and sometimes forming the greater part of the mass. In one place I saw in the gravel concretionary nodules (not rounded) of crystallised gypsum, some as large as a man's head. I traced this bed for forty-five miles inland, and was assured that it extended far into the interior. As the surface of the calcareo- argillaceous plain of Pampean formation, on the northern side of the wide valley of the Colorado, stands at about the same height with the mortar- like cemented gravel capping the sandstone on the southern side, it is probable, considering the apparent equability of the subterranean movements along this side of America, that this gravel of the Rio Negro and the upper beds of the Pampean formation northward of the Colorado, are of nearly contemporaneous origin, and that the calcareous matter has been derived from the same source.

Southward of the Rio Negro, the cliffs along the great bay of S. Antonio are capped with gravel: at San Josef, I found that the pebbles closely resembled those on the plain of the Rio Negro, but that they were not cemented by calcareous matter. Between San Josef and Port Desire, I was assured by the Officers of the Survey that the whole face of the country is coated with gravel. At Port Desire and over a space of twenty-five miles inland, on the three step-formed plains and in the valleys, I everywhere passed over gravel which, where thickest, was between thirty and forty feet. Here, as in other parts of Patagonia, the gravel, or its sandy covering, was, as we have seen, often strewed with recent marine shells. The sandy covering sometimes fills up furrows in the gravel, as does the gravel in the underlying tertiary formations. The pebbles are frequently whitewashed and even cemented together by a peculiar, white, friable, aluminous, fusible substance, which I believe is decomposed feldspar. At Port Desire, the gravel rested sometimes on the basal formation of porphyry, and sometimes on the upper or the lower denuded tertiary strata. It is remarkable that most of the porphyritic pebbles differ from those varieties of porphyry which occur here abundantly in situ. The peculiar gallstone-yellow variety was common, but less numerous than at Port S. Julian, where it formed nearly one-third of the mass of the gravel; the remaining part there consisting of pale grey and greenish porphyries with many crystals of feldspar. At Port S. Julian, I ascended one of the flat- topped hills, the denuded remnant of the highest plain, and found it, at the height of 950 feet, capped with the usual bed of gravel.

Near the mouth of the Santa Cruz, the bed of gravel on the 355 feet plain is from twenty to about thirty-five feet in thickness. The pebbles vary from minute ones to the size of a hen's egg, and even to that of half a man's head; they consist of paler varieties of porphyry than those found further northward, and there are fewer of the gallstone-yellow kind; pebbles of compact black clay-slate were here first observed. The gravel, as we have seen, covers the step-formed plains at the mouth, head, and on the sides of the great valley of the Santa Cruz. At a distance of 110 miles from the coast, the plain has risen to the height of 1,416 feet above the sea; and the gravel, with the associated great boulder formation, has attained a thickness of 212 feet. The plain, apparently with its usual gravel covering, slopes up to the foot of the Cordillera to the height of between 3,200 and 3,300 feet. In ascending the valley, the gravel gradually becomes entirely altered in character: high up, we have pebbles of crystalline feldspathic rocks, compact clay-slate, quartzose schists, and pale-coloured porphyries; these rocks, judging both from the gigantic boulders in the surface and from some small pebbles embedded beneath 700 feet in thickness of the old tertiary strata, are the prevailing kinds in this part of the Cordillera; pebbles of basalt from the neighbouring streams of basaltic lava are also numerous; there are few or none of the reddish or of the gallstone-yellow porphyries so common near the coast. Hence the pebbles on the 350 feet plain at the mouth of the Santa Cruz cannot have been derived (with the exception of those of compact clay-slate, which, however, may equally well have come from the south) from the Cordillera in this latitude; but probably, in chief part, from farther north.

Southward of the Santa Cruz, the gravel may be seen continuously capping the great 840 feet plain: at the Rio Gallegos, where this plain is succeeded by a lower one, there is, as I am informed by Captain Sulivan, an irregular covering of gravel from ten to twelve feet in thickness over the whole country. The district on each side of the Strait of Magellan is covered up either with gravel or the boulder formation: it was interesting to observe the marked difference between the perfectly rounded state of the pebbles in the great shingle formation of Patagonia, and the more or less angular fragments in the boulder formation. The pebbles and fragments near the Strait of Magellan nearly all belong to rocks known to occur in Fuegia. I was therefore much surprised in dredging south of the Strait to find, in latitude 54 degrees 10' south, many pebbles of the gallstone-yellow siliceous porphyry; I procured others from a great depth off Staten Island, and others were brought me from the western extremity of the Falkland Islands. (At my request, Mr. Kent collected for me a bag of pebbles from the beach of White Rock harbour, in the northern part of the sound, between the two Falkland Islands. Out of these well-rounded pebbles, varying in size from a walnut to a hen's egg, with some larger, thirty-eight evidently belonged to the rocks of these islands; twenty-six were similar to the pebbles of porphyry found on the Patagonian plains, which rocks do not exist in situ in the Falklands; one pebble belonged to the peculiar yellow siliceous porphyry; thirty were of doubtful origin.) The distribution of the pebbles of this peculiar porphyry, which I venture to affirm is not found in situ either in Fuegia, the Falkland Islands, or on the coast of Patagonia, is very remarkable, for they are found over a space of 840 miles in a north and south line, and at the Falklands, 300 miles eastward of the coast of Patagonia. Their occurrence in Fuegia and the Falklands may, however, perhaps be due to the same ice-agency by which the boulders have been there transported.

We have seen that porphyritic pebbles of a small size are first met with on the northern side of the Rio Colorado, the bed becoming well developed near the Rio Negro: from this latter point I have every reason to believe that the gravel extends uninterruptedly over the plains and valleys of Patagonia for at least 630 nautical miles southward to the Rio Gallegos. From the slope of the plains, from the nature of the pebbles, from their extension at the Rio Negro far into the interior, and at the Santa Cruz close up to the Cordillera, I think it highly probable that the whole breadth of Patagonia is thus covered. If so, the average width of the bed must be about two hundred miles. Near the coast the gravel is generally from ten to thirty feet in thickness; and as in the valley of Santa Cruz it attains, at some distance from the

Cordillera, a thickness of 214 feet, we may, I think, safely assume its average thickness over the whole area of 630 by 200 miles, at fifty feet!

The transportal and origin of this vast bed of pebbles is an interesting problem. From the manner in which they cap the step-formed plains, worn by the sea within the period of existing shells, their deposition, at least on the plains up to a height of 400 feet, must have been a recent geological event. From the form of the continent, we may feel sure that they have come from the westward, probably, in chief part from the Cordillera, but, perhaps, partly from unknown rocky ridges in the central districts of Patagonia. That the pebbles have not been transported by rivers, from the interior towards the coast, we may conclude from the fewness and smallness of the streams of Patagonia: moreover, in the case of the one great and rapid river of Santa Cruz, we have good evidence that its transporting power is very trifling. This river is from two to three hundred yards in width, about seventeen feet deep in its middle, and runs with a singular degree of uniformity five knots an hour, with no lakes and scarcely any still reaches: nevertheless, to give one instance of its small transporting power, upon careful examination, pebbles of compact basalt could not be found in the bed of the river at a greater distance than ten miles below the point where the stream rushes over the debris of the great basaltic cliffs forming its shore: fragments of the CELLULAR varieties have been washed down twice or thrice as far. That the pebbles in Central and Northern Patagonia have not been transported by ice-agency, as seems to have been the case to a considerable extent farther south, and likewise in the northern hemisphere, we may conclude, from the absence of all angular fragments in the gravel, and from the complete contrast in many other respects between the shingle and neighbouring boulder formation.

Looking to the gravel on any one of the step-formed plains, I cannot doubt, from the several reasons assigned in this chapter, that it has been spread out and leveled by the long-continued action of the sea, probably during the slow rise of the land. The smooth and perfectly rounded condition of the innumerable pebbles alone would prove long-continued action. But how the whole mass of shingle on the coast-plains has been transported from the mountains of the interior, is another and more difficult question. The following considerations, however, show that the sea by its ordinary action has considerable power in distributing pebbles. Table 3 above shows how very uniformly and gradually the pebbles decrease in size with the gradually seaward increasing depth and distance. (I may mention, that at the distance of 150 miles from the Patagonian shore I carefully examined the minute rounded particles in the sand, and found them to be fusible like the porphyries of the great shingle bed. I could even distinguish particles of the gallstone-yellow porphyry. It was interesting to notice how gradually the particles of white quartz increased, as we approached the Falkland Islands, which are thus constituted. In the whole line of soundings between these islands and the coast of Patagonia dead or living organic remains were most rare. On the relations between the depth of water and the nature of the bottom, see Martin White on "Soundings in the Channel"; also Captain Beechey's "Voyage to the Pacific" chapter 18.) A series of this kind irresistibly leads to the conclusion, that the sea has the power of sifting and distributing the loose matter on its bottom. According to Martin White, the bed of the British Channel is disturbed during gales at depths of sixty-three and sixty-seven fathoms, and at thirty fathoms, shingle and fragments of shells are often deposited, afterwards to be carried away again. ("Soundings in the Channel" M. Siau states ("Edinburgh New Philosophical Journal" volume 31, that he found the sediment, at a depth of 188 metres, arranged in ripples of different degrees of fineness. There are some excellent discussions on this and allied subjects in Sir H. De la Beche's "Theoretical Researches.") Groundswells, which are believed to be caused by distant gales, seem especially to affect the bottom: at such times, according to Sir R. Schomburgk, the sea to a great distance round the West Indian Islands, at depths from five to fifteen fathoms, becomes discoloured, and even the anchors of vessels have been moved. ("Journal of Royal Geographical Society" volume 5. It appears from Mr. Scott Russell's investigations (see Mr. Murchison's "Anniversary Address Geological Society" 1843, that in waves of translation the motion of the particles of water is nearly as great at the bottom as at the top.) There are, however, some difficulties in understanding how the sea can transport pebbles lying at the bottom, for, from experiments instituted on the power of running water, it would appear that the currents of the sea have not sufficient velocity to move stones of even moderate size: moreover, I have repeatedly found in the most exposed situations that the pebbles which lie at the bottom are encrusted with full-grown living corallines, furnished with the most delicate, yet unbroken spines: for instance, in ten fathoms water off the mouth of the Santa Cruz, many pebbles, under half an inch in diameter, were thus coated with Flustracean zoophytes. (A pebble, one and a half inch square and half an inch thick, was given me, dredged up from twenty-seven fathoms depth off the western end of the Falkland Islands, where the sea is remarkably stormy, and subject to violent tides. This pebble was encrusted on all sides by a delicate living coralline. I have seen many pebbles from depths between forty and seventy fathoms thus encrusted; one from the latter depth off Cape Horn.) Hence we must conclude that these pebbles are not often violently disturbed: it should, however, be borne in mind that the growth of corallines is rapid. The view, propounded by Professor Playfair, will, I believe, explain this apparent difficulty,—namely, that from the undulations of the sea TENDING to lift up and down pebbles or other loose bodies at the bottom, such are liable, when thus quite or partially raised, to be moved even by a very small force, a little onwards. We can thus understand how oceanic or tidal currents of no great strength, or that recoil movement of the bottom-water near the land, called by sailors the "undertow" (which I presume must extend out seaward as far as the BREAKING waves impel the surface-water towards the beach), may gain the power during storms of sifting and distributing pebbles even of considerable size, and yet without so violently disturbing them as to injure the encrusting corallines. (I may take this opportunity of remarking on a singular, but very common character in the form of the bottom, in the creeks which deeply penetrate the western shores of Tierra del Fuego; namely, that they are almost invariably much shallower close to the open sea at their mouths than inland. Thus, Cook, in entering Christmas Sound, first had soundings in thirty-seven fathoms, then in fifty, then in sixty, and a little farther in no bottom with 170 fathoms. The sealers are so familiar with this fact, that they always look out for anchorage near the entrances of the creeks. See, also, on this subject, the "Voyages of the 'Adventure' and 'Beagle'" volume 1 and "Appendix". This Shoalness of the sea- channels near their entrances probably results from the quantity of sediment formed by the wear and tear of the outer rocks exposed to the full force of the open sea. I have no doubt that many lakes, for instance in Scotland, which are very deep within, and are separated from the sea apparently only by a tract of detritus, were originally sea-channels with banks of this nature near their mouths, which have since been upheaved.)

The sea acts in another and distinct manner in the distribution of pebbles, namely by the waves on the beach. Mr. Palmer, in his excellent memoir on this subject, has shown that vast masses of shingle travel with surprising quickness along lines of coast, according to the direction with which the waves break on the beach and that this is determined by the prevailing direction of the winds. ("Philosophical Transactions" 1834.) This agency must be powerful in mingling together and disseminating pebbles derived from different sources: we may, perhaps, thus understand the wide distribution of the gallstone-yellow porphyry; and likewise, perhaps, the great difference in the nature of the pebbles at the mouth of the Santa Cruz from those in the same latitude at the head of the valley.

I will not pretend to assign to these several and complicated agencies their shares in the distribution of the Patagonian shingle: but from the several considerations given in this chapter, and I may add, from the frequency of a capping of gravel on tertiary deposits in all parts of the world, as I have myself observed and seen stated in the works of various authors, I cannot doubt that the power of widely dispersing gravel is an ordinary contingent on the action of the sea; and that even in the case of the great Patagonian shingle-bed we have no occasion to call in the aid of debacles. I at one time imagined that perhaps an immense accumulation of shingle had originally been collected at the foot of the Cordillera;

and that this accumulation, when upraised above the level of the sea, had been eaten into and partially spread out (as off the present line of coast); and that the newly-spread out bed had in its turn been upraised, eaten into, and re-spread out; and so onwards, until the shingle, which was first accumulated in great thickness at the foot of the Cordillera, had reached in thinner beds its present extension. By whatever means the gravel formation of Patagonia may have been distributed, the vastness of its area, its thickness, its superficial position, its recent origin, and the great degree of similarity in the nature of its pebbles, all appear to me well deserving the attention of geologists, in relation to the origin of the widely-spread beds of conglomerate belonging to past epochs.

FORMATION OF CLIFFS.
(DIAGRAM 7.—SECTION OF COAST-CLIFFS AND BOTTOM OF SEA, OFF THE ISLAND OF ST. HELENA.
Height in feet above sea level.
Depths in fathoms.
Vertical and horizontal scale, two inches to a nautical mile. The point marked 1,600 feet is at the foot of High Knoll; point marked 510 feet is on the edge of Ladder Hill. The strata consist of basaltic streams.
Section left to right:
Height at the foot of High Knoll: 1,600 at top of strata.
Height on the edge of Ladder Hill: 510 at top of strata.
Bottom at coast rocky only to a depth of five or six fathoms.
30 fathoms: bottom mud and sand.
100 fathoms sloping more sharply to 250 fathoms.)

When viewing the sea-worn cliffs of Patagonia, in some parts between eight hundred and nine hundred feet in height, and formed of horizontal tertiary strata, which must once have extended far seaward—or again, when viewing the lofty cliffs round many volcanic islands, in which the gentle inclination of the lava-streams indicates the former extension of the land, a difficulty often occurred to me, namely, how the strata could possibly have been removed by the action of the sea at a considerable depth beneath its surface. The section in Diagram 7, which represents the general form of the land on the northern and leeward side of St. Helena (taken from Mr. Seale's large model and various measurements), and of the bottom of the adjoining sea (taken chiefly from Captain Austin's survey and some old charts), will show the nature of this difficulty.

If, as seems probable, the basaltic streams were originally prolonged with nearly their present inclination, they must, as shown by the dotted line in the section, once have extended at least to a point, now covered by the sea to a depth of nearly thirty fathoms: but I have every reason to believe they extended considerably further, for the inclination of the streams is less near the coast than further inland. It should also be observed, that other sections on the coast of this island would have given far more striking results, but I had not the exact measurements; thus, on the windward side, the cliffs are about two thousand feet in height and the cut-off lava streams very gently inclined, and the bottom of the sea has nearly a similar slope all round the island. How, then, has all the hard basaltic rock, which once extended beneath the surface of the sea, been worn away? According to Captain Austin, the bottom is uneven and rocky only to that very small distance from the beach within which the depth is from five to six fathoms; outside this line, to a depth of about one hundred fathoms, the bottom is smooth, gently inclined, and formed of mud and sand; outside the one hundred fathoms, it plunges suddenly into unfathomable depths, as is so very commonly the case on all coasts where sediment is accumulating. At greater depths than the five or six fathoms, it seems impossible, under existing circumstances, that the sea can both have worn away hard rock, in parts to a thickness of at least 150 feet, and have deposited a smooth bed of fine sediment. Now, if we had any reason to suppose that St. Helena had, during a long period, gone on slowly subsiding, every difficulty would be removed: for looking at the diagram, and imagining a fresh amount of subsidence, we can see that the waves would then act on the coast-cliffs with fresh and unimpaired vigour, whilst the rocky ledge near the beach would be carried down to that depth, at which sand and mud would be deposited on its bare and uneven surface: after the formation near the shore of a new rocky shoal, fresh subsidence would carry it down and allow it to be smoothly covered up. But in the case of the many cliff-bounded islands, for instance in some of the Canary Islands and of Madeira, round which the inclination of the strata shows that the land once extended far into the depths of the sea, where there is no apparent means of hard rock being worn away—are we to suppose that all these islands have slowly subsided? Madeira, I may remark, has, according to Mr. Smith of Jordan Hill, subsided. Are we to extend this conclusion to the high, cliff- bound, horizontally stratified shores of Patagonia, off which, though the water is not deep even at the distance of several miles, yet the smooth bottom of pebbles gradually decreasing in size with the increasing depth, and derived from a foreign source, seem to declare that the sea is now a depositing and not a corroding agent? I am much inclined to suspect, that we shall hereafter find in all such cases, that the land with the adjoining bed of the sea has in truth subsided: the time will, I believe, come, when geologists will consider it as improbable, that the land should have retained the same level during a whole geological period, as that the atmosphere should have remained absolutely calm during an entire season.

CHAPTER II. ON THE ELEVATION OF THE WESTERN COAST OF SOUTH AMERICA.

Chonos Archipelago.

Chiloe, recent and gradual elevation of, traditions of the inhabitants on this subject.

Concepcion, earthquake and elevation of.

VALPARAISO, great elevation of, upraised shells, earth of marine origin, gradual rise of the land within the historical period.

COQUIMBO, elevation of, in recent times; terraces of marine origin, their inclination, their escarpments not horizontal.

Guasco, gravel terraces of.

Copiapo.

PERU.

Upraised shells of Cobija, Iquique, and Arica.

Lima, shell-beds and sea-beach on San Lorenzo, human remains, fossil earthenware, earthquake debacle, recent subsidence.

On the decay of upraised shells.

General summary.

Commencing at the south and proceeding northward, the first place at which I landed, was at Cape Tres Montes, in latitude 46 degrees 35'. Here, on the shores of Christmas Cove, I observed in several places a beach of pebbles with recent shells, about twenty feet above high-water mark. Southward of Tres Montes (between latitude 47 and 48 degrees), Byron remarks, "We thought it very strange, that upon the summits of the highest hills were found beds of shells, a foot or two thick." ("Narrative of the Loss of the 'Wager'.") In the Chonos Archipelago, the island of Lemus (latitude 44 degrees 30') was, according to M. Coste, suddenly elevated eight feet, during the earthquake of 1829: he adds, "Des roches jadis toujours couvertes par la mer, restant aujourd'hui constamment decouvertes." ("Comptes Rendus" October 1838.) In other parts of this archipelago, I observed two terraces of gravel, abutting to the foot of each other: at Lowe's Harbour (43 degrees 48'), under a great mass of the boulder formation, about three hundred feet in thickness, I found a layer of sand, with numerous comminuted fragments of sea-shells, having a fresh aspect, but too small to be identified.

THE ISLAND OF CHILOE.

The evidence of recent elevation is here more satisfactory. The bay of San Carlos is in most parts bounded by precipitous cliffs from about ten to forty feet in height, their bases being separated from the present line of tidal action by a talus, a few feet in height, covered with vegetation. In one sheltered creek (west of P. Arena), instead of a loose talus, there was a bare sloping bank of tertiary mudstone, perforated, above the line of the highest tides, by numerous shells of a Pholas now common in the harbour. The upper extremities of these shells, standing upright in their holes with grass growing out of them, were abraded about a quarter of an inch, to the same level with the surrounding worn strata. In other parts, I observed (as at Pudeto) a great beach, formed of comminuted shells, twenty feet above the present shore. In other parts again, there were small caves worn into the foot of the low cliffs, and protected from the waves by the talus with its vegetation: one such cave, which I examined, had its mouth about twenty feet, and its bottom, which was filled with sand containing fragments of shells and legs of crabs, from eight to ten feet above high-water mark. From these several facts, and from the appearance of the upraised shells, I inferred that the elevation had been quite recent; and on inquiring from Mr. Williams, the Portmaster, he told me he was convinced that the land had risen, or the sea fallen, four feet within the last four years. During this period, there had been one severe earthquake, but no particular change of level was then observed; from the habits of the people who all keep boats in the protected creeks, it is absolutely impossible that a rise of four feet could have taken place suddenly and been unperceived. Mr. Williams believes that the change has been quite gradual. Without the elevatory movement continues at a quick rate, there can be no doubt that the sea will soon destroy the talus of earth at the foot of the cliffs round the bay, and will then reach its former lateral extension, but not of course its former level: some of the inhabitants assured me that one such talus, with a footpath on it, was even already sensibly decreasing in width.

I received several accounts of beds of shells, existing at considerable heights in the inland parts of Chiloe; and to one of these, near Catiman, I was guided by a countryman. Here, on the south side of the peninsula of Lacuy, there was an immense bed of the Venus costellata and of an oyster, lying on the summit-edge of a piece of tableland, 350 feet (by the barometer) above the level of the sea. The shells were closely packed together, embedded in and covered by a very black, damp, peaty mould, two or three feet in thickness, out of which a forest of great trees was growing. Considering the nature and dampness of this peaty soil, it is surprising that the fine ridges on the outside of the Venus are perfectly preserved, though all the shells have a blackened appearance. I did not doubt that the black soil, which when dry, cakes hard, was entirely of terrestrial origin, but on examining it under the microscope, I found many very minute rounded fragments of shells, amongst which I could distinguish bits of Serpulae and mussels. The Venus costellata, and the Ostrea (O. edulis, according to Captain King) are now the commonest shells in the adjoining bays. In a bed of shells, a few feet below the 350 feet bed, I found a horn of the little Cervus humilis, which now inhabits Chiloe.

The eastern or inland side of Chiloe, with its many adjacent islets, consists of tertiary and boulder deposits, worn into irregular plains capped by gravel. Near Castro, and for ten miles southward, and on the islet of Lemuy, I found the surface of the ground to a height of between twenty and thirty feet above high-water mark, and in several places apparently up to fifty feet, thickly coated by much comminuted shells, chiefly of the Venus costellata and Mytilus Chiloensis; the species now most abundant on this line of coast. As the inhabitants carry immense numbers of these shells inland, the continuity of the bed at the same height was often the only means of recognising its natural origin. Near Castro, on each side of the creek and rivulet of the Gamboa, three distinct terraces are seen: the lowest was estimated at about one hundred and fifty feet in height, and the highest at about five hundred feet, with the country irregularly rising behind it; obscure traces, also, of these same terraces could be seen along other parts of the coast. There can be no doubt that their three escarpments record pauses in the elevation of the island. I may remark that several promontories have the word Huapi, which signifies in the Indian tongue, island, appended to them, such as Huapilinao, Huapilacuy, Caucahuapi, etc.; and these, according to Indian traditions, once existed as islands. In the same manner the term Pulo in Sumatra is appended to the names of promontories, traditionally said to have been islands (Marsden's "Sumatra".); in Sumatra, as in Chiloe, there are upraised recent shells. The Bay of Carelmapu, on the mainland north of Chiloe, according to Aguerros, was in 1643 a good harbour ("Descripcion Hist. de la Provincia de Chiloe". From the account given by the old Spanish writers, it would appear that several other harbours, between this point and Concepcion, were formerly much deeper than they now are.); it is now quite useless, except for boats.

VALDIVIA.
I did not observe here any distinct proofs of recent elevation; but in a bed of very soft sandstone, forming a fringe-like plain, about sixty feet in height, round the hills of mica-slate, there are shells of Mytilus, Crepidula, Solen, Novaculina, and Cytheraea, too imperfect to be specifically recognised. At Imperial, seventy miles north of Valdivia, Aguerros states that there are large beds of shells, at a considerable distance from the coast, which are burnt for lime. (Ibid) The island of Mocha, lying a little north of Imperial, was uplifted two feet, during the earthquake of 1835. ("Voyages of 'Adventure' and 'Beagle'" volume 2

CONCEPCION.
I cannot add anything to the excellent account by Captain Fitzroy of the elevation of the land at this place, which accompanied the earthquake of 1835. (Ibid volume 2 et seq. In volume 5 of the "Geological Transactions" I have given an account of the remarkable volcanic phenomena, which accompanied this earthquake. These phenomena appear to me to prove that the action, by which large tracts of land are uplifted, and by which volcanic eruptions are produced, is in every respect identical.) I will only recall to the recollection of geologists, that the southern end of the island of St. Mary was uplifted eight feet, the central part nine, and the northern end ten feet; and the whole island more than the surrounding districts. Great beds of mussels, patellae, and chitons still adhering to the rocks were upraised above high-water mark; and some acres of a rocky flat, which was formerly always covered by the sea, was left standing dry, and exhaled an offensive smell, from the many attached and putrefying shells. It appears from the researches of Captain Fitzroy that both the island of St. Mary and Concepcion (which was uplifted only four or five feet) in the course of some weeks subsided, and lost part of their first elevation. I will only add as a lesson of caution, that round the sandy shores of the great Bay of Concepcion, it was most difficult, owing to the obliterating effects of the great accompanying wave, to recognise any distinct evidence of this considerable upheaval; one spot must be excepted, where there was a detached rock which before the earthquake had always been covered by the sea, but afterwards was left uncovered.
On the island of Quiriquina (in the Bay of Concepcion), I found, at an estimated height of four hundred feet, extensive layers of shells, mostly comminuted, but some perfectly preserved and closely packed in black vegetable mould; they consisted of Concholepas, Fissurella, Mytilus, Trochus, and Balanus. Some of these layers of shells rested on a thick bed of bright-red, dry, friable earth, capping the surface of the tertiary sandstone, and extending, as I observed whilst sailing along the coast, for 150 miles southward: at Valparaiso, we shall presently see that a similar red earthy mass, though quite like terrestrial mould, is really in chief part of recent marine origin. On the flanks of this island of Quiriquina, at a less height than the 400 feet, there were spaces several feet square, thickly strewed with fragments of similar shells. During a subsequent visit of the "Beagle" to Concepcion, Mr. Kent, the assistant-surgeon, was so kind as to make for me some measurements with the barometer: he found many marine remains along the shores of the whole bay, at a height of about twenty feet; and from the hill of Sentinella behind Talcahuano, at the height of 160 feet, he collected numerous shells, packed together close beneath the surface in black earth,

consisting of two species of Mytilus, two of Crepidula, one of Concholepas, of Fissurella, Venus, Mactra, Turbo, Monoceros, and the Balanus psittacus. These shells were bleached, and within some of the Balani other Balani were growing, showing that they must have long lain dead in the sea. The above species I compared with living ones from the bay, and found them identical; but having since lost the specimens, I cannot give their names: this is of little importance, as Mr. Broderip has examined a similar collection, made during Captain Beechey's expedition, and ascertained that they consisted of ten recent species, associated with fragments of Echini, crabs, and Flustrae; some of these remains were estimated by Lieutenant Belcher to lie at the height of nearly a thousand feet above the level of the sea. ("Zoology of Captain Beechey's Voyage" In some places round the bay, Mr. Kent observed that there were beds formed exclusively of the Mytilus Chiloensis: this species now lives in parts never uncovered by the tides. At considerable heights, Mr. Kent found only a few shells; but from the summit of one hill, 625 feet high, he brought me specimens of the Concholepas, Mytilus Chiloensis, and a Turbo. These shells were softer and more brittle than those from the height of 164 feet; and these latter had obviously a much more ancient appearance than the same species from the height of only twenty feet.

COAST NORTH OF CONCEPCION.

The first point examined was at the mouth of the Rapel (160 miles north of Concepcion and sixty miles south of Valparaiso), where I observed a few shells at the height of 100 feet, and some barnacles adhering to the rocks three or four feet above the highest tides: M. Gay found here recent shells at the distance of two leagues from the shore. ("Annales des Scienc. Nat." Avril 1833.) Inland there are some wide, gravel-capped plains, intersected by many broad, flat-bottomed valleys (now carrying insignificant streamlets), with their sides cut into successive wall-like escarpments, rising one above another, and in many places, according to M. Gay, worn into caves. The one cave (C. del Obispo) which I examined, resembled those formed on many sea-coasts, with its bottom filled with shingle. These inland plains, instead of sloping towards the coast, are inclined in an opposite direction towards the Cordillera, like the successively rising terraces on the inland or eastern side of Chiloe: some points of granite, which project through the plains near the coast, no doubt once formed a chain of outlying islands, on the inland shores of which the plains were accumulated. At Bucalemu, a few miles northward of the Rapel, I observed at the foot, and on the summit-edge of a plain, ten miles from the coast, many recent shells, mostly comminuted, but some perfect. There were, also, many at the bottom of the great valley of the Maypu. At San Antonio, shells are said to be collected and burnt for lime. At the bottom of a great ravine (Quebrada Onda, on the road to Casa Blanca), at the distance of several miles from the coast, I noticed a considerable bed, composed exclusively of Mesodesma donaciforme, Desh., lying on a bed of muddy sand: this shell now lives associated together in great numbers, on tidal-flats on the coast of Chile.

VALPARAISO.

During two successive years I carefully examined, part of the time in company with Mr. Alison, into all the facts connected with the recent elevation of this neighbourhood. In very many parts a beach of broken shells, about fourteen or fifteen feet above high-water mark, may be observed; and at this level the coast-rocks, where precipitous, are corroded in a band. At one spot, Mr. Alison, by removing some birds' dung, found at this same level barnacles adhering to the rocks. For several miles southward of the bay, almost every flat little headland, between the heights of 60 and 230 feet (measured by the barometer), is smoothly coated by a thick mass of comminuted shells, of the same species, and apparently in the same proportional numbers with those existing in the adjoining sea. The Concholepas is much the most abundant, and the best preserved shell; but I extracted perfectly preserved specimens of the Fissurella biradiata, a Trochus and Balanus (both well-known, but according to Mr. Sowerby yet unnamed) and parts of the Mytilus Chiloensis. Most of these shells, as well as an encrusting Nullipora, partially retain their colour; but they are brittle, and often stained red from the underlying brecciated mass of primary rocks; some are packed together, either in black or reddish moulds; some lie loose on the bare rocky surfaces. The total number of these shells is immense; they are less numerous, though still far from rare, up a height of 1,000 feet above the sea. On the summit of a hill, measured 557 feet, there was a small horizontal band of comminuted shells, of which MANY consisted (and likewise from lesser heights) of very young and small specimens of the still living Concholepas, Trochus, Patellae, Crepidulae, and of Mytilus Magellanicus (?) (Mr. Cuming informs me that he does not think this species identical with, though closely resembling, the true M. Magellanicus of the southern and eastern coast of South America; it lives abundantly on the coast of Chile.): several of these shells were under a quarter of an inch in their greatest diameter. My attention was called to this circumstance by a native fisherman, whom I took to look at these shell-beds; and he ridiculed the notion of such small shells having been brought up for food; nor could some of the species have adhered when alive to other larger shells. On another hill, some miles distant, and 648 feet high, I found shells of the Concholepas and Trochus, perfect, though very old, with fragments of Mytilus Chiloensis, all embedded in reddish-brown mould: I also found these same species, with fragments of an Echinus and of Balanus psittacus, on a hill 1,000 feet high. Above this height, shells became very rare, though on a hill 1,300 feet high (Measured by the barometer: the highest point in the range behind Valparaiso I found to be 1,626 feet above the level of the sea.), I collected the Concholepas, Trochus, Fissurella, and a Patella. At these greater heights the shells are almost invariably embedded in mould, and sometimes are exposed only by tearing up bushes. These shells obviously had a very much more ancient appearance than those from the lesser heights; the apices of the Trochi were often worn down; the little holes made by burrowing animals were greatly enlarged; and the Concholepas was often perforated quite through, owing to the inner plates of shell having scaled off.

Many of these shells, as I have said, were packed in, and were quite filled with, blackish or reddish-brown earth, resting on the granitic detritus. I did not doubt until lately that this mould was of purely terrestrial origin, when with a microscope examining some of it from the inside of a Concholepas from the height of about one hundred feet, I found that it was in considerable part composed of minute fragments of the spines, mouth-bones, and shells of Echini, and of minute fragments, of chiefly very young Patellae, Mytili, and other species. I found similar microscopical fragments in earth filling up the central orifices of some large Fissurellae. This earth when crushed emits a sickly smell, precisely like that from garden-mould mixed with guano. The earth accidentally preserved within the shells, from the greater heights, has the same general appearance, but it is a little redder; it emits the same smell when rubbed, but I was unable to detect with certainty any marine remains in it. This earth resembles in general appearance, as before remarked, that capping the rocks of Quiriquina in the Bay of Concepcion, on which beds of sea-shells lay. I have, also, shown that the black, peaty soil, in which the shells at the height of 350 feet at Chiloe were packed, contained many minute fragments of marine animals. These facts appear to me interesting, as they show that soils, which would naturally be considered of purely terrestrial nature, may owe their origin in chief part to the sea.

Being well aware from what I have seen at Chiloe and in Tierra del Fuego, that vast quantities of shells are carried, during successive ages, far inland, where the inhabitants chiefly subsist on these productions, I am bound to state that at greater heights than 557 feet, where the number of very young and small shells proved that they had not been carried up for food, the only evidence of the shells having been naturally left by the sea, consists in their invariable and uniform appearance of extreme antiquity—in the distance of some of the places from the coast, in others being inaccessible from the nearest part of the beach, and in the absence of fresh water for men to drink—in the shells NOT LYING IN HEAPS,—and, lastly, in the close similarity of the soil in which they are embedded, to that which lower down can be unequivocally shown to be in great part formed from the debris of the sea animals. (In the "Proceedings of the Geological Society" volume 2, I have given a brief account of the upraised shells on the coast of Chile, and have there stated that the proofs of elevation are not satisfactory above the height of 230 feet. I had at that time unfortunately overlooked a separate page written during my second visit to Valparaiso, describing the shells now in my possession from the 557 feet hill; I had not then unpacked my collections, and had not reconsidered the obvious appearance of greater antiquity of the shells from the greater heights, nor had I at that time discovered the marine origin of the earth in which many of the shells are packed. Considering these facts, I do not now feel a shadow of doubt that the shells, at the height of 1,300 feet, have been upraised by natural causes into their present position.)

With respect to the position in which the shells lie, I was repeatedly struck here, at Concepcion, and at other places, with the frequency of their occurrence on the summits and edges either of separate hills, or of little flat headlands often terminating precipitously over the sea. The several above-enumerated species of mollusca, which are found strewed on the surface of the land from a few feet above the level of the sea up to the height of 1,300 feet, all now live either on the beach, or at only a few fathoms' depth: Mr. Edmondston, in a letter to Professor E. Forbes, states that in dredging in the Bay of Valparaiso, he found the common species of Concholepas, Fissurella, Trochus, Monoceros, Chitons, etc., living in abundance from the beach to a depth of seven fathoms; and dead shells occurred only a few fathoms deeper. The common Turritella cingulata was dredged up living at even from ten to fifteen fathoms; but this is a species which I did not find here amongst the upraised shells. Considering this fact of the species being all littoral or sub-littoral, considering their occurrence at various heights, their vast numbers, and their generally comminuted state, there can be little doubt that they were left on successive beach-lines during a gradual elevation of the land. The presence, however, of so many whole and perfectly preserved shells appears at first a difficulty on this view, considering that the coast is exposed to the full force of an open ocean: but we may suppose, either that these shells were thrown during gales on flat ledges of rock just above the level of high-water mark, and that during the elevation of the land they are never again touched by the waves, or, that during earthquakes, such as those of 1822, 1835, and 1837, rocky reefs covered with marine-animals were it one blow uplifted above the future reach of the sea. This latter explanation is, perhaps, the most probable one with respect to the beds at Concepcion entirely composed of the Mytilus Chiloensis, a species which lives below the lowest tides; and likewise with respect to the great beds occurring both north and south of Valparaiso, of the Mesodesma donaciforme,—a shell which, as I am informed by Mr. Cuming, inhabits sandbanks at the level of the lowest tides. But even in the case of shells having the habits of this Mytilus and Mesodesma, beds of them, wherever the sea gently throws up sand or mud, and thus protects its own accumulations, might be upraised by the slowest movement, and yet remain undisturbed by the waves of each new beach-line.

It is worthy of remark, that nowhere near Valparaiso above the height of twenty feet, or rarely of fifty feet, I saw any lines of erosion on the solid rocks, or any beds of pebbles; this, I believe, may be accounted for by the disintegrating tendency of most of the rocks in this neighbourhood. Nor is the land here modelled into terraces: Mr. Alison, however, informs me, that on both sides of one narrow ravine, at the height of 300 feet above the sea, he found a succession of rather indistinct step-formed beaches, composed of broken shells, which together covered a space of about eighty feet vertical.

I can add nothing to the accounts already published of the elevation of the land at Valparaiso, which accompanied the earthquake of 1822 (Dr. Meyen "Reise um Erde" Th. 1 s. 221, found in 1831 seaweed and other bodies still adhering to some rocks which during the shock of 1822 were lifted above the sea.): but I heard it confidently asserted, that a sentinel on duty, immediately after the shock, saw a part of a fort, which previously was not within the line of his vision, and this would indicate that the uplifting was not horizontal: it would even appear from some facts collected by Mr. Alison, that only the eastern half of the bay was then elevated. Through the kindness of this same gentleman, I am able to give an interesting account of the changes of level, which have supervened here within historical periods: about the year 1680 a long sea-wall (or Prefil) was built, of which only a few fragments now remain; up to the year 1817, the sea often broke over it, and washed the houses on the opposite side of the road (where the prison now stands); and even in 1819, Mr. J. Martin remembers walking at the foot of this wall, and being often obliged to climb over it to escape the waves. There now stands (1834) on the seaward side of this wall, and between it and the beach, in one part a single row of houses, and in another part two rows with a street between them. This great extension of the beach in so short a time cannot be attributed simply to the accumulation of detritus; for a resident engineer measured for me the height between the lowest part of the wall visible, and the present beach-line at spring-tides, and the difference was eleven feet six inches. The church of S. Augustin is believed to have been built in 1614, and there is a tradition that the sea formerly flowed very near it; by levelling, its foundations were found to stand nineteen feet six inches above the highest beach-line; so that we see in a period of 220 years, the elevation cannot have been as much as nineteen feet six inches. From the facts given with respect to the sea-wall, and from the testimony of the elder inhabitants, it appears certain that the change in level began to be manifest about the year 1817. The only sudden elevation of which there is any record occurred in 1822, and this seems to have been less than three feet. Since that year, I was assured by several competent observers, that part of an old wreck, which is firmly embedded near the beach, has sensibly emerged; hence here, as at Chiloe, a slow rise of the land appears to be now in progress. It seems highly probable that the rocks which are corroded in a band at the height of fourteen feet above the sea were acted on during the period, when by tradition the base of S. Augustin church, now nineteen feet six inches above the highest water-mark, was occasionally washed by the waves.

VALPARAISO TO COQUIMBO.

For the first seventy-five miles north of Valparaiso I followed the coast-road, and throughout this space I observed innumerable masses of upraised shells. About Quintero there are immense accumulations (worked for lime) of the Mesodesma donaciforme, packed in sandy earth; they abound chiefly about fifteen feet above high-water, but shells are here found, according to Mr. Miers, to a height of 500 feet, and at a distance of three leagues from the coast ("Travels in Chile" volume 1. I received several similar accounts from the inhabitants, and was assured that there are many shells on the plain of Casa Blanca, between Valparaiso and Santiago, at the height of 800 feet.): I here noticed barnacles adhering to the rocks three or four feet above the highest tides. In the neighbourhood of Plazilla and Catapilco, at heights of between two hundred and three hundred feet, the number of comminuted shells, with some perfect ones, especially of the Mesodesma, packed in layers, was truly immense: the land at Plazilla had evidently existed as a bay, with abrupt rocky masses rising out of it, precisely like the islets in the broken bays now indenting this coast. On both sides of the rivers Ligua, Longotomo, Guachen, and Quilimari, there are plains of gravel about two hundred feet in height, in many parts absolutely covered with shells. Close to Conchalee, a gravel-plain is fronted by a lower and similar plain about sixty feet in height, and this again is separated from the beach by a wide tract of low land: the surfaces of all three plains or terraces were strewed with vast numbers of the Concholepas, Mesodesma, an existing Venus, and other still existing littoral shells. The two upper terraces closely resemble in miniature the plains of Patagonia; and like them are furrowed by dry, flat-bottomed, winding valleys. Northward of this place I turned inward; and therefore found no more shells: but the valleys of Chuapa, Illapel, and Limari, are bounded by gravel-capped plains, often including a lower terrace within. These plains send bay-like arms between and into the surrounding hills; and they are continuously united with other extensive gravel-capped plains, separating the coast mountain-ranges from the Cordillera.

COQUIMBO.

A narrow fringe-like plain, gently inclined towards the sea, here extends for eleven miles along the coast, with arms stretching up between the coast-mountains, and likewise up the valley of Coquimbo: at its southern extremity it is directly connected with the plain of Limari, out of which hills abruptly rise like islets, and other hills project like headlands on a coast. The surface of the fringe-like plain appears level, but differs insensibly in height, and greatly in composition, in different parts.

At the mouth of the valley of Coquimbo, the surface consists wholly of gravel, and stands from 300 to 350 feet above the level of the sea, being about one hundred feet higher than in other parts. In these other and lower parts the superficial beds consist of calcareous matter, and rest on ancient tertiary deposits hereafter to be described. The uppermost calcareous layer is cream-coloured, compact, smooth-fractured, sub- stalactiform, and contains some sand, earthy matter, and recent shells. It lies on, and sends wedge-like veins into, a much more friable, calcareous, tuff-like variety; and both rest on a mass about twenty feet in thickness, formed of fragments of recent shells, with a few whole ones, and with small pebbles firmly cemented together. (In many respects this upper hard, and the underlying more friable, varieties, resemble the great superficial beds at King George's Sound in Australia, which I have described in my "Geological Observations on Volcanic Islands." There could be little doubt that the upper layers there have been hardened by the action of rain on the friable, calcareous matter, and that the whole mass has originated in the decay of minutely comminuted sea-shells and corals.) This latter rock is called by the inhabitants losa, and is used for building: in many parts it is divided into strata, which dip at an angle of ten degrees seaward, and appear as if they had originally been heaped in successive layers (as may be seen on coral-reefs) on a steep beach. This stone is remarkable from being in parts entirely formed of empty, pellucid capsules or cells of calcareous matter, of the size of small seeds: a series of specimens unequivocally showed that all these capsules once contained minute rounded fragments of shells which have since been gradually dissolved by water percolating through the mass. (I have incidentally described this rock in the above work on Volcanic Islands.)

The shells embedded in the calcareous beds forming the surface of this fringe-like plain, at the height of from 200 to 250 feet above the sea, consist of:—

1. Venus opaca. 2. Mulinia Byronensis. 3. Pecten purpuratus. 4. Mesodesma donaciforme. 5. Turritella cingulata. 6. Monoceros costatum. 7. Concholepas Peruviana. 8. Trochus (common Valparaiso species). 9. Calyptraea Byronensis.

Although these species are all recent, and are all found in the neighbouring sea, yet I was particularly struck with the difference in the proportional numbers of the several species, and of those now cast up on the present beach. I found only one specimen of the Concholepas, and the Pecten was very rare, though both these shells are now the commonest kinds, with the exception, perhaps, of the Calyptraea radians, of which I did not find one in the calcareous beds. I will not pretend to determine how far this difference in the proportional numbers depends on the age of the deposit, and how far on the difference in nature between the present sandy beaches and the calcareous bottom, on which the embedded shells must have lived.

(DIAGRAM 8.—SECTION OF PLAIN OF COQUIMBO.
Section through Plain B-B and Ravine A.
Surface of plain 252 feet above sea.
A. Stratified sand, with recent shells in same proportions as on the beach, half filling up a ravine.
B. Surface of plain, with scattered shells in nearly same proportions as on the beach.
C. Upper calcareous bed, and D. Lower calcareous sandy bed (Losa), both with recent shells, but not in same proportions as on the beach.
E. Upper ferrugino-sandy old tertiary stratum, and F. Lower old tertiary stratum, both with all, or nearly all, extinct shells.)

On the bare surface of the calcareous plain, or in a thin covering of sand, there were lying, at a height from 200 to 252 feet, many recent shells, which had a much fresher appearance than the embedded ones: fragments of the Concholepas, and of the common Mytilus, still retaining a tinge of its colour, were numerous, and altogether there was manifestly a closer approach in proportional numbers to those now lying on the beach. In a mass of stratified, slightly agglutinated sand, which in some places covers up the lower half of the seaward escarpment of the plain, the included shells appeared to be in exactly the same proportional numbers with those on the beach. On one side of a steep-sided ravine, cutting through the plain behind Herradura Bay, I observed a narrow strip of stratified sand, containing similar shells in similar proportional numbers; a section of the ravine is represented in Diagram 8, which serves also to show the general composition of the plain. I mention this case of the ravine chiefly because without the evidence of the marine shells in the sand, any one would have supposed that it had been hollowed out by simple alluvial action.

The escarpment of the fringe-like plain, which stretches for eleven miles along the coast, is in some parts fronted by two or three narrow, step- formed terraces, one of which at Herradura Bay expands into a small plain. Its surface was there formed of gravel, cemented together by calcareous matter; and out of it I extracted the following recent shells, which are in a more perfect condition than those from the upper plain:—

1. Calyptraea radians. 2. Turritella cingulata. 3. Oliva Peruviana. 4. Murex labiosus, var. 5. Nassa (identical with a living species). 6. Solen Dombeiana. 7. Pecten purpuratus. 8. Venus Chilensis. 9. Amphidesma rugulosum. The small irregular wrinkles of the posterior part of this shell are rather stronger than in the recent specimens of this species from Coquimbo. (G.B. Sowerby.) 10. Balanus (identical with living species).

On the syenitic ridge, which forms the southern boundary of Herradura Bay and Plain, I found the Concholepas and Turritella cingulata (mostly in fragments), at the height of 242 feet above the sea. I could not have told that these shells had not formerly been brought up by man, if I had not found one very small mass of them cemented together in a friable calcareous tuff. I mention this fact more particularly, because I carefully looked, in many apparently favourable spots, at lesser heights on the side of this ridge, and could not find even the smallest fragment of a shell. This is only one instance out of many, proving that the absence of sea-shells on the surface, though in many respects inexplicable, is an argument of very little weight in opposition to other evidence on the recent elevation of the land. The highest point in this neighbourhood at which I found upraised shells of existing species was on an inland calcareous plain, at the height of 252 feet above the sea.

It would appear from Mr. Caldcleugh's researches, that a rise has taken place here within the last century and a half ("Proceedings of the Geological Society" volume 2.); and as no sudden change of level has been observed during the not very severe earthquakes, which have occasionally occurred here, the rising has probably been slow, like that now, or quite lately, in progress at Chiloe and at Valparaiso: there are three well-known rocks, called the Pelicans, which in 1710, according to Feuillee, were a fleur d'eau, but now are said to stand twelve feet above low-water mark: the spring-tides rise here only five feet. There is another rock, now nine feet above high-water mark, which in the time of Frezier and Feuillee rose only five or six feet out of water. Mr. Caldcleugh, I may add, also shows (and I received similar accounts) that there has been a considerable decrease in the soundings during the last twelve years in the Bays of Coquimbo, Concepcion, Valparaiso, and Guasco; but as in these cases it is nearly impossible to distinguish between the accumulation of sediment and the upheavement of the bottom, I have not entered into any details.

VALLEY OF COQUIMBO.

(FIGURE 9. EAST AND WEST SECTION THROUGH THE TERRACES AT COQUIMBO, WHERE THEY DEBOUCH FROM THE VALLEY, AND FRONT THE SEA.

Vertical scale 1/10 of inch to 100 feet: horizontal scale much contracted.

Height of terrace in feet from east (high) to west (low):
Terrace F. 364
Terrace E. 302
Terrace D. shown dotted, height not given.
Terrace C. 120
Terrace B. 70
Terrace A. 25 sloping down to level of sea at Town of Coquimbo.)

The narrow coast-plain sends, as before stated, an arm, or more correctly a fringe, on both sides, but chiefly on the southern side, several miles up the valley. These fringes are worn into steps or terraces, which present a most remarkable appearance, and have been compared (though not very correctly) by Captain Basil Hall, to the parallel roads of Glen Roy in Scotland: their origin has been ably discussed by Mr. Lyell. ("Principles of Geology" 1st edition volume 3.) The first section which I will give (Figure 9), is not drawn across the valley, but in an east and west line at its mouth, where the step-formed terraces debouch and present their very gently inclined surfaces towards the Pacific.

The bottom plain (A) is about a mile in width, and rises quite insensibly from the beach to a height of twenty-five feet at the foot of the next plain; it is sandy, and abundantly strewed with shells.

Plain or terrace B is of small extent, and is almost concealed by the houses of the town, as is likewise the escarpment of terrace C. On both sides of a ravine, two miles south of the town, there are two little terraces, one above the other, evidently corresponding with B and C; and on them marine remains of the species already enumerated were plentiful. Terrace E is very narrow, but quite distinct and level; a little southward of the town there were traces of a terrace D intermediate between E and C. Terrace F is part of the fringe-like plain, which stretches for the eleven miles along the coast; it is here composed of shingle, and is 100 feet higher than where composed of calcareous matter. This greater height is obviously due to the quantity of shingle, which at some former period has been brought down the great valley of Coquimbo.

Considering the many shells strewed over the terraces A, B, and C, and a few miles southward on the calcareous plain, which is continuously united with the upper step-like plain F, there cannot, I apprehend, be any doubt, that these six terraces have been formed by the action of the sea; and that their five escarpments mark so many periods of comparative rest in the elevatory movement, during which the sea wore into the land. The elevation between these periods may have been sudden and on AN AVERAGE not more than seventy-two feet each time, or it may have been gradual and insensibly slow. From the shells on the three lower terraces, and on the upper one, and I may add on the three gravel-capped terraces at Conchalee, being all littoral and sub-littoral species, and from the analogical facts given at Valparaiso, and lastly from the evidence of a slow rising lately or still in progress here, it appears to me far more probable that the movement has been slow. The existence of these successive escarpments, or old cliff-lines, is in another respect highly instructive, for they show periods of comparative rest in the elevatory movement, and of denudation, which would never even have been suspected from a close examination of many miles of coast southward of Coquimbo.

(FIGURE 10. NORTH AND SOUTH SECTION ACROSS THE VALLEY OF COQUIMBO.
From north F (high) through E?, D, C, B, A (low), B?, C, D?, E, F (high).
Vertical scale 1/10 of inch to 100 feet: horizontal scale much contracted.

Terraces marked with ? do not occur on that side of the valley, and are introduced only to make the diagram more intelligible. A river and bottom-plain of valley C, E, and F, on the south side of valley, are respectively, 197, 377, and 420 feet above the level of the sea.

AA. The bottom of the valley, believed to be 100 feet above the sea: it is continuously united with the lowest plain A of Figure 9.

B. This terrace higher up the valley expands considerably; seaward it is soon lost, its escarpment being united with that of C: it is not developed at all on the south side of the valley.

C. This terrace, like the last, is considerably expanded higher up the valley. These two terraces apparently correspond with B and C of Figure 9.

D is not well developed in the line of this section; but seaward it expands into a plain: it is not present on the south side of the valley; but it is met with, as stated under the former section, a little south of the town.

E is well developed on the south side, but absent on the north side of the valley: though not continuously united with E of Figure 9, it apparently corresponds with it.

F. This is the surface-plain, and is continuously united with that which stretches like a fringe along the coast. In ascending the valley it gradually becomes narrower, and is at last, at the distance of about ten miles from the sea, reduced to a row of flat-topped patches on the sides of the mountains. None of the lower terraces extend so far up the valley.)

We come now to the terraces on the opposite sides of the east and west valley of Coquimbo: the section in Figure 10 is taken in a north and south line across the valley at a point about three miles from the sea. The valley measured from the edges of the escarpments of the upper plain FF is about a mile in width; but from the bases of the bounding mountains it is from three to four miles wide. The terraces marked with an interrogative do not exist on that side of the valley, but are introduced merely to render the diagram more intelligible.

These five terraces are formed of shingle and sand; three of them, as marked by Captain B. Hall (namely, B, C, and F), are much more conspicuous than the others. From the marine remains copiously strewed at the mouth of the valley on the lower terraces, and southward of the town on the upper one, they are, as before remarked, undoubtedly of marine origin; but within the valley, and this fact well deserves notice, at a distance of from only a mile and a half to three or four miles from the sea, I could not find even a fragment of a shell.

ON THE INCLINATION OF THE TERRACES OF COQUIMBO, AND ON THE UPPER AND BASAL EDGES OF THEIR ESCARPMENTS NOT BEING HORIZONTAL.

The surfaces of these terraces slope in a slight degree, as shown by the sections in Figures 9 and 10 taken conjointly, both towards the centre of the valley, and seawards towards its mouth. This double or diagonal inclination, which is not the same in the several terraces, is, as we shall immediately see, of simple explanation. There are, however, some other points which at first appear by no means obvious,—namely, first, that each terrace, taken in its whole breadth from the summit-edge of one escarpment to the base of that above it, and followed up the valley, is not horizontal; nor have the several terraces, when followed up the valley, all the same inclination; thus I found the terraces C, E, and F, measured at a point about two miles from the mouth of the valley, stood severally between fifty-six to seventy-seven feet higher than at the mouth. Again, if we look to any one line of cliff or escarpment, neither its summit-edge nor its base is horizontal. On the theory of the terraces having been formed during a slow and equable rise of the land, with as many intervals of rest as there are escarpments, it appears at first very surprising that horizontal lines of some kind should not have been left on the land.

The direction of the diagonal inclination in the different terraces being different,—in some being directed more towards the middle of the valley, in others more towards its mouth,—naturally follows on the view of each terrace, being an accumulation of successive beach-lines round bays, which must have been of different forms and sizes when the land stood at different levels: for if we look to the actual beach of a narrow creek, its slope is directed towards the middle; whereas, in an open bay, or slight concavity on a coast, the slope is towards the mouth, that is, almost directly seaward; hence as a bay alters in form and size, so will the direction of the inclination of its successive beaches become changed.

(FIGURE 11. DIAGRAM OF A BAY IN A DISTRICT WHICH HAS BEGUN SLOWLY RISING)

If it were possible to trace any one of the many beach-lines, composing each sloping terrace, it would of course be horizontal; but the only lines of demarcation are the summit and basal edges of the escarpments. Now the summit-edge of one of these escarpments marks the furthest line or point to which the sea has cut into a mass of gravel sloping seaward; and as the sea will generally have greater power at the mouth than at the protected head of the bay, so will the escarpment at the mouth be cut deeper into the land, and its summit-edge be higher; consequently it will not be horizontal. With respect to the basal or lower edges of the escarpments, from picturing in one's mind ancient bays ENTIRELY surrounded at successive periods by cliff-formed shores, one's first impression is that they at least necessarily must be horizontal, if the elevation has been horizontal. But here is a fallacy: for after the sea has, during a cessation of the elevation, worn cliffs all round the shores of a bay, when the movement recommences, and especially if it recommences slowly, it might well happen that, at the exposed mouth of the bay, the waves might continue for some time wearing into the land, whilst in the protected and upper parts successive beach-lines might be accumulating in a sloping surface or terrace at the foot of the cliffs which had been lately reached: hence, supposing the whole line of escarpment to be finally uplifted above the reach of the sea, its basal line or foot near the mouth will run at a lower level than in the upper and protected parts of the bay; consequently this basal line will not be horizontal. And it has already been shown that the summit-edges of each escarpment will generally be higher near the mouth (from the seaward sloping land being there most exposed and cut into) than near the head of the bay; therefore the total height of the escarpments will be greatest near the mouth; and further up the old bay or valley they will on both sides generally thin out and die away: I have observed this thinning out of the successive escarpment at other places besides Coquimbo; and for a long time I was quite unable to understand its meaning. The rude diagram in Figure 11 will perhaps render what I mean more intelligible; it represents a bay in a district which has begun slowly rising. Before the movement commenced, it is supposed that the waves had been enabled to eat into the land and form cliffs, as far up, but with gradually diminishing power, as the points AA: after the movement had commenced and gone on for a little time, the sea is supposed still to have retained the power, at the exposed mouth of the bay, of cutting down and into the land as it slowly emerged; but in the upper parts of the bay it is supposed soon to have lost this power, owing to the more protected situation and to the quantity of detritus brought down by the river; consequently low land was there accumulated. As this low land was formed during a slow elevatory movement, its surface will gently slope upwards from the beach on all sides. Now, let us imagine the bay, not to make the diagram more complicated, suddenly converted into a valley: the basal line of the cliffs will of course be horizontal, as far as the beach is now seen extending in the diagram; but in the upper part of the valley, this line will be higher, the level of the district having been raised whilst the low land was accumulating at the foot of the inland cliffs. If, instead of the bay in the diagram being suddenly converted into a valley, we suppose with much more probability it to be upraised slowly, then the waves in the upper parts of the bay will continue very gradually to fail to reach the cliffs, which are now in the diagram represented as washed by the sea, and which, consequently, will be left standing higher and higher above its level; whilst at the still exposed mouth, it might well happen that the waves might be enabled to cut deeper and deeper, both down and into the cliffs, as the land slowly rose.

The greater or lesser destroying power of the waves at the mouths of successive bays, comparatively with this same power in their upper and protected parts, will vary as the bays become changed in form and size, and therefore at different levels, at their mouths and heads, more or less of the surfaces between the escarpments (that is, the accumulated beach-lines or terraces) will be left undestroyed: from what has gone before we can see that, according as the elevatory movements after each cessation recommence more or less slowly, according to the amount of detritus delivered by the river at the heads of the successive bays, and according to the degree of protection afforded by their altered forms, so will a greater or less extent of terrace be accumulated in the upper part, to which there will be no surface at a corresponding level at the mouth: hence we can perceive why no one terrace, taken in its whole breadth and followed up the valley, is horizontal, though each separate beach-line must have been so; and why the inclination of the several terraces, both transversely, and longitudinally up the valley, is not alike.

I have entered into this case in some detail, for I was long perplexed (and others have felt the same difficulty) in understanding how, on the idea of an equable elevation with the sea at intervals eating into the land, it came that neither the terraces nor the upper nor lower edges of the escarpments were horizontal. Along lines of coast, even of great lengths, such as that of Patagonia, if they are nearly uniformly exposed, the corroding power of the waves will be checked and conquered by the elevatory movement, as often as it recommences, at about the same period; and hence the terraces, or accumulated beach-lines, will commence being formed at nearly the same levels: at each succeeding period of rest, they will, also, be eaten into at nearly the same rate, and consequently there will be a much closer coincidence in their levels and inclinations, than in the terraces and escarpments formed round bays with their different parts very differently exposed to the action of the sea. It is only where the waves are enabled, after a long lapse of time, slowly to corrode hard rocks, or to throw up, owing to the supply of sediment being small and to the surface being steeply inclined, a narrow beach or mound, that we can expect, as at Glen Roy in Scotland ("Philosophical Transactions" 1839 a distinct line marking an old sea-level, and which will be strictly horizontal, if the subsequent elevatory movements have been so: for in these cases no discernible effects will be produced, except during the long intervening periods of rest; whereas in the case of step-formed coasts, such as those described in this and the preceding chapter, the terraces themselves are accumulated during the slow elevatory process, the accumulation commencing sooner in protected than in exposed situations, and sooner where there is copious supply of detritus than where there is little; on the other hand, the steps or escarpments are formed during the stationary periods, and are more deeply cut down and into the coast-land in exposed than in protected situations;—the cutting action, moreover, being prolonged in the most exposed parts, both during the beginning and ending, if slow, of the upward movement.

Although in the foregoing discussion I have assumed the elevation to have been horizontal, it may be suspected, from the considerable seaward slope of the terraces, both up the valley of S. Cruz and up that of Coquimbo, that the rising has been greater inland than nearer the coast. There is reason to believe (Mr. Place in the "Quarterly Journal of Science" 1824 volume 17, from the effects produced on the water-course of a mill during the earthquake of 1822 in Chile, that the upheaval one mile inland was nearly double, namely, between five and seven feet, to what it was on the Pacific. We know, also, from the admirable researches of M. Bravais, that in Scandinavia the ancient sea-beaches gently slope from the interior mountain-ranges towards the coast, and that they are not parallel one to the other ("Voyages de la Comm. du Nord" etc. also "Comptes Rendus" October 1842.), showing that the proportional difference in the amount of elevation on the coast and in the interior, varied at different periods.

COQUIMBO TO GUASCO.

In this distance of ninety miles, I found in almost every part marine shells up to a height of apparently from two hundred to three hundred feet. The desert plain near Choros is thus covered; it is bounded by the escarpment of a higher plain, consisting of pale-coloured, earthy, calcareous stone, like that of Coquimbo, with the same recent shells embedded in it. In the valley of Chaneral, a similar bed occurs in which, differently from that of Coquimbo, I observed many shells of the Concholepas: near Guasco the same calcareous bed is likewise met with.

In the valley of Guasco, the step-formed terraces of gravel are displaced in a more striking manner than at any other point. I followed the valley for thirty-seven miles (as reckoned by the inhabitants) from the coast to Ballenar; in nearly the whole of this distance, five grand terraces, running at corresponding heights on both sides of the broad valley, are more conspicuous than the three best-developed ones at Coquimbo. They give to the landscape the most singular and formal aspect; and when the clouds hung low, hiding the neighbouring mountains, the valley resembled in the most striking manner that of Santa Cruz. The whole thickness of these terraces or plains seems composed of gravel, rather firmly aggregated together, with occasional parting seams of clay: the pebbles on the upper plain are often whitewashed with an aluminous substance, as in Patagonia. Near the coast I observed many sea-shells on the lower plains. At Freyrina (twelve miles up the valley), there are six terraces beside the bottom- surface of the valley: the two lower ones are here only from two hundred to three hundred yards in width, but higher up the valley they expand into plains; the third terrace is generally narrow; the fourth I saw only in one place, but there it was distinct for the length of a mile; the fifth is very broad; the sixth is the summit-plain, which expands inland into a great basin. Not having a barometer with me, I did not ascertain the height of these plains, but they appeared considerably higher than those at Coquimbo. Their width varies much, sometimes being very broad, and sometimes contracting into mere fringes of separate flat-topped projections, and then quite disappearing: at the one spot, where the fourth terrace was visible, the whole six terraces were cut off for a short space by one single bold escarpment. Near Ballenar (thirty-seven miles from the mouth of the river), the valley between the summit-edges of the highest escarpments is several miles in width, and the five terraces on both sides are broadly developed: the highest cannot be less than six hundred feet above the bed of the river, which itself must, I conceive, be some hundred feet above the sea.

A north and south section across the valley in this part is represented in
Figure 12.

(FIGURE 12. NORTH AND SOUTH SECTION ACROSS THE VALLEY OF GUASCO, AND OF A PLAIN NORTH OF IT.

From left (north, high) to right (south, high) through plains B and A and the River of Guasco at the Town of Ballenar.)

On the northern side of the valley the summit-plain of gravel, A, has two escarpments, one facing the valley, and the other a great basin-like plain, B, which stretches for several leagues northward. This narrow plain, A, with the double escarpment, evidently once formed a spit or promontory of gravel, projecting into and dividing two great bays, and subsequently was worn on both sides into steep cliffs. Whether the several escarpments in this valley were formed during the same stationary periods with those of Coquimbo, I will not pretend to conjecture; but if so the intervening and subsequent elevatory movements must have been here much more energetic, for these plains certainly stand at a much higher level than do those of Coquimbo.

COPIAPO.

From Guasco to Copiapo, I followed the road near the foot of the Cordillera, and therefore saw no upraised remains. At the mouth, however, of the valley of Copiapo there is a plain, estimated by Meyen ("Reise um die Erde" th. 1 s. 372 et seq.) between fifty and seventy feet in height, of which the upper part consists chiefly of gravel, abounding with recent shells, chiefly of the Concholepas, Venus Dombeyi, and Calyptraea trochiformis. A little inland, on a plain estimated by myself at nearly three hundred feet, the upper stratum was formed of broken shells and sand cemented by white calcareous matter, and abounding with embedded recent shells, of which the Mulinia Byronensis and Pecten purpuratus were the most numerous. The lower plain stretches for some miles southward, and for an unknown distance northward, but not far up the valley; its seaward face, according to Meyen, is worn into caves above the level of the present beach. The valley of Copiapo is much less steeply inclined and less direct in its course than any other valley which I saw in Chile; and its bottom does not generally consist of gravel: there are no step-formed terraces in it, except at one spot near the mouth of the great lateral valley of the Despoblado where there are only two, one above the other: lower down the valley, in one place I observed that the solid rock had been cut into the shape of a beach, and was smoothed over with shingle.

Northward of Copiapo, in latitude 26 degrees S., the old voyager Wafer found immense numbers of sea-shells some miles from the coast. (Burnett's "Collection of Voyages" volume 4 At Cobija (latitude 22 degrees 34') M. d'Orbigny observed beds of gravel and broken shells, containing ten species of recent shells; he also found, on projecting points of porphyry, at a height of 300 feet, shells of Concholepas, Chiton, Calyptraea, Fissurella, and Patella, still attached to the spots on which they had lived. M. d'Orbigny argues from this fact, that the elevation must have been great and sudden ("Voyage, Part Geolog.". M. d'Orbigny , in summing up, says: "S'il est certain (as he believes) que tous les terrains en pente, compris entre la mer et les montagnes sont l'ancien rivage de la mer, on doit supposer, pour l'ensemble, un exhaussement que ce ne serait pas moindre de deux cent metres; il faudrait supposer encore que ce soulevement n'a point ete graduel;...mais qu'il resulterait d'une seule et meme cause fortuite," etc. Now, on this view, when the sea was forming the beach at the foot of the mountains, many shells of Concholepas, Chiton, Calyptraea, Fissurella, and Patella (which are known to live close to the beach), were attached to rocks at a depth of 300 feet, and at a depth of 600 feet several of these same shells were accumulating in great numbers in horizontal beds. From what I have myself seen in dredging, I believe this to be improbable in the highest degree, if not impossible; and I think everyone who has read Professor E. Forbes's excellent researches on the subject, will without hesitation agree in this conclusion.): to me it appears far more probable that the movement was gradual, with small starts as during the earthquakes of 1822 and 1835, by which whole beds of shells attached to the rocks were lifted above the subsequent reach of the waves. M. d'Orbigny also found rolled pebbles extending up the mountain to a height of at least six hundred feet. At Iquique (latitude 20 degrees 12' S.), in a great accumulation of sand, at a height estimated between one hundred and fifty and two hundred feet, I observed many large sea-shells which I thought could not have been blown up by the wind to that height. Mr. J.H. Blake has lately described these shells: he states that "inland toward the mountains they form a compact uniform bed, scarcely a trace of the original shells being discernible; but as we approach the shore, the forms become gradually more distinct till we meet with the living shells on the coast." ("Silliman's American Journal of Science" volume 44.) This interesting observation, showing by the gradual decay of the shells how slowly and gradually the coast must have been uplifted, we shall presently see fully confirmed at Lima. At Arica (latitude 18 degrees 28'), M. d'Orbigny found a great range of sand-dunes, fourteen leagues in length, stretching towards Tacna, including recent shells and bones of Cetacea, and reaching up to a height of 300 feet above the sea. ("Voyage" etc.) Lieutenant Freyer has given some more precise facts: he states (In a letter to Mr. Lyell "Geological Proceedings" volume 2.) that the Morro of Arica is about four hundred feet high; it is worn into obscure terraces, on the bare rock of which he found Balini and Milleporae adhering. At the height of between twenty and thirty feet the shells and corals were in a quite fresh state, but at fifty feet they were much abraded; there were, however, traces of organic remains at greater heights. On the road from Tacna to Arequipa, between Loquimbo and Moquegua, Mr. M. Hamilton found numerous recent sea shells in sand, at a considerable distance from the sea. ("Edinburgh New Philosophical Journal" volume 30)

LIMA.

Northward of Arica, I know nothing of the coast for about a space of five degrees of latitude; but near Callao, the port of Lima, there is abundant and very curious evidence of the elevation of the land. The island of San Lorenzo is upwards of one thousand feet high; the basset edges of the strata composing the lower part are worn into three obscure, narrow, sloping steps or ledges, which can be seen only when standing on them: they probably resemble those described by Lieutenant Freyer at Arica. The surface of the lower ledge, which extends from a low cliff overhanging the sea to the foot of the next upper escarpment, is covered by an enormous accumulation of recent shells. (M. Chevalier, in the "Voyage of the 'Bonite'" observed these shells; but his specimens were lost.—"L'Institut" 1838) The bed is level, and in some parts more than two feet in thickness; I traced it over a space of one mile in length, and heard of it in other places: the uppermost part is eighty-five feet by the barometer above high-water mark. The shells are packed together, but not stratified: they are mingled with earth and stones, and are generally covered by a few inches of detritus; they rest on a mass of nearly angular fragments of the underlying sandstone, sometimes cemented together by common salt. I collected eighteen species of shells of all ages and sizes. Several of the univalves had evidently long lain dead at the bottom of the sea, for their INSIDES were incrusted with Balani and Serpulae. All, according to Mr. G.B. Sowerby, are recent species: they consist of:—

1. Mytilus Magellanicus: same as that found at Valparaiso, and there stated to be probably distinct from the true M. Magellanicus of the east coast.
2. Venus costellata, Sowerby "Zoological Proceedings."
3. Pecten purpuratus, Lam.
4. Chama, probably echinulata, Brod.
5. Calyptraea Byronensis, Gray.
6. Calyptraea radians (Trochus, Lam.)
7. Fissurella affinis, Gray.
8. Fissurella biradiata, Trembly.
9. Purpura chocolatta, Duclos.
10. Purpura Peruviana, Gray.
11. Purpura labiata, Gray.
12. Purpura buxea (Murex, Brod.).
13. Concholepas Peruviana.
14. Nassa, related to reticulata.
15. Triton rudis, Brod.
16. Trochus, not yet described, but well-known and very common.
17 and 18. Balanus, two species, both common on the coast.

These upraised shells appear to be nearly in the same proportional numbers- -with the exception of the Crepidulae being more numerous—with those on the existing beach. The state of preservation of the different species differed much; but most of them were much corroded, brittle, and bleached: the upper and lower surfaces of the Concholepas had generally quite scaled off: some of the Trochi and Fissurellae still partially retain their colours. It is remarkable that these shells, taken all together, have fully as ancient an appearance, although the extremely arid climate appears highly favourable for their preservation, as those from 1,300 feet at Valparaiso, and certainly a more ancient appearance than those from five to six hundred feet from Valparaiso and Concepcion; at which places I have seen grass and other vegetables actually growing out of the shells. Many of the univalves here at San Lorenzo were filled with, and united together by, pure salt, probably left by the evaporation of the sea-spray, as the land slowly emerged. (The underlying sandstone contains true layers of salt; so that the salt may possibly have come from the beds in the higher parts of the island; but I think more probably from the sea-spray. It is generally asserted that rain never falls on the coast of Peru; but this is not quite accurate; for, on several days, during our visit, the so-called Peruvian dew fell in sufficient quantity to make the streets muddy, and it would certainly have washed so deliquescent a substance as salt into the soil. I state this because M. d'Orbigny, in discussing an analogous subject, supposes that I had forgotten that it never rains on this whole line of coast. See Ulloa's "Voyage" volume 2 English Translation for an account of the muddy streets of Lima, and on the continuance of the mists during the whole winter. Rain, also, falls at rare intervals even in the driest districts, as, for instance, during forty days, in 1726, at Chocope (7 degrees 46'); this rain entirely ruined ("Ulloa" etc.) the mud houses of the inhabitants.) On the highest parts of the ledge, small fragments of the shells were mingled with, and evidently in process of reduction into, a yellowish-white, soft, calcareous powder, tasting strongly of salt, and in some places as fine as prepared medicinal chalk.

FOSSIL-REMAINS OF HUMAN ART.

In the midst of these shells on San Lorenzo, I found light corallines, the horny ovule-cases of Mollusca, roots of seaweed (Mr. Smith of Jordan Hill found pieces of seaweed in an upraised pleistocene deposit in Scotland. See his admirable Paper in the "Edinburgh New Philosophical Journal" volume 25), bones of birds, the heads of Indian corn and other vegetable matter, a piece of woven rushes, and another of nearly decayed COTTON string. I extracted these remains by digging a hole, on a level spot; and they had all indisputably been embedded with the shells. I compared the plaited rush, the COTTON string, and Indian corn, at the house of an antiquary, with similar objects, taken from the Huacas or burial-grounds of the ancient Peruvians, and they were undistinguishable; it should be observed that the Peruvians used string only of cotton. The small quantity of sand or gravel with the shells, the absence of large stones, the width and thickness of the bed, and the time requisite for a ledge to be cut into the sandstone, all show that these remains were not thrown high up by an earthquake-wave: on the other hand, these facts, together with the number of dead shells, and of floating objects, both marine and terrestrial, both natural and human, render it almost certain that they were accumulated on a true beach, since upraised eighty-five feet, and upraised this much since INDIAN MAN INHABITED PERU. The elevation may have been, either by several small sudden starts, or quite gradual; in this latter case the unrolled shells having been thrown up during gales beyond the reach of the waves which afterwards broke on the slowly emerging land. I have made these remarks, chiefly because I was at first surprised at the complete difference in nature, between this broad, smooth, upraised bed of shells, and the present shingle-beach at the foot of the low sandstone-cliffs; but a beach formed, when the sea is cutting into the land, as is shown now to be the case by the low bare sandstone-cliffs, ought not to be compared with a beach accumulated on a gently inclined rocky surface, at a period when the sea (probably owing to the elevatory movement in process) was not able to eat into the land. With respect to the mass of nearly angular, salt-cemented fragments of sandstone, which lie under the shells, and which are so unlike the materials of an ordinary sea-beach; I think it probable after having seen the remarkable effects of the earthquake of 1835 (I have described this in my "Journal of Researches" 2nd edition.), in absolutely shattering as if by gunpowder the SURFACE of the primary rocks near Concepcion, that a smooth bare surface of stone was left by the sea covered by the shelly mass, and that afterwards when upraised, it was superficially shattered by the severe shocks so often experienced here.

The very low land surrounding the town of Callao, is to the south joined by an obscure escarpment to a higher plain (south of Bella Vista), which stretches along the coast for a length of about eight miles. This plain appears to the eye quite level; but the sea-cliffs show that its height varies (as far as I could estimate) from seventy to one hundred and twenty feet. It is composed of thin, sometimes waving, beds of clay, often of bright red and yellow colours, of layers of impure sand, and in one part with a great stratified mass of granitic pebbles. These beds are capped by a remarkable mass, varying from two to six feet in thickness, of reddish loam or mud, containing many scattered and broken fragments of recent marine shells, sometimes though rarely single large round pebble, more frequently short irregular layers of fine gravel, and very many pieces of red coarse earthenware, which from their curvatures must once have formed parts of large vessels. The earthenware is of Indian manufacture; and I found exactly similar pieces accidentally included within the bricks, of which the neighbouring ancient Peruvian burial-mounds are built. These fragments abounded in such numbers in certain spots, that it appeared as if waggon-loads of earthenware had been smashed to pieces. The broken sea-shells and pottery are strewed both on the surface, and throughout the whole thickness of this upper loamy mass. I found them wherever I examined the cliffs, for a space of between two and three miles, and for half a mile inland; and there can be little doubt that this same bed extends with a smooth surface several miles further over the entire plain. Besides the little included irregular layers of small pebbles, there are occasionally very obscure traces of stratification.

At one of the highest parts of the cliff, estimated 120 feet above the sea, where a little ravine came down, there were two sections, at right angles to each other, of the floor of a shed or building. In both sections or faces, two rows, one over the other, of large round stones could be distinctly seen; they were packed close together on an artificial layer of sand two inches thick, which had been placed on the natural clay-beds; the round stones were covered by three feet in thickness of the loam with broken sea-shells and pottery. Hence, before this widely spread-out bed of loam was deposited, it is certain that the plain was inhabited; and it is probable, from the broken vessels being so much more abundant in certain spots than in others, and from the underlying clay being fitted for their manufacture, that the kilns stood here.

The smoothness and wide extent of the plain, the bulk of matter deposited, and the obscure traces of stratification seem to indicate that the loam was deposited under water; on the other hand, the presence of sea-shells, their broken state, the pebbles of various sizes, and the artificial floor of round stones, almost prove that it must have originated in a rush of water from the sea over the land. The height of the plain, namely, 120 feet, renders it improbable that an earthquake-wave, vast as some have here been, could have broken over the surface at its present level; but when the land stood eighty-five feet lower, at the period when the shells were thrown up on the ledge at S. Lorenzo, and when as we know man inhabited this district, such an event might well have occurred; and if we may further suppose, that the plain was at that time converted into a temporary lake, as actually occurred, during the earthquakes of 1713 and 1746, in the case of the low land round Callao owing to its being encircled by a high shingle-beach, all the appearances above described will be perfectly explained. I must add, that at a lower level near the point where the present low land round Callao joins the higher plain, there are appearances of two distinct deposits both apparently formed by debacles: in the upper one, a horse's tooth and a dog's jaw were embedded; so that both must have been formed after the settlement of the Spaniards: according to Acosta, the earthquake-wave of 1586 rose eighty-four feet.

The inhabitants of Callao do not believe, as far as I could ascertain, that any change in level is now in progress. The great fragments of brickwork, which it is asserted can be seen at the bottom of the sea, and which have been adduced as a proof of a late subsidence, are, as I am informed by Mr. Gill, a resident engineer, loose fragments; this is probable, for I found on the beach, and not near the remains of any building, masses of brickwork, three and four feet square, which had been washed into their present places, and smoothed over with shingle during the earthquake of 1746. The spit of land, on which the ruins of OLD Callao stand, is so extremely low and narrow, that it is improbable in the highest degree that a town should have been founded on it in its present state; and I have lately heard that M. Tschudi has come to the conclusion, from a comparison of old with modern charts, that the coast both south and north of Callao has subsided. (I am indebted for this fact to Dr. E. Dieffenbach. I may add that there is a tradition, that the islands of San Lorenzo and Fronton were once joined, and that the channel between San Lorenzo and the mainland, now above two miles in width, was so narrow that cattle used to swim over.) I have shown that the island of San Lorenzo has been upraised eighty-five feet since the Peruvians inhabited this country; and whatever may have been the amount of recent subsidence, by so much more must the elevation have exceeded the eighty-five feet. In several places in this neighbourhood, marks of sea-action have been observed: Ulloa gives a detailed account of such appearances at a point five leagues northward of Callao: Mr. Cruikshank found near Lima successive lines of sea-cliffs, with rounded blocks at their bases, at a height of 700 feet above the present level of the sea. ("Observaciones sobre el Clima del Lima" par Dr. H.

Unanue.—Ulloa's "Voyage" volume 2 English Translation.—For Mr. Cruikshank's observations, see Mr. Lyell's "Principles of Geology" 1st edition volume 3) ON

THE DECAY OF UPRAISED SEA-SHELLS.

I have stated that many of the shells on the lower inclined ledge or terrace of San Lorenzo are corroded in a peculiar manner, and that they have a much more ancient appearance than the same species at considerably greater heights on the coast of Chile. I have, also, stated that these shells in the upper part of the ledge, at the height of eighty-five feet above the sea, are falling, and in some parts are quite changed into a fine, soft, saline, calcareous powder. The finest part of this powder has been analysed for me, at the request of Sir H. De la Beche, by the kindness of Mr. Trenham Reeks of the Museum of Economic Geology; it consists of carbonate of lime in abundance, of sulphate and muriate of lime, and of muriate and sulphate of soda. The carbonate of lime is obviously derived from the shells; and common salt is so abundant in parts of the bed, that, as before remarked, the univalves are often filled with it. The sulphate of lime may have been derived, as has probably the common salt, from the evaporation of the sea-spray, during the emergence of the land; for sulphate of lime is now copiously deposited from the spray on the shores of Ascension. (See "Volcanic Islands" etc. by the Author.) The other saline bodies may perhaps have been partially thus derived, but chiefly, as I conclude from the following facts, through a different means.

On most parts of the second ledge or old sea-beach, at a height of 170 feet, there is a layer of white powder of variable thickness, as much in some parts as two inches, lying on the angular, salt-cemented fragments of sandstone and under about four inches of earth, which powder, from its close resemblance in nature to the upper and most decayed parts of the shelly mass, I can hardly doubt originally existed as a bed of shells, now much collapsed and quite disintegrated. I could not discover with the microscope a trace of organic structure in it; but its chemical constituents, according to Mr. Reeks, are the same as in the powder extracted from amongst the decaying shells on the lower ledge, with the marked exception that the carbonate of lime is present in only very small quantity. On the third and highest ledge, I observed some of this powder in a similar position, and likewise occasionally in small patches at considerably greater heights near the summit of the island. At Iquique, where the whole face of the country is covered by a highly saliferous alluvium, and where the climate is extremely dry, we have seen that, according to Mr. Blake, the shells which are perfect near the beach become, in ascending, gradually less and less perfect, until scarcely a trace of their original structure can be discovered. It is known that carbonate of lime and common salt left in a mass together, and slightly moistened, partially decompose each other (I am informed by Dr. Kane, through Mr. Reeks, that a manufactory was established on this principle in France, but failed from the small quantity of carbonate of soda produced. Sprengel "Gardeners' Chronicle" 1845, states, that salt and carbonate of lime are liable to mutual decomposition in the soil. Sir H. De la Beche informs me, that calcareous rocks washed by the spray of the sea, are often corroded in a peculiar manner; see also on this latter subject "Gardeners' Chronicle" 1844.): now we have at San Lorenzo and at Iquique, in the shells and salt packed together, and occasionally moistened by the so- called Peruvian dew, the proper elements for this action. We can thus understand the peculiar corroded appearance of the shells on San Lorenzo, and the great decrease of quantity in the carbonate of lime in the powder on the upper ledge. There is, however, a great difficulty on this view, for the resultant salts should be carbonate of soda and muriate of lime; the latter is present, but not the carbonate of soda. Hence I am led to the perhaps unauthorised conjecture (which I shall hereafter have to refer to) that the carbonate of soda, by some unexplained means, becomes converted into a sulphate.

If the above remarks be just, we are led to the very unexpected conclusion, that a dry climate, by leaving the salt from the sea-spray undissolved, is much less favourable to the preservation of upraised shells than a humid climate. However this may be, it is interesting to know the manner in which masses of shells, gradually upraised above the sea-level, decay and finally disappear.

SUMMARY ON THE RECENT ELEVATION OF THE WEST COAST OF SOUTH AMERICA.

We have seen that upraised marine remains occur at intervals, and in some parts almost continuously, from latitude 45 degrees 35' to 12 degrees S., along the shores of the Pacific. This is a distance, in a north and south line, of 2,075 geographical miles. From Byron's observations, the elevation has no doubt extended sixty miles further south; and from the similarity in the form of the country near Lima, it has probably extended many leagues further north. (I may take this opportunity of stating that in a MS. in the Geological Society by Mr. Weaver, it is stated that beds of oysters and other recent shells are found thirty feet above the level of the sea, in many parts of Tampico, in the Gulf of Mexico.) Along this great line of coast, besides the organic remains, there are in very many parts, marks of erosion, caves, ancient beaches, sand-dunes, and successive terraces of gravel, all above the present level of the sea. From the steepness of the land on this side of the continent, shells have rarely been found at greater distances inland than from two to three leagues; but the marks of sea-action are evident farther from the coast; for instance, in the valley of Guasco, at a distance of between thirty and forty miles. Judging from the upraised shells alone, the elevation in Chiloe has been 350 feet, at Concepcion certainly 625 feet; and by estimation 1,000 feet; at Valparaiso 1,300 feet; at Coquimbo 252 feet; northward of this place, sea-shells have not, I believe, been found above 300 feet; and at Lima they were falling into decay (hastened probably by the salt) at 85 feet. Not only has this amount of elevation taken place within the period of existing Mollusca and Cirripedes; but their proportional numbers in the neighbouring sea have in most cases remained the same. Near Lima, however, a small change in this respect between the living and the upraised was observed: at Coquimbo this was more evident, all the shells being existing species, but with those embedded in the uppermost calcareous plain not approximating so closely in proportional numbers, as do those that lie loose on its surface at the height of 252 feet, and still less closely than those which are strewed on the lower plains, which latter are identical in proportional numbers with those now cast up on the beach. From this circumstance, and from not finding, upon careful examination, near Coquimbo any shells at a greater height than 252 feet, I believe that the recent elevation there has been much less than at Valparaiso, where it has been 1,300 feet, and I may add, than at Concepcion. This considerable inequality in the amount of elevation at Coquimbo and Valparaiso, places only 200 miles apart, is not improbable, considering, first, the difference in the force and number of the shocks now yearly affecting different parts of this coast; and, secondly, the fact of single areas, such as that of the province of Concepcion, having been uplifted very unequally during the same earthquake. It would, in most cases, be very hazardous to infer an inequality of elevation, from shells being found on the surface or in superficial beds at different heights; for we do not know on what their rate of decay depends; and at Coquimbo one instance out of many has been given, of a promontory, which, from the occurrence of one very small collection of lime-cemented shells, has indisputably been elevated 242 feet, and yet on which, not even a fragment of shell could be found on careful examination between this height and the beach, although many sites appeared very favourable for the preservation of organic remains: the absence, also, of shells on the gravel-terraces a short distance up the valley of Coquimbo, though abundant on the corresponding terraces at its mouth, should be borne in mind.

There are other epochs, besides that of the existence of recent Mollusca, by which to judge of the changes of level on this coast. At Lima, as we have just seen, the elevation has been at least eighty-five feet, within the Indo-human period; and since the arrival of the Spaniards in 1530, there has apparently been a sinking of the surface. At Valparaiso, in the course of 220 years, the rise must have been less than nineteen feet; but it has been as much as from ten to eleven feet in the seventeen years subsequently to 1817, and of this rise only a part can be attributed to the earthquake of 1822, the remainder having been insensible and apparently still, in 1834, in progress. At Chiloe the elevation has been gradual, and about four feet during four years. At Coquimbo, also, it has been gradual, and in the course of 150 years has amounted to several feet. The sudden small upheavals, accompanied by earthquakes, as in 1822 at Valparaiso, in 1835 at Concepcion, and in 1837 in the Chonos Archipelago, are familiar to most geologists, but the gradual rising of the coast of Chile has been hardly noticed; it is, however, very important, as connecting together these two orders of events.

The rise of Lima, having been eighty-five feet within the period of man, is the more surprising if we refer to the eastern coast of the continent, for at Port S. Julian, in Patagonia, there is good evidence (as we shall hereafter see) that when the land stood ninety feet lower, the Macrauchenia, a mammiferous beast, was alive; and at Bahia Blanca, when it stood only a few feet lower than it now does, many gigantic quadrupeds ranged over the adjoining country. But the coast of Patagonia is some way distant from the Cordillera, and the movement at Bahia Blanca is perhaps noways connected with this great range, but rather with the tertiary volcanic rocks of Banda Oriental, and therefore the elevation at these places may have been infinitely slower than on the coast of Peru. All such speculations, however, must be vague, for as we know with certainty that the elevation of the whole coast of Patagonia has been interrupted by many and long pauses, who will pretend to say that, in such cases, many and long periods of subsidence may not also have been intercalated?

In many parts of the coast of Chile and Peru there are marks of the action of the sea at successive heights on the land, showing that the elevation has been interrupted by periods of comparative rest in the upward movement, and of denudation in the action of the sea. These are plainest at Chiloe, where, in a height of about five hundred feet, there are three escarpments,—at Coquimbo, where in a height of 364 feet, there are five,— at Guasco, where there are six, of which five may perhaps correspond with those at Coquimbo, but if so, the subsequent and intervening elevatory movements have been here much more energetic,—at Lima, where, in a height of about 250 feet there are three terraces, and others, as it is asserted, at considerably greater heights. The almost entire absence of ancient marks of sea-action at defined levels along considerable spaces of coast, as near Valparaiso and Concepcion, is highly instructive, for as it is improbable that the elevation at these places alone should have been continuous, we must attribute the absence of such marks to the nature and form of the coast-rocks. Seeing over how many hundred miles of the coast of Patagonia, and on how many places on the shores of the Pacific, the elevatory process has been interrupted by periods of comparative rest, we may conclude, conjointly with the evidence drawn from other quarters of the world, that the elevation of the land is generally an intermittent action. From the quantity of matter removed in the formation of the escarpments, especially of those of Patagonia, it appears that the periods of rest in the movement, and of denudation of the land, have generally been very long. In Patagonia, we have seen that the elevation has been equable, and the periods of denudation synchronous over very wide spaces of coast; on the shores of the Pacific, owing to the terraces chiefly occurring in the valleys, we have not equal means of judging on this point; and the very different heights of the upraised shells at Coquimbo, Valparaiso, and Concepcion seem directly opposed to such a conclusion.

Whether on this side of the continent the elevation, between the periods of comparative rest when the escarpments were formed, has been by small sudden starts, such as those accompanying recent earthquakes, or, as is most probable, by such starts conjointly with a gradual upward movement, or by great and sudden upheavals, I have no direct evidence. But as on the eastern coast, I was led to think, from the analogy of the last hundred feet of elevation in La Plata, and from the nearly equal size of the pebbles over the entire width of the terraces, and from the upraised shells being all littoral species, that the elevation had been gradual; so do I on this western coast, from the analogy of the movements now in progress, and from the vast numbers of shells now living exclusively on or close to the beach, which are strewed over the whole surface of the land up to very considerable heights, conclude, that the movement here also has been slow and gradual, aided probably by small occasional starts. We know at least that at Coquimbo, where five escarpments occur in a height of 364 feet, the successive elevations, if they have been sudden, cannot have been very great. It has, I think, been shown that the occasional preservation of shells, unrolled and unbroken, is not improbable even during a quite gradual rising of the land; and their preservation, if the movement has been aided by small starts, is quite conformable with what actually takes place during recent earthquakes.

Judging from the present action of the sea, along the shores of the Pacific, on the deposits of its own accumulation, the present time seems in most places to be one of comparative rest in the elevatory movement, and of denudation of the land. Undoubtedly this is the case along the whole great length of Patagonia. At Chiloe, however, we have seen that a narrow sloping fringe, covered with vegetation, separates the present sea-beach from a line of low cliffs, which the waves lately reached; here, then, the land is gaining in breadth and height, and the present period is not one of rest in the elevation and of contingent denudation; but if the rising be not prolonged at a quick rate, there is every probability that the sea will soon regain its former horizontal limits. I observed similar low sloping fringes on several parts of the coast, both northward of Valparaiso and near Coquimbo; but at this latter place, from the change in form which the coast has undergone since the old escarpments were worn, it may be doubted whether the sea, acting for any length of time at its present level, would eat into the land; for it now rather tends to throw up great masses of sand. It is from facts such as these that I have generally used the term COMPARATIVE rest, as applied to the elevation of the land; the rest or cessation in the movement being comparative both with what has preceded it and followed it, and with the sea's power of corrosion at each spot and at each level. Near Lima, the cliff-formed shores of San Lorenzo, and on the mainland south of Callao, show that the sea is gaining on the land; and as we have here some evidence that its surface has lately subsided or is still sinking, the periods of comparative rest in the elevation and of contingent denudation, may probably in many cases include periods of subsidence. It is only, as was shown in detail when discussing the terraces of Coquimbo, when the sea with difficulty and after a long lapse of time has either corroded a narrow ledge into solid rock, or has heaped up on a steep surface a NARROW mound of detritus, that we can confidently assert that the land at that level and at that period long remained absolutely stationary. In the case of terraces formed of gravel or sand, although the elevation may have been strictly horizontal, it may well happen that no one level beach-line may be traceable, and that neither the terraces themselves nor the summit nor basal edges of their escarpments may be horizontal.

Finally, comparing the extent of the elevated area, as deduced from the upraised recent organic remains, on the two sides of the continent, we have seen that on the Atlantic, shells have been found at intervals from Eastern Tierra del Fuego for 1,180 miles northward, and on the Pacific for a space of 2,075 miles. For a length of 775 miles, they occur in the same latitudes on both sides of the continent. Without taking this circumstance into consideration, it is probable from the reasons assigned in the last chapter, that the entire breadth of the continent in Central Patagonia has been uplifted in mass; but from other reasons there given, it would be hazardous to extend this conclusion to La Plata. From the continent being narrow in the southern-most parts of Patagonia, and from the shells found at the Inner Narrows of the Strait of Magellan, and likewise far up the valley of the Santa Cruz, it is probable that the southern part of the western coast, which was not visited by me, has been elevated within the period of recent Mollusca: if so, the shores of the Pacific have been continuously, recently, and in a geological sense synchronously upraised, from Lima for a length of 2,480 nautical miles southward,—a distance equal to that from the Red Sea to the North Cape of Scandinavia!

CHAPTER III. ON THE PLAINS AND VALLEYS OF CHILE:—SALIFEROUS SUPERFICIAL DEPOSITS.

Basin-like plains of Chile; their drainage, their marine origin.

Marks of sea-action on the eastern flanks of the Cordillera.
Sloping terrace-like fringes of stratified shingle within the valleys of the Cordillera; their marine origin.
Boulders in the valley of Cachapual.
Horizontal elevation of the Cordillera.
Formation of valleys.
Boulders moved by earthquake-waves.
Saline superficial deposits.
Bed of nitrate of soda at Iquique.
Saline incrustations.
Salt-lakes of La Plata and Patagonia; purity of the salt; its origin.

The space between the Cordillera and the coast of Chile is on a rude average from eighty to above one hundred miles in width; it is formed, either of an almost continuous mass of mountains, or more commonly of several nearly parallel ranges, separated by plains; in the more southern parts of this province the mountains are quite subordinate to the plains; in the northern part the mountains predominate.

The basin-like plains at the foot of the Cordillera are in several respects remarkable; that on which the capital of Chile stands is fifteen miles in width, in an east and west line, and of much greater length in a north and south line; it stands 1,750 feet above the sea; its surface appears smooth, but really falls and rises in wide gentle undulations, the hollows corresponding with the main valleys of the Cordillera: the striking manner in which it abruptly comes up to the foot of this great range has been remarked by every author since the time of Molina. (This plain is partially separated into two basins by a range of hills; the southern half, according to Meyen ("Reise um Erde" Th. 1 s. 274), falls in height, by an abrupt step, of between fifteen and twenty feet.) Near the Cordillera it is composed of a stratified mass of pebbles of all sizes, occasionally including rounded boulders: near its western boundary, it consists of reddish sandy clay, containing some pebbles and numerous fragments of pumice, and sometimes passes into pure sand or into volcanic ashes. At Podaguel, on this western side of the plain, beds of sand are capped by a calcareous tuff, the uppermost layers being generally hard and substalagmitic, and the lower ones white and friable, both together precisely resembling the beds at Coquimbo, which contain recent marine shells. Abrupt, but rounded, hummocks of rock rise out of this plain: those of Sta. Lucia and S. Cristoval are formed of greenstone-porphyry almost entirely denuded of its original covering of porphyritic claystone breccia; on their summits, many fragments of rock (some of them kinds not found in situ) are coated and united together by a white, friable, calcareous tuff, like that found at Podaguel. When this matter was deposited on the summit of S. Cristoval, the water must have stood 946 feet above the surface of the surrounding plain. (Or 2,690 feet above the sea, as measured barometrically by Mr. Eck. This tuff appears to the eye nearly pure; but when placed in acid it leaves a considerable residue of sand and broken crystals, apparently of feldspar. Dr. Meyen ("Reise" Th. 1 s. 269) says he found a similar substance on the neighbouring hill of Dominico (and I found it also on the Cerro Blanco), and he attributes it to the weathering of the stone. In some places which I examined, its bulk put this view of its origin quite out of the question; and I should much doubt whether the decomposition of a porphyry would, in any case, leave a crust chiefly composed of carbonate of lime. The white crust, which is commonly seen on weathered feldspathic rocks, does not appear to contain any free carbonate of lime.)

To the south this basin-like plain contracts, and rising scarcely perceptibly with a smooth surface, passes through a remarkable level gap in the mountains, forming a true land-strait, and called the Angostura. It then immediately expands into a second basin-formed plain: this again to the south contracts into another land-strait, and expands into a third basin, which, however, falls suddenly in level about forty feet. This third basin, to the south, likewise contracts into a strait, and then again opens into the great plain of San Fernando, stretching so far south that the snowy peaks of the distant Cordillera are seen rising above its horizon as above the sea. These plains, near the Cordillera, are generally formed of a thick stratified mass of shingle (The plain of San Fernando has, according to MM. Meyen and Gay "Reise" etc. Th. 1 ss. 295 and 298, near the Cordillera, an upper step-formed plain of clay, on the surface of which they found numerous blocks of rocks, from two to three feet long, either lying single or piled in heaps, but all arranged in nearly straight lines.); in other parts, of a red sandy clay, often with an admixture of pumiceous matter. Although these basins are connected together like a necklace, in a north and south line, by smooth land-straits, the streams which drain them do not all flow north and south, but mostly westward, through breaches worn in the bounding mountains; and in the case of the second basin, or that of Rancagua, there are two distinct breaches. Each basin, moreover, is not drained singly; thus, to give the most striking instance, but not the only one, in proceeding southward over the plain of Rancagua, we first find the water flowing northward to and through the northern land-strait; then, without crossing any marked ridge or watershed, we see it flowing south-westward towards the northern one of the two breaches in the western mountainous boundary; and lastly, again without any ridge, it flows towards the southern breach in these same mountains. Hence the surface of this one basin-like plain, appearing to the eye so level, has been modelled with great nicety, so that the drainage, without any conspicuous watersheds, is directed towards three openings in the encircling mountains. ((It appears from Captain Herbert's account of the Diluvium of the Himalaya, "Gleanings of Science" Calcutta volume 2, that precisely similar remarks apply to the drainage of the plains or valleys between those great mountains.) The streams flowing from the southern basin-like plains, after passing through the breaches to the west, unite and form the river Rapel, which enters the Pacific near Navidad. I followed the southernmost branch of this river, and found that the basin or plain of San Fernando is continuously and smoothly united with those plains, which were described in the Second Chapter, as being worn near the coast into successive cave-eaten escarpments, and still nearer to the coast, as being strewed with upraised recent marine remains.

I might have given descriptions of numerous other plains of the same general form, some at the foot of the Cordillera, some near the coast, and some halfway between these points. I will allude only to one other, namely, the plain of Uspallata, lying on the eastern or opposite side of the Cordillera, between that great range and the parallel lower range of Uspallata. According to Miers, its surface is 6,000 feet above the level of the sea: it is from ten to fifteen miles in width, and is said to extend with an unbroken surface for 180 miles northwards: it is drained by two rivers passing through breaches in the mountains to the east. On the banks of the River Mendoza it is seen to be composed of a great accumulation of stratified shingle, estimated at 400 feet in thickness. In general appearance, and in numerous points of structure, this plain closely resembles those of Chile.

The origin and manner of formation of the thick beds of gravel, sandy clay, volcanic detritus, and calcareous tuff, composing these basin-like plains, is very important; because, as we shall presently show, they send arms or fringes far up the main valleys of the Cordillera. Many of the inhabitants believe that these plains were once occupied by lakes, suddenly drained; but I conceive that the number of the separate breaches at nearly the same level in the mountains surrounding them quite precludes this idea. Had not such distinguished naturalists as MM. Meyen and Gay stated their belief that these deposits were left by great debacles rushing down from the Cordillera, I should not have noticed a view, which appears to me from many reasons improbable in the highest degree—namely, from the vast accumulation of WELL-ROUNDED PEBBLES—their frequent stratification with layers of sand—the overlying beds of calcareous tuff—this same substance coating and uniting the fragments of rock on the hummocks in the plain of Santiago—and lastly even from the worn, rounded, and much denuded state of these hummocks, and of the headlands which project from the surrounding mountains. On the other hand, these several circumstances, as well as the continuous union of the basins at the foot of the Cordillera, with the great plain of the Rio Rapel which still retains the marks of sea-action at various levels, and their general similarity in form and composition with the many plains near the coast, which are either similarly marked or are strewed with upraised marine remains, fully convince me that the mountains bounding these basin-plains were breached, their islet-like projecting rocks worn, and the loose stratified detritus forming their now level surfaces deposited, by the sea, as the land slowly emerged. It is hardly possible to state too strongly the perfect resemblance in outline between these basin-like, long, and narrow plains of Chile (especially when in the early morning the mists hanging low represented water), and the creeks and fiords now intersecting the southern and western shores of the continent. We can on this view of the sea, when the land stood lower, having long and tranquilly occupied the spaces between the mountain-ranges, understand how the boundaries of the separate basins were breached in more than one place; for we see that this is the general character of the inland bays and channels of Tierra del Fuego; we there, also, see in the sawing action of the tides, which flow with great force in the cross channels, a power sufficient to keep the breaches open as the land emerged. We can further see that the waves would naturally leave the smooth bottom of each great bay or channel, as it became slowly converted into land, gently inclined to as many points as there were mouths, through which the sea finally retreated, thus forming so many watersheds, without any marked ridges, on a nearly level surface. The absence of marine remains in these high inland plains cannot be properly adduced as an objection to their marine origin: for we may conclude, from shells not being found in the great shingle beds of Patagonia, though copiously strewed on their surfaces, and from many other analogous facts, that such deposits are eminently unfavourable for the embedment of such remains; and with respect to shells not being found strewed on the surface of these basin-like plains, it was shown in the last chapter that remains thus exposed in time decay and disappear.

(FIGURE 13. SECTION OF THE PLAIN AT THE EASTERN FOOT OF THE CHILEAN CORDILLERA.
From Cordillera (left) through Talus-plain and Level surface, 2,700 feet above sea, to Gravel terraces (right).)

I observed some appearances on the plains at the eastern and opposite foot of the Cordillera which are worth notice, as showing that the sea there long acted at nearly the same level as on the basin-plains of Chile. The mountains on this eastern side are exceedingly abrupt; they rise out of a smooth, talus-like, very gentle, slope, from five to ten miles in width (as represented in Figure 13), entirely composed of perfectly rounded pebbles, often white-washed with an aluminous substance like decomposed feldspar. This sloping plain or talus blends into a perfectly flat space a few miles in width, composed of reddish impure clay, with small calcareous concretions as in the Pampean deposit,—of fine white sand with small pebbles in layers,—and of the above-mentioned white aluminous earth, all interstratified together. This flat space runs as far as Mendoza, thirty miles northward, and stands probably at about the same height, namely, 2,700 feet (Pentland and Miers) above the sea. To the east it is bounded by an escarpment, eighty feet in height, running for many miles north and south, and composed of perfectly round pebbles, and loose, white-washed, or embedded in the aluminous earth: behind this escarpment there is a second and similar one of gravel. Northward of Mendoza, these escarpments become broken and quite obliterated; and it does not appear that they ever enclosed a lake-like area: I conclude, therefore, that they were formed by the sea, when it reached the foot of the Cordillera, like the similar escarpments occurring at so many points on the coasts of Chile and Patagonia.

The talus-like plain slopes up with a smooth surface into the great dry valleys of the Cordillera. On each hand of the Portillo valley, the mountains are formed of red granite, mica-slate, and basalt, which all have suffered a truly astonishing amount of denudation; the gravel in the valley, as well as on the talus-like plain in front of it, is composed of these rocks; but at the mouth of the valley, in the middle (height probably about three thousand five hundred feet above the sea), a few small isolated hillocks of several varieties of porphyry project, round which, on all sides, smooth and often white-washed pebbles of these same porphyries, to the exclusion of all others, extend to a circumscribed distance. Now, it is difficult to conceive any other agency, except the quiet and long-continued action of the sea on these hillocks, which could have rounded and whitewashed the fragments of porphyry, and caused them to radiate from such small and quite insignificant centres, in the midst of that vast stream of stones which has descended from the main Cordillera.

SLOPING TERRACES OF GRAVEL IN THE VALLEYS OF THE CORDILLERA.
(FIGURE 14. GROUND-PLAN OF A BIFURCATING VALLEY IN THE CORDILLERA, bordered by smooth, sloping gravel-fringes (AA), worn along the course of the river into cliffs.)

All the main valleys on both flanks of the Chilean Cordillera have formerly had, or still have, their bottoms filled up to a considerable thickness by a mass of rudely stratified shingle. In Central Chile the greater part of this mass has been removed by the torrents; cliff-bounded fringes, more or less continuous, being left at corresponding heights on both sides of the valleys. These fringes, or as they may be called terraces, have a smooth surface, and as the valleys rise, they gently rise with them: hence they are easily irrigated, and afford great facilities for the construction of the roads. From their uniformity, they give a remarkable character to the scenery of these grand, wild, broken valleys. In width, the fringes vary much, sometimes being only broad enough for the roads, and sometimes expanding into narrow plains. Their surfaces, besides gently rising up the valley, are slightly inclined towards its centre in such a manner as to show that the whole bottom must once have been filled up with a smooth and slightly concave mass, as still are the dry unfurrowed valleys of Northern Chile. Where two valleys unite into one, these terraces are particularly well exhibited, as is represented in Figure 14. The thickness of the gravel forming these fringes, on a rude average, may be said to vary from thirty to sixty or eighty feet; but near the mouths of the valleys it was in several places from two to three hundred feet. The amount of matter removed by the torrents has been immense; yet in the lower parts of the valleys the terraces have seldom been entirely worn away on either side, nor has the solid underlying rock been reached: higher up the valleys, the terraces have frequently been removed on one or the other side, and sometimes on both sides; but in this latter case they reappear after a short interval on the line, which they would have held had they been unbroken. Where the solid rock has been reached, it has been cut into deep and narrow gorges. Still higher up the valleys, the terraces gradually become more and more broken, narrower, and less thick, until, at a height of from seven to nine thousand feet, they become lost, and blended with the piles of fallen detritus.

I carefully examined in many places the state of the gravel, and almost everywhere found the pebbles equally and perfectly rounded, occasionally with great blocks of rock, and generally distinctly stratified, often with parting seams of sand. The pebbles were sometimes coated with a white aluminous, and less frequently with a calcareous, crust. At great heights up the valleys the pebbles become less rounded; and as the terraces become obliterated, the whole mass passes into the nature of ordinary detritus. I was repeatedly struck with the great difference between this detritus high up the valleys, and the gravel of the terraces low down, namely, in the greater number of the quite angular fragments in the detritus,—in the unequal degree to which the other fragments have been rounded,—in the quantity of associated earth,—in the absence of stratification,—and in the irregularity of the upper surfaces. This difference was likewise well shown at points low down the valleys, where precipitous ravines, cutting through mountains of highly coloured rock, have thrown down wide, fan-shaped accumulations of detritus on the terraces: in such cases, the line of separation between the detritus and the terrace could be pointed out to within an inch or two; the detritus consisting entirely of angular and only partially rounded fragments of the adjoining coloured rocks; the stratified shingle (as I ascertained by close inspection, especially in one case, in the valley of the River Mendoza) containing only a small proportion of these fragments, and those few well rounded.

I particularly attended to the appearance of the terraces where the valleys made abrupt and considerable bends, but I could perceive no difference in their structure: they followed the bends with their usual nearly equable inclination. I observed, also, in several valleys, that wherever large blocks of any rock became numerous, either on the surface of the terrace or embedded in it, this rock soon appeared higher up in situ: thus I have noticed blocks of porphyry, of andesitic syenite, of porphyry and of syenite, alternately becoming numerous, and in each case succeeded by mountains thus constituted. There is, however, one remarkable exception to this rule; for along the valley of the Cachapual, M. Gay found numerous large blocks of white granite, which does not occur in the neighbourhood. I observed these blocks, as well as others of andesitic syenite (not occurring here in situ), near the baths of Cauquenes at a height of between two and three hundred feet above the river, and therefore quite above the terrace or fringe which borders that river; some miles up the valleys there were other blocks at about the same height. I also noticed, at a less height, just above the terrace, blocks of porphyries (apparently not found in the immediately impending mountains), arranged in rude lines, as on a sea-beach. All these blocks were rounded, and though large, not gigantic, like the true erratic boulders of Patagonia and Fuegia. M. Gay states that the granite does not occur in situ within a distance of twenty leagues ("Annales des Science Nat. " 1 series tome 28. M. Gay, as I was informed, penetrated the Cordillera by the great oblique valley of Los Cupressos, and not by the most direct line.); I suspect, for several reasons, that it will ultimately be found at a much less distance, though certainly not in the immediate neighbourhood. The boulders found by MM. Meyen and Gay on the upper plain of San Fernando (mentioned in a previous note) probably belong to this same class of phenomena.

These fringes of stratified gravel occur along all the great valleys of the Cordillera, as well as along their main branches; they are strikingly developed in the valleys of the Maypu, Mendoza, Aconcagua, Cachapual, and according to Meyen, in the Tinguirica. ("Reise" etc. Th. 1 s. 302.) In the valleys, however, of Northern Chile, and in some on the eastern flank of the Cordillera, as in the Portillo Valley, where streams have never flowed, or are quite insignificant in volume, the presence of a mass of stratified gravel can be inferred only from the smooth slightly concave form of the bottom. One naturally seeks for some explanation of so general and striking a phenomenon; that the matter forming the fringes along the valleys, or still filling up their entire beds, has not fallen from the adjoining mountains like common detritus, is evident from the complete contrast in every respect between the gravel and the piles of detritus, whether seen high up the valleys on their sides, or low down in front of the more precipitous ravines; that the matter has not been deposited by debacles, even if we could believe in debacles having rushed down EVERY valley, and all their branches, eastward and westward from the central pinnacles of the Cordillera, we must admit from the following reasons,—from the distinct stratification of the mass,—its smooth upper surface,—the well-rounded and sometimes encrusted state of the pebbles, so different from the loose debris on the mountains,—and especially from the terraces preserving their uniform inclination round the most abrupt bends. To suppose that as the land now stands, the rivers deposited the shingle along the course of every valley, and all their main branches, appears to me preposterous, seeing that these same rivers not only are now removing and have removed much of this deposit, but are everywhere tending to cut deep and narrow gorges in the hard underlying rocks.

I have stated that these fringes of gravel, the origin of which are inexplicable on the notion of debacles or of ordinary alluvial action, are directly continuous with the similarly-composed basin-like plains at the foot of the Cordillera, which, from the several reasons before assigned, I cannot doubt were modelled by the agency of the sea. Now if we suppose that the sea formerly occupied the valleys of the Chilean Cordillera, in precisely the same manner as it now does in the more southern parts of the continent, where deep winding creeks penetrate into the very heart of, and in the case of Obstruction Sound quite through, this great range; and if we suppose that the mountains were upraised in the same slow manner as the eastern and western coasts have been upraised within the recent period, then the origin and formation of these sloping, terrace-like fringes of gravel can be simply explained. For every part of the bottom of each valley will, on this view, have long stood at the head of a sea creek, into which the then existing torrents will have delivered fragments of rocks, where, by the action of the tides, they will have been rolled, sometimes encrusted, rudely stratified, and the whole surface levelled by the blending together of the successive beach lines. (Sloping terraces of precisely similar structure have been described by me "Philosophical Transactions" 1839, in the valleys of Lochaber in Scotland, where, at higher levels, the parallel roads of Glen Roy show the marks of the long and quiet residence of the sea. I have no doubt that these sloping terraces would have been present in the valleys of most of the European ranges, had not every trace of them, and all wrecks of sea-action, been swept away by the glaciers which have since occupied them. I have shown that this is the case with the mountains ("London and Edinburgh Philosophical Journal" volume 21) of North Wales.) As the land rose, the torrents in every valley will have tended to have removed the matter which just before had been arrested on, or near, the beach-lines; the torrents, also, having continued to gain in force by the continued elevation increasing their total descent from their sources to the sea. This slow rising of the Cordillera, which explains so well the otherwise inexplicable origin and structure of the terraces, judging from all known analogies, will probably have been interrupted by many periods of rest; but we ought not to expect to find any evidence of these periods in the structure of the gravel-terraces: for, as the waves at the heads of deep creeks have little erosive power, so the only effect of the sea having long remained at the same level will be that the upper parts of the creeks will have become filled up at such periods to the level of the water with gravel and sand; and that afterwards the rivers will have thrown down on the filled-up parts a talus of similar matter, of which the inclination (as at the head of a partially filled-up lake) will have been determined by the supply of detritus, and the force of the stream. (I have attempted to explain this process in a more detailed manner, in a letter to Mr. Maclaren, published in the "Edinburgh New Philosophical Journal" volume 35) Hence, after the final conversion of the creeks into valleys, almost the only difference in the terraces at those points at which the sea stood long, will be a somewhat more gentle inclination, with river-worn instead of sea-worn detritus on the surface.

I know of only one difficulty on the foregoing view, namely, the far-transported blocks of rock high on the sides of the valley of the Cachapual: I will not attempt any explanation of this phenomenon, but I may state my belief that a mountain-ridge near the Baths of Cauquenes has been upraised long subsequently to all the other ranges in the neighbourhood, and that when this was effected the whole face of the country must have been greatly altered. In the course of ages, moreover, in this and other valleys, events may have occurred like, but even on a grander scale than, that described by Molina, when a slip during the earthquake of 1762 banked up for ten days the great River Lontue, which then bursting its barrier "inundated the whole country," and doubtless transported many great fragments of rock. ("Compendio de la Hist." etc. etc. tome 1. M. Brongniart, in his report on M. Gay's labours "Annales des Sciences" 1833, considers that the blocks in the Cachapual belong to the same class with the erratic boulders of Europe. As the blocks which I saw are not gigantic, and especially as they are not angular, and as they have not been transported fairly across low spaces or wide valleys, I am unwilling to class them with those which, both in the northern and southern hemisphere "Geological Transactions" volume 6, have been transported by ice. It is to be hoped that when M. Gay's long-continued and admirable labours in Chile are published, more light will be thrown on this subject. However, the boulders may have been primarily transported; the final position of those of porphyry, which have been described as arranged at the foot of the mountain in rude lines, I cannot doubt, has been due to the action of waves on a beach. The valley of the Cachapual, in the part where the boulders occur, bursts through the high ridge of Cauquenes, which runs parallel to, but at some distance from, the Cordillera. This ridge has been subjected to excessive violence; trachytic lava has burst from it, and hot springs yet flow at its base. Seeing the enormous amount of denudation of solid rock in the upper and much broader parts of this valley where it enters the Cordillera, and seeing to what extent the ridge of Cauquenes now protects the great range, I could not help believing (as alluded to in the text) that this ridge with its trachytic eruptions had been thrown up at a much later period than the Cordillera. If this has been the case, the boulders, after having been transported to a low level by the torrents (which exhibit in every valley proofs of their power of moving great fragments), may have been raised up to their present height, with the land on which they rested.) Finally, notwithstanding this one case of difficulty, I cannot entertain any doubt, that these terrace-like fringes, which are continuously united with the basin-shaped plains at the foot of the Cordillera, have been formed by the arrestment of river-borne detritus at successive levels, in the same manner as we see now taking place at the heads of all those many, deep, winding fiords intersecting the southern coasts. To my mind, this has been one of the most important conclusions to which my observations on the geology of South America have led me; for we thus learn that one of the grandest and most symmetrical mountain-chains in the world, with its several parallel lines, has been together uplifted in mass between seven and nine thousand feet, in the same gradual manner as have the eastern and western coasts within the recent period. (I do not wish to affirm that all the lines have been uplifted quite equally; slight differences in the elevation would leave no perceptible effect on the terraces. It may, however, be inferred, perhaps with one exception, that since the period when the sea occupied these valleys, the several ranges have not been dislocated by GREAT and ABRUPT faults or upheavals; for if such had occurred, the terraces of gravel at these points would not have been continuous. The one exception is at the lower end of a plain in the Valle del Yeso (a branch of the Maypu), where, at a great height, the terraces and valley appear to have been broken through by a line of upheaval, of which the evidence is plain in the adjoining mountains; this dislocation, perhaps, occurred AFTER THE ELEVATION of this part of the valley above the level of the sea. The valley here is almost blocked up by a pile about one thousand feet in thickness, formed, as far as I could judge, from three sides, entirely, or at least in chief part, of gravel and detritus. On the south side, the river has cut quite through this mass; on the northern side, and on the very summit, deep ravines, parallel to the line of the valley, are worn, as if the drainage from the valley above had passed by these two lines before following its present course.)

FORMATION OF VALLEYS.

The bulk of solid rock which has been removed in the lower parts of the valleys of the Cordillera has been enormous. It is only by reflecting on such cases as that of the gravel beds of Patagonia, covering so many thousand square leagues of surface, and which, if heaped into a ridge, would form a mountain-range almost equal to the Cordillera, that the amount of denudation becomes credible. The valleys within this range often follow anticlinal but rarely synclinal lines; that is, the strata on the two sides more often dip from the line of valley than towards it. On the flanks of the range, the valleys most frequently run neither along anticlinal nor synclinal axes, but along lines of flexure or faults: that is, the strata on both sides dip in the same direction, but with different, though often only slightly different, inclinations. As most of the nearly parallel ridges which together form the Cordillera run approximately north and south, the east and west valleys cross them in zig-zag lines, bursting through the points where the strata have been least inclined. No doubt the greater part of the denudation was affected at the periods when tidal- creeks occupied the valleys, and when the outer flanks of the mountains were exposed to the full force of an open ocean. I have already alluded to the power of the tidal action in the channels connecting great bays; and I may here mention that one of the surveying vessels in a channel of this kind, though under sail, was whirled round and round by the force of the current. We shall hereafter see, that of the two main ridges forming the Chilean Cordillera, the eastern and loftiest one owes the greater part of its ANGULAR upheaval to a period subsequent to the elevation of the western ridge; and it is likewise probable that many of the other parallel ridges have been angularly upheaved at different periods; consequently many parts of the surfaces of these mountains must formerly have been exposed to the full force of the waves, which, if the Cordillera were now sunk into the sea, would be protected by parallel chains of islands. The torrents in the valleys certainly have great power in wearing the rocks; as could be told by the dull rattling sound of the many fragments night and day hurrying downwards; and as was attested by the vast size of certain fragments, which I was assured had been carried onwards during floods; yet we have seen in the lower parts of the valleys, that the torrents have seldom removed all the sea-checked shingle forming the terraces, and have had time since the last elevation in mass only to cut in the underlying rocks, gorges, deep and narrow, but quite insignificant in dimensions compared with the entire width and depth of the valleys.

Along the shores of the Pacific, I never ceased during my many and long excursions to feel astonished at seeing every valley, ravine, and even little inequality of surface, both in the hard granitic and soft tertiary districts, retaining the exact outline, which they had when the sea left their surfaces coated with organic remains. When these remains shall have decayed, there will be scarcely any difference in appearance between this line of coast-land and most other countries, which we are accustomed to believe have assumed their present features chiefly through the agency of the weather and fresh-water streams. In the old granitic districts, no doubt it would be rash to attribute all the modifications of outline exclusively to the sea-action; for who can say how often this lately submerged coast may not previously have existed as land, worn by running streams and washed by rain? This source of doubt, however, does not apply to the districts superficially formed of the modern tertiary deposits. The valleys worn by the sea, through the softer formations, both on the Atlantic and Pacific sides of the continent, are generally broad, winding, and flat-bottomed: the only district of this nature now penetrated by arms of the sea, is the island of Chiloe.

Finally, the conclusion at which I have arrived, with respect to the relative powers of rain and sea water on the land, is, that the latter is far the most efficient agent, and that its chief tendency is to widen the valleys; whilst torrents and rivers tend to deepen them, and to remove the wreck of the sea's destroying action. As the waves have more power, the more open and exposed the space may be, so will they always tend to widen more and more the mouths of valleys compared with their upper parts: hence, doubtless, it is, that most valleys expand at their mouths,—that part, at which the rivers flowing in them, generally have the least wearing power.

When reflecting on the action of the sea on the land at former levels, the effect of the great waves, which generally accompany earthquakes, must not be overlooked: few years pass without a severe earthquake occurring on some part of the west coast of South America; and the waves thus caused have great power. At Concepcion, after the shock of 1835, I saw large slabs of sandstone, one of which was six feet long, three in breadth, and two in thickness, thrown high up on the beach; and from the nature of the marine animals still adhering to it, it must have been torn up from a considerable depth. On the other hand, at Callao, the recoil-wave of the earthquake of 1746 carried great masses of brickwork, between three and four feet square, some way out seaward. During the course of ages, the effect thus produced at each successive level, cannot have been small; and in some of the tertiary deposits on this line of coast, I observed great boulders of granite and other neighbouring rocks, embedded in fine sedimentary layers, the transportal of which, except by the means of earthquake-waves, always appeared to me inexplicable.

SUPERFICIAL SALINE DEPOSITS.

This subject may be here conveniently treated of: I will begin with the most interesting case, namely, the superficial saline beds near Iquique in Peru. The porphyritic mountains on the coast rise abruptly to a height of between one thousand nine hundred and three thousand feet: between their summits and an inland plain, on which the celebrated deposit of nitrate of soda lies, there is a high undulatory district, covered by a remarkable superficial saliferous crust, chiefly composed of common salt, either in white, hard, opaque nodules, or mingled with sand, in this latter case forming a compact sandstone. This saliferous superficial crust extends from the edge of the coast-escarpment, over the whole face of the country; but never attains, as I am assured by Mr. Bollaert (long resident here) any great thickness. Although a very slight shower falls only at intervals of many years, yet small funnel-shaped cavities show that the salt has been in some parts dissolved. (It is singular how slowly, according to the observations of M. Cordier on the salt-mountain of Cardona in Spain "Ann. des Mines, Translation of Geolog. Mem." by De la Beche, salt is dissolved, where the amount of rain is supposed to be as much as 31.4 of an inch in the year. It is calculated that only five feet in thickness is dissolved in the course of a century.) In several places I saw large patches of sand, quite moist, owing to the quantity of muriate of lime (as ascertained by Mr. T. Reeks) contained in them. From the compact salt- cemented sand being either red, purplish, or yellow, according to the colour of the rocky strata on which it rested, I imagined that this substance had probably been derived through common alluvial action from the layers of salt which occur interstratified in the surrounding mountains ("Journal of Researches" first edition.): but from the interesting details given by M. d'Orbigny, and from finding on a fresh examination of this agglomerated sand, that it is not irregularly cemented, but consists of thin layers of sand of different tints of colour, alternating with excessively fine parallel layers of salt, I conclude that it is not of alluvial origin. M. d'Orbigny observed analogous saline beds extending from Cobija for five degrees of latitude northward, and at heights varying from six hundred to nine hundred feet ("Voyage" etc.. M. d'Orbigny found this deposit intersected, in many places, by deep ravines, in which there was no salt. Streams must once, though historically unknown, have flowed in them; and M. d'Orbigny argues from the presence of undissolved salt over the whole surrounding country, that the streams must have arisen from rain or snow having fallen, not in the adjoining country, but on the now arid Cordillera. I may remark, that from having observed ruins of Indian buildings in absolutely sterile parts of the Chilian Cordillera ("Journal" 2nd edition), I am led to believe that the climate, at a time when Indian man inhabited this part of the continent, was in some slight degree more humid than it is at present.): from finding recent sea- shells strewed on these saliferous beds, and under them, great well-rounded blocks, exactly like those on the existing beach, he believes that the salt, which is invariably superficial, has been left by the evaporation of the sea-water. This same conclusion must, I now believe, be extended to the superficial saliferous beds of Iquique, though they stand about three thousand feet above the level of the sea.

Associated with the salt in the superficial beds, there are numerous, thin, horizontal layers of impure, dirty-white, friable, gypseous and calcareous tuffs. The gypseous beds are very remarkable, from abounding with, so as sometimes to be almost composed of, irregular concretions, from the size of an egg to that of a man's head, of very hard, compact, heavy gypsum, in the form of anhydrite. This gypsum contains some foreign particles of stone; it is stained, judging from its action with borax, with iron, and it exhales a strong aluminous odour. The surfaces of the concretions are marked by sharp, radiating, or bifurcating ridges, as if they had been (but not really) corroded: internally they are penetrated by branching veins (like those of calcareous spar in the septaria of the London clay) of pure white anhydrite. These veins might naturally have been thought to have been formed by subsequent infiltration, had not each little embedded fragment of rock been likewise edged in a very remarkable manner by a narrow border of the same white anhydrite: this shows that the veins must have been formed by a process of segregation, and not of infiltration. Some of the little included and CRACKED fragments of foreign rock are penetrated by the anhydrite, and portions have evidently been thus mechanically displaced: at St. Helena, I observed that calcareous matter, deposited by rain water, also had the power to separate small fragments of rock from the larger masses. ("Volcanic Islands" etc.) I believe the superficial gypseous deposit is widely extended: I received specimens of it from Pisagua, forty miles north of Iquique, and likewise from Arica, where it coats a layer of pure salt. M. d'Orbigny found at Cobija a bed of clay, lying above a mass of upraised recent shells, which was saturated with sulphate of soda, and included thin layers of fibrous gypsum. ("Voyage Geolog." etc. These widely extended, superficial, beds of salt and gypsum, appear to me an interesting geological phenomenon, which could be presented only under a very dry climate.

The plain or basin, on the borders of which the famous bed of nitrate of soda lies, is situated at the distance of about thirty miles from the sea, being separated from it by the saliferous district just described. It stands at a height of 3,300 feet; its surface is level, and some leagues in width; it extends forty miles northward, and has a total length (as I was informed by Mr. Belford Wilson, the Consul-General at Lima) of 420 miles. In a well near the works, thirty-six yards in depth, sand, earth, and a little gravel were found: in another well, near Almonte, fifty yards deep, the whole consisted, according to Mr. Blake, of clay, including a layer of sand two feet thick, which rested on fine gravel, and this on coarse gravel, with large rounded fragments of rock. (See an admirable paper "Geological and Miscellaneous Notices of Tarapaca" in "Silliman's American Journal" volume 44.) In many parts of this now utterly desert plain, rushes and large prostrate trees in a hardened state, apparently Mimosas, are found buried, at a depth from three to six feet; according to Mr. Blake, they have all fallen to the south-west. The bed of nitrate of soda is said to extend for forty to fifty leagues along the western margin of the plain, but is not found in its central parts: it is from two to three feet in thickness, and is so hard that it is generally blasted with gunpowder; it slopes gently upwards from the edge of the plain to between ten and thirty feet above its level. It rests on sand in which, it is said, vegetable remains and broken shells have been found; shells have also been found, according to Mr. Blake, both on and in the nitrate of soda. It is covered by a superficial mass of sand, containing nodules of common salt, and, as I was assured by a miner, much soft gypseous matter, precisely like that in the superficial crust already described: certainly this crust, with its characteristic concretions of anhydrite, comes close down to the edge of the plain.

The nitrate of soda varies in purity in different parts, and often contains nodules of common salt. According to Mr. Blake, the proportion of nitrate of soda varies from 20 to 75 per cent. An analysis by Mr. A. Hayes, of an average specimen, gave:—

Nitrate of Soda.... 64.98
Sulphate of Soda.... 3.00
Chloride of Soda... 28.69
Iodic Salts......... 0.63
Shells and Marl..... 2.60
    99.90

The "mother-water" at some of the refineries is very rich in iodic salts, and is supposed to contain much muriate of lime. ("Literary Gazette" 1841) In an unrefined specimen brought home by myself, Mr. T. Reeks has ascertained that the muriate of lime is very abundant. With respect to the origin of this saline mass, from the manner in which the gently inclined, compact bed follows for so many miles the sinuous margin of the plain, there can be no doubt that it was deposited from a sheet of water: from the fragments of embedded shells, from the abundant iodic salts, from the superficial saliferous crust occurring at a higher level and being probably of marine origin, and from the plain resembling in form those of Chile and that of Uspallata, there can be little doubt that this sheet of water was, at least originally, connected with the sea. (From an official document, shown me by Mr. Belford Wilson, it appears that the first export of nitrate of soda to Europe was in July 1830, on French account, in a British ship:—

In year, the entire export was in Quintals.
1830........................... 17,300
1831........................... 40,885
1832........................... 51,400

1833............................ 91,335
1834............................ 149,538
The Spanish quintal nearly equals 100 English pounds.)

THIN, SUPERFICIAL, SALINE INCRUSTATIONS.

These saline incrustations are common in many parts of America: Humboldt met with them on the tableland of Mexico, and the Jesuit Falkner and other authors state that they occur at intervals over the vast plains extending from the mouth of the Plata to Rioja and Catamarca. (Azara "Travels" volume 1, considers that the Parana is the eastern boundary of the saliferous region; but I heard of "salitrales" in the Province of Entre Rios.) Hence it is that during droughts, most of the streams in the Pampas are saline. I nowhere met with these incrustations so abundantly as near Bahia Blanca: square miles of the mud-flats, which near that place are raised only a few feet above the sea, just enough to protect them from being overflowed, appear, after dry weather, whiter than the ground after the thickest hoar-frost. After rain the salts disappear, and every puddle of water becomes highly saline; as the surface dries, the capillary action draws the moisture up pieces of broken earth, dead sticks, and tufts of grass, where the salt effloresces. The incrustation, where thickest, does not exceed a quarter of an inch. M. Parchappe has analysed it (M. d'Orbigny "Voyage" etc. Part. Hist. tome 1); and finds that the specimens collected at the extreme head of the low plain, near the River Manuello, consist of 93 per cent of sulphate of soda, and 7 of common salt; whilst the specimens taken close to the coast contain only 63 per cent of the sulphate, and 37 of the muriate of soda. This remarkable fact, together with our knowledge that the whole of this low muddy plain has been covered by the sea within the recent period, must lead to the suspicion that the common salt, by some unknown process, becomes in time changed into the sulphate. Friable, calcareous matter is here abundant, and the case of the apparent double decomposition of the shells and salt on San Lorenzo, should not be forgotten.

The saline incrustations, near Bahia Blanca, are not confined to, though most abundant on, the low muddy flats; for I noticed some on a calcareous plain between thirty and forty feet above the sea, and even a little occurs in still higher valleys. Low alluvial tracts in the valleys of the Rivers Negro and Colorado are also encrusted, and in the latter valley such spaces appeared to be occasionally overflowed by the river. I observed saline incrustations in some of the valleys of Southern Patagonia. At Port Desire a low, flat, muddy valley was thickly incrusted by salts, which on analysis by Mr. T. Reeks, are found to consist of a mixture of sulphate and muriate of soda, with carbonate of lime and earthy matter. On the western side of the continent, the southern coasts are much too humid for this phenomenon; but in Northern Chile I again met with similar incrustations. On the hardened mud, in parts of the broad, flat-bottomed valley of Copiapo, the saline matter encrusts the ground to the thickness of some inches: specimens, sent by Mr. Bingley to Apothecaries' Hall for analysis, were said to consist of carbonate and sulphate of soda. Much sulphate of soda is found in the desert of Atacama. In all parts of South America, the saline incrustations occur most frequently on low damp surfaces of mud, where the climate is rather dry; and these low surfaces have, in almost every case, been upraised above the level of the sea, within the recent period.

SALT-LAKES OF PATAGONIA AND LA PLATA.

Salinas, or natural salt-lakes, occur in various formations on the eastern side of the continent,—in the argillaceo-calcareous deposit of the Pampas, in the sandstone of the Rio Negro, where they are very numerous, in the pumiceous and other beds of the Patagonian tertiary formation, and in small primary districts in the midst of this latter formation. Port S. Julian is the most southerly point (latitude 49 degrees to 50 degrees) at which salinas are known to occur. (According to Azara "Travels" volume 1, there are salt-lakes as far north as Chaco (latitude 25 degrees), on the banks of the Vermejo. The salt-lakes of Siberia appear (Pallas "Travels" English Translation volume 1) to occur in very similar depressions to those of Patagonia.) The depressions, in which these salt-lakes lie, are from a few feet to sixty metres, as asserted by M. d'Orbigny, below the surface of the surrounding plains ("Voyage Geolog."); and, according to this same author, near the Rio Negro they all trend, either in the N.E. and S.W. or in E. and W. lines, coincident with the general slope of the plain. These depressions in the plain generally have one side lower than the others, but there are no outlets for drainage. Under a less dry climate, an outlet would soon have been formed, and the salt washed away. The salinas occur at different elevations above the sea; they are often several leagues in diameter; they are generally very shallow, but there is a deep one in a quartz-rock formation near C. Blanco. In the wet season, the whole, or a part, of the salt is dissolved, being redeposited during the succeeding dry season. At this period the appearance of the snow-white expanse of salt crystallised in great cubes, is very striking. In a large salina, northward of the Rio Negro, the salt at the bottom, during the whole year, is between two and three feet in thickness.

The salt rests almost always on a thick bed of black muddy sand, which is fetid, probably from the decay of the burrowing worms inhabiting it. (Professor Ehrenberg examined some of this muddy sand, but was unable to find in it any infusoria.) In a salina, situated about fifteen miles above the town of El Carmen on the Rio Negro, and three or four miles from the banks of that river, I observed that this black mud rested on gravel with a calcareous matrix, similar to that spread over the whole surrounding plains: at Port S. Julian the mud, also, rested on the gravel: hence the depressions must have been formed anteriorly to, or contemporaneously with, the spreading out of the gravel. I was informed that one small salina occurs in an alluvial plain within the valley of the Rio Negro, and therefore its origin must be subsequent to the excavation of that valley. When I visited the salina, fifteen miles above the town, the salt was beginning to crystallise, and on the muddy bottom there were lying many crystals, generally placed crossways of sulphate of soda (as ascertained by Mr. Reeks), and embedded in the mud, numerous crystals of sulphate of lime, from one to three inches in length: M. d'Orbigny states that some of these crystals are acicular and more than even nine inches in length ("Voyage Geolog."); others are macled and of great purity: those I found all contained some sand in their centres. As the black and fetid sand overlies the gravel, and that overlies the regular tertiary strata, I think there can be no doubt that these remarkable crystals of sulphate of lime have been deposited from the waters of the lake. The inhabitants call the crystals of selenite, the padre del sal, and those of the sulphate of soda, the madre del sal; they assured me that both are found under the same circumstances in several of the neighbouring salinas; and that the sulphate of soda is annually dissolved, and is always crystallised before the common salt on the muddy bottom. (This is what might have been expected; for M. Ballard asserts "Acad. des Sciences" October 7, 1844, that sulphate of soda is precipitated from solution more readily from water containing muriate of soda in excess, than from pure water.) The association of gypsum and salt in this case, as well as in the superficial deposits of Iquique, appears to me interesting, considering how generally these substances are associated in the older stratified formations.

Mr. Reeks has analysed for me some of the salt from the salina near the Rio Negro; he finds it composed entirely of chloride of sodium, with the exception of 0.26 of sulphate of lime and of 0.22 of earthy matter: there are no traces of iodic salts. Some salt from the salina Chiquitos, in the Pampean formation, is equally pure. It is a singular fact, that the salt from these salinas does not serve so well for preserving meat, as sea-salt from the Cape de Verde Islands; and a merchant at Buenos Ayres told me that he considered it as 50 per cent less valuable. The purity of the Patagonian salt, or absence from it of those other saline bodies found in all sea- water, is the only assignable cause for this inferiority; a conclusion which is supported by the fact lately ascertained, that those salts answer best for preserving cheese which contain most of the deliquescent chlorides. ("Horticultural and Agricultural Gazette" 1845) (It would probably well answer for the merchants of Buenos Ayres (considering the great consumption there of salt for preserving meat) to import the deliquescent chlorides to mix with the salt from the salinas: I may call attention to the fact, that at Iquique, a large quantity of muriate of lime, left in the MOTHER-WATER during the refinement of the nitrate of soda, is annually thrown away.)

With respect to the origin of the salt in the salinas, the foregoing analysis seems opposed to the view entertained by M. d'Orbigny and others, and which seems so probable considering the recent elevation of this line of coast, namely, that it is due to the evaporation of sea-water and to the drainage from the surrounding strata impregnated with sea-salt. I was informed (I know not whether accurately) that on the northern side of the salina on the Rio Negro, there is a small brine spring which flows at all times of the year: if this be so, the salt in this case at least, probably is of subterranean origin. It at first appears very singular that fresh water can often be procured in wells, and is sometimes found in small lakes, quite close to these salinas. (Sir W. Parish states "Buenos Ayres" etc., that this is the case near the great salinas westward of the S. Ventana. I have seen similar statements in an ancient MS. Journal lately published by S. Angelis. At Iquique, where the surface is so thickly encrusted with saline matter, I tasted water only slightly brackish, procured in a well thirty-six yards deep; but here one feels less surprise at its presence, as pure water might percolate under ground from the not very distant Cordillera.) I am not aware that this fact bears particularly on the origin of the salt; but perhaps it is rather opposed to the view of the salt having been washed out of the surrounding superficial strata, but not to its having been the residue of sea-water, left in depressions as the land was slowly elevated.

CHAPTER IV. ON THE FORMATIONS OF THE PAMPAS.
Mineralogical constitution.
Microscopical structure.
Buenos Ayres, shells embedded in tosca-rock.
Buenos Ayres to the Colorado.
San Ventana.
Bahia Blanca; M. Hermoso, bones and infusoria of; P. Alta, shells, bones, and infusoria of; co-existence of the recent shells and extinct mammifers.
Buenos Ayres to Santa Fe.
Skeletons of Mastodon.
Infusoria.
Inferior marine tertiary strata, their age.
Horse's tooth.

BANDA ORIENTAL.
Superficial Pampean formation.
Inferior tertiary strata, variation of, connected with volcanic action;
Macrauchenia Patachonica at San Julian in Patagonia, age of, subsequent to living mollusca and to the erratic block period.
SUMMARY.
Area of Pampean formation.
Theories of origin.
Source of sediment.
Estuary origin.
Contemporaneous with existing mollusca.
Relations to underlying tertiary strata.
Ancient deposit of estuary origin.
Elevation and successive deposition of the Pampean formation.
Number and state of the remains of mammifers; their habitation, food, extinction, and range.
Conclusion.
Localities in Pampas at which mammiferous remains have been found.
The Pampean formation is highly interesting from its vast extent, its disputed origin, and

from the number of extinct gigantic mammifers embedded in it. It has upon the whole a very uniform character: consisting of a more or less dull reddish, slightly indurated, argillaceous earth or mud, often, but not always, including in horizontal lines concretions of marl, and frequently passing into a compact marly rock. The mud, wherever I examined it, even close to the concretions, did not contain any carbonate of lime. The concretions are generally nodular, sometimes rough externally, sometimes stalactiformed; they are of a compact structure, but often penetrated (as well as the mud) by hair-like serpentine cavities, and occasionally with irregular fissures in their centres, lined with minute crystals of carbonate of lime; they are of white, brown, or pale pinkish tints, often marked by black dendritic manganese or iron; they are either darker or lighter tinted than the surrounding mass; they contain much carbonate of lime, but exhale a strong aluminous odour, and leave, when dissolved in acids, a large but varying residue, of which the greater part consists of sand. These concretions often unite into irregular strata; and over very large tracts of country, the entire mass consists of a hard, but generally cavernous marly rock: some of the varieties might be called calcareous tuffs.

Dr. Carpenter has kindly examined under the microscope, sliced and polished specimens of these concretions, and of the solid marl-rock, collected in various places between the Colorado and Santa Fe Bajada. In the greater number, Dr. Carpenter finds that the whole substance presents a tolerably uniform amorphous character, but with traces of incipient crystalline metamorphosis; in other specimens he finds microscopically minute rounded concretions of an amorphous substance (resembling in size those in oolitic rocks, but not having a concentric structure), united by a cement which is often crystalline. In some, Dr. Carpenter can perceive distinct traces of shells, corals, Polythalamia, and rarely of spongoid bodies. For the sake of comparison, I sent Dr. Carpenter specimens of the calcareous rock, formed chiefly of fragments of recent shells, from Coquimbo in Chile: in one of these specimens, Dr. Carpenter finds, besides the larger fragments, microscopical particles of shells, and a varying quantity of opaque amorphous matter; in another specimen from the same bed, he finds the whole composed of the amorphous matter, with layers showing indications of an incipient crystalline metamorphosis: hence these latter specimens, both in external appearance and in microscopical structure, closely resemble those of the Pampas. Dr. Carpenter informs me that it is well known that chemical precipitation throws down carbonate of lime in the opaque amorphous state; and he is inclined to believe that the long-continued attrition of a calcareous body in a state of crystalline or semi-crystalline aggregation (as, for instance, in the ordinary shells of Mollusca, which, when sliced, are transparent) may yield the same result. From the intimate relations between all the Coquimbo specimens, I can hardly doubt that the amorphous carbonate of lime in them has resulted from the attrition and decay of the larger fragments of shell: whether the amorphous matter in the marly rocks of the Pampas has likewise thus originated, it would be hazardous to conjecture.

For convenience' sake, I will call the marly rock by the name given to it by the inhabitants, namely, Tosca-rock; and the reddish argillaceous earth, Pampean mud. This latter substance, I may mention, has been examined for me by Professor Ehrenberg, and the result of his examination will be given under the proper localities.

I will commence my descriptions at a central spot, namely, at Buenos Ayres, and thence proceed first southward to the extreme limit of the deposit, and afterwards northward. The plain on which Buenos Ayres stands is from thirty to forty feet in height. The Pampean mud is here of a rather pale colour, and includes small nearly white nodules, and other irregular strata of an unusually arenaceous variety of tosca-rock. In a well at the depth of seventy feet, according to Ignatio Nunez, much tosca-rock was met with, and at several points, at one hundred feet deep, beds of sand have been found. I have already given a list of the recent marine and estuary shells found in many parts on the surface near Buenos Ayres, as far as three or four leagues from the Plata. Specimens from near Ensenada, given me by Sir W. Parish, where the rock is quarried just beneath the surface of the plain, consist of broken bivalves, cemented by and converted into white crystalline carbonate of lime. I have already alluded, in the first chapter, to a specimen (also given me by Sir W. Parish) from the A. del Tristan, in which shells, resembling in every respect the Azara labiata, d'Orbigny, as far as their worn condition permits of comparison, are embedded in a reddish, softish, somewhat arenaceous marly rock: after careful comparison, with the aid of a microscope and acids, I can perceive no difference between the basis of this rock and the specimens collected by me in many parts of the Pampas. I have also stated, on the authority of Sir W. Parish, that northward of Buenos Ayres, on the highest parts of the plain, about forty feet above the Plata, and two or three miles from it, numerous shells of the Azara labiata (and I believe of Venus sinuosa) occur embedded in a stratified earthy mass, including small marly concretions, and said to be precisely like the great Pampean deposit. Hence we may conclude that the mud of the Pampas continued to be deposited to within the period of this existing estuary shell. Although this formation is of such immense extent, I know of no other instance of the presence of shells in it.

BUENOS AYRES TO THE RIO COLORADO.

With the exception of a few metamorphic ridges, the country between these two points, a distance of 400 geographical miles, belongs to the Pampean formation, and in the southern part is generally formed of the harder and more calcareous varieties. I will briefly describe my route: about twenty- five miles S.S.W. of the capital, in a well forty yards in depth, the upper part, and, as I was assured, the entire thickness, was formed of dark red Pampean mud without concretions. North of the River Salado, there are many lakes; and on the banks of one (near the Guardia) there was a little cliff similarly composed, but including many nodular and stalactiform concretions: I found here a large piece of tessellated armour, like that of the Glyptodon, and many fragments of bones. The cliffs on the Salado consist of pale-coloured Pampean mud, including and passing into great masses of tosca-rock: here a skeleton of the Megatherium and the bones of other extinct quadrupeds (see the list at the end of this chapter) were found. Large quantities of crystallised gypsum (of which specimens were given me) occur in the cliffs of this river; and likewise (as I was assured by Mr. Lumb) in the Pampean mud on the River Chuelo, seven leagues from Buenos Ayres: I mention this because M. d'Orbigny lays some stress on the supposed absence of this mineral in the Pampean formation.

Southward of the Salado the country is low and swampy, with tosca-rock appearing at long intervals at the surface. On the banks, however, of the Tapalguen (sixty miles south of the Salado) there is a large extent of tosca-rock, some highly compact and even semi-crystalline, overlying pale Pampean mud with the usual concretions. Thirty miles further south, the small quartz-ridge of Tapalguen is fringed on its northern and southern flank, by little, narrow, flat-topped hills of tosca-rock, which stand higher than the surrounding plain. Between this ridge and the Sierra of Guitru-gueyu, a distance of sixty miles, the country is swampy, with the tosca-rock appearing only in four or five spots: this sierra, precisely like that of Tapalguen, is bordered by horizontal, often cliff-bounded, little hills of tosca-rock, higher than the surrounding plain. Here, also, a new appearance was presented in some extensive and level banks of alluvium or detritus of the neighbouring metamorphic rocks; but I neglected to observe whether it was stratified or not. Between Guitru-gueyu and the Sierra Ventana, I crossed a dry plain of tosca-rock higher than the country hitherto passed over, and with small pieces of denuded tableland of the same formation, standing still higher.

The marly or calcareous beds not only come up nearly horizontally to the northern and southern foot of the great quartzose mountains of the Sierra Ventana, but interfold between the parallel ranges. The superficial beds (for I nowhere obtained sections more than twenty feet deep) retain, even close to the mountains, their usual character: the uppermost layer, however, in one place included pebbles of quartz, and rested on a mass of detritus of the same rock. At the very foot of the mountains, there were some few piles of quartz and tosca-rock detritus, including land-shells; but at the distance of only half a mile from these lofty, jagged, and battered mountains, I could not, to my great surprise, find on the boundless surface of the calcareous plain even a single pebble. Quartz- pebbles, however, of considerable size have at some period been transported to a distance of between forty and fifty miles to the shores of Bahia Blanca. (Schmidtmeyer "Travels in Chile", states that he first noticed on the Pampas, very small bits of red granite, when fifty miles distant from the southern extremity of the mountains of Cordova, which project on the plain, like a reef into the sea.)

The highest peak of the St. Ventana is, by Captain Fitzroy's measurement, 3,340 feet, and the calcareous plain at its foot (from observations taken by some Spanish officers) 840 feet above the sea-level. ("La Plata" etc. by Sir W. Parish) On the flanks of the mountains, at a height of three hundred or four hundred feet above the plain, there were a few small patches of conglomerate and breccia, firmly cemented by ferruginous matter to the abrupt and battered face of the quartz—traces being thus exhibited of ancient sea-action. The high plain round this range sinks quite insensibly to the eye on all sides, except to the north, where its surface is broken into low cliffs. Round the Sierras Tapalguen, Guitru-gueyu, and between the latter and the Ventana we have seen (and shall hereafter see round some hills in Banda Oriental), that the tosca-rock forms low, flat- topped, cliff-bounded hills, higher than the surrounding plains of similar composition. From the horizontal stratification and from the appearance of the broken cliffs, the greater height of the Pampean formation round these primary hills ought not to be altogether or in chief part attributed to these several points having been uplifted more energetically than the surrounding country, but to the argillaceo-calcareous mud having collected round them, when they existed as islets or submarine rocks, at a greater height, than at the bottom of the adjoining open sea;—the cliffs having been subsequently worn during the elevation of the whole country in mass.

Southward of the Ventana, the plain extends farther than the eye can range; its surface is not very level, having slight depressions with no drainage exits; it is generally covered by a few feet in thickness of sandy earth; and in some places, according to M. Parchappe, by beds of clay two yards thick. (M. d'Orbigny "Voyage" Part Geolog.) On the banks of the Sauce, four leagues S.E. of the Ventana, there is an imperfect section about two hundred feet in height, displaying in the upper part tosca-rock and in the lower part red Pampean mud. At the settlement of Bahia Blanca, the uppermost plain is composed of very compact, stratified tosca-rock, containing rounded grains of quartz distinguishable by the naked eye: the lower plain, on which the fortress stands, is described by M. Parchappe as composed of solid tosca-rock (Ibid.); but the sections which I examined appeared more like a redeposited mass of this rock, with small pebbles and fragments of quartz. I shall immediately return to the important sections on the shores of Bahia Blanca. Twenty miles southward of this place, there is a remarkable ridge extending W. by N. and E. by S., formed of small, separate, flat-topped, steep-sided hills, rising between one hundred and two hundred feet above the Pampean plain at its southern base, which plain is a little lower than that to the north. The uppermost stratum in this ridge consists of pale, highly calcareous, compact tosca-rock, resting (as seen in one place) on reddish Pampean mud, and this again on a paler kind: at the foot of the ridge, there is a well in reddish clay or mud. I have seen no other instance of a chain of hills belonging to the Pampean formation; and as the strata show no signs of disturbance, and as the direction of the ridge is the same with that common to all the metamorphic lines in this whole area, I suspect that the Pampean sediment has in this instance been accumulated on and over a ridge of hard rocks, instead of, as in the case of the above-mentioned Sierras, round their submarine flanks. South of this little chain of tosca-rock, a plain of Pampean mud declines towards the banks of the Colorado: in the middle a well has been dug in red Pampean mud, covered by two feet of white, softish, highly calcareous tosca-rock, over which lies sand with small pebbles three feet in thickness—the first appearance of that vast shingle formation described in the First Chapter. In the first section after crossing the Colorado, an old tertiary formation, namely, the Rio Negro sandstone (to be described in the next chapter), is met with: but from the accounts given me by the Gauchos, I believe that at the mouth of the Colorado the Pampean formation extends a little further southwards.

BAHIA BLANCA.

To return to the shores of this bay. At Monte Hermoso there is a good section, about one hundred feet in height, of four distinct strata, appearing to the eye horizontal, but thickening a little towards the N.W. The uppermost bed, about twenty feet in thickness, consists of obliquely laminated, soft sandstone, including many pebbles of quartz, and falling at the surface into loose sand. The second bed, only six inches thick, is a hard, dark-coloured sandstone. The third bed is pale-coloured Pampean mud; and the fourth is of the same nature, but darker coloured, including in its lower part horizontal layers and lines of concretions of not very compact pinkish tosca-rock. The bottom of the sea, I may remark, to a distance of several miles from the shore, and to a depth of between sixty and one hundred feet, was found by the anchors to be composed of tosca-rock and reddish Pampean mud. Professor Ehrenberg has examined for me specimens of the two lower beds, and finds in them three Polygastrica and six Phytolitharia.

(The following list is given in the "Monatsberichten der konig. Akad. zu Berlin" April 1845:—

POLYGASTRICA.

Fragilaria rhabdosoma.

Gallionella distans.

Pinnularia?

PHYTOLITHARIA.

Lithodontium Bursa.

Lithodontium furcatum.

Lithostylidium exesum.

Lithostylidium rude.

Lithostylidium Serra.

Spongolithis Fustis?)

Of these, only one (Spongolithis Fustis?) is a marine form; five of them are identical with microscopical structures of brackish-water origin, hereafter to be mentioned, which form a central point in the Pampean formation. In these two beds, especially in the lower one, bones of extinct mammifers, some embedded in their proper relative positions and others single, are very numerous in a small extent of the cliffs. These remains consist of, first, the head of Ctenomys antiquus, allied to the living Ctenomys Braziliensis; secondly, a fragment of the remains of a rodent; thirdly, molar teeth and other bones of a large rodent, closely allied to, but distinct from, the existing species of Hydrochoerus, and therefore probably an inhabitant of fresh water; fourth and fifthly, portions of vertebrae, limbs, ribs, and other bones of two rodents; sixthly, bones of the extremities of some great megatheroid quadruped. (See "Fossil Mammalia" by Professor Owen, in the "Zoology of the Voyage of the 'Beagle';" and Catalogue of Fossil Remains in Museum of Royal College of Surgeons.) The number of the remains of rodents gives to this collection a peculiar character, compared with those found in any other locality. All these bones are compact and heavy; many of them are stained red, with their surfaces polished; some of the smaller ones are as black as jet.

Monte Hermoso is between fifty and sixty miles distant in a S.E. line from the Ventana, with the intermediate country gently rising towards it, and all consisting of the Pampean formation. What relation, then, do these beds, at the level of the sea and under it, bear to those on the flanks of the Ventana, at the height of 840 feet, and on the flanks of the other neighbouring sierras, which, from the reasons already assigned, do not appear to owe their greater height to unequal elevation? When the tosca- rock was accumulating round the Ventana, and when, with the exception of a few small rugged primary islands, the whole wide surrounding plains must have been under water, were the strata at Monte Hermoso depositing at the bottom of a great open sea, between eight hundred and one thousand feet in depth? I much doubt this; for if so, the almost perfect carcasses of the several small rodents, the remains of which are so very numerous in so limited a space, must have been drifted to this spot from the distance of many hundred miles. It appears to me far more probable, that during the Pampean period this whole area had commenced slowly rising (and in the cliffs, at several different heights we have proofs of the land having been exposed to sea-action at several levels), and that tracts of land had thus been formed of Pampean sediment round the Ventana and the other primary ranges, on which the several rodents and other quadrupeds lived, and that a stream (in which perhaps the extinct aquatic Hydrochoerus lived) drifted their bodies into the adjoining sea, into which the Pampean mud continued to be poured from the north. As the land continued to rise, it appears that this source of sediment was cut off; and in its place sand and pebbles were borne down by stronger currents, and conformably deposited over the Pampean strata.

(FIGURE 15. SECTION OF BEDS WITH RECENT SHELLS AND EXTINCT MAMMIFERS, AT
PUNTA ALTA IN BAHIA BLANCA. (Showing beds from bottom to top: A, B, C, D.))

Punta Alta is situated about thirty miles higher up on the northern side of this same bay: it consists of a small plain, between twenty and thirty feet in height, cut off on the shore by a line of low cliffs about a mile in length, represented in Figure 15 with its vertical scale necessarily exaggerated. The lower bed (A) is more extensive than the upper ones; it consists of stratified gravel or conglomerate, cemented by calcareo- arenaceous matter, and is divided by curvilinear layers of pinkish marl, of which some are precisely like tosca-rock, and some more sandy. The beds are curvilinear, owing to the action of currents, and dip in different directions; they include an extraordinary number of bones of gigantic mammifers and many shells. The pebbles are of considerable size, and are of hard sandstone, and of quartz, like that of the Ventana: there are also a few well-rounded masses of tosca-rock.

The second bed B is about fifteen feet in thickness, but towards both extremities of the cliff (not included in the diagram) it either thins out and dies away, or passes insensibly into an overlying bed of gravel. It consists of red, tough clayey mud, with minute linear cavities; it is marked with faint horizontal shades of colour; it includes a few pebbles, and rarely a minute particle of shell: in one spot, the dermal armour and a few bones of a Dasypoid quadruped were embedded in it: it fills up furrows in the underlying gravel. With the exception of the few pebbles and particles of shells, this bed resembles the true Pampean mud; but it still more closely resembles the clayey flats (mentioned in the First Chapter) separating the successively rising parallel ranges of sand-dunes.

The bed C is of stratified gravel, like the lowest one; it fills up furrows in the underlying red mud, and is sometimes interstratified with it, and sometimes insensibly passes into it; as the red mud thins out, this upper gravel thickens. Shells are more numerous in it than in the lower gravel; but the bones, though some are still present, are less numerous. In one part, however, where this gravel and the red mud passed into each other, I found several bones and a tolerably perfect head of the Megatherium. Some of the large Volutas, though embedded in the gravel-bed C, were filled with the red mud, including great numbers of the little recent Paludestrina australis. These three lower beds are covered by an unconformable mantle D of stratified sandy earth, including many pebbles of quartz, pumice and phonolite, land and sea-shells.

M. d'Orbigny has been so obliging as to name for me the twenty species of
Mollusca embedded in the two gravel beds: they consist of:—
1. Volutella angulata, d'Orbigny, "Voyage" Mollusq. and Pal. 2. Voluta Braziliana, Sol 3. Olicancilleria Braziliensis d'Orbigny. 4. Olicancilleria auricularia, d'Orbigny. 5. Olivina puelchana, d'Orbigny. 6. Buccinanops cochlidium, d'Orbigny. 7. Buccinanops globulosum, d'Orbigny. 8. Colombella sertulariarum, d'Orbigny. 9. Trochus Patagonicus, and var. of ditto, d'Orbigny. 10. Paludestrina Australis, d'Orbigny. 11. Fissurella Patagonica, d'Orbigny. 12. Crepidula muricata, Lam. 13. Venus purpurata, Lam. 14. Venus rostrata, Phillippi. 15. Mytilus Darwinianus, d'Orbigny. 16. Nucula semiornata, d'Orbigny. 17. Cardita Patagonica,

d'Orbigny. 18. Corbula Patagonica (?), d'Orbigny. 19. Pecten tethuelchus, d'Orbigny. 20. Ostrea puelchana, d'Orbigny. 21. A living species of Balanus. 22 and 23. An Astrae and encrusting Flustra, apparently identical with species now living in the bay.

All these shells now live on this coast, and most of them in this same bay. I was also struck with the fact, that the proportional numbers of the different kinds appeared to be the same with those now cast up on the beach: in both cases specimens of Voluta, Crepidula, Venus, and Trochus are the most abundant. Four or five of the species are the same with the upraised shells on the Pampas near Buenos Ayres. All the specimens have a very ancient and bleached appearance, and do not emit, when heated, an animal odour: some of them are changed throughout into a white, soft, fibrous substance; others have the space between the external walls, either hollow, or filled up with crystalline carbonate of lime. (A Bulinus, mentioned in the Introduction to the "Fossil Mammalia" in the "Zoology of the Voyage of the 'Beagle'" has so much fresher an appearance, than the marine species, that I suspect it must have fallen amongst the others, and been collected by mistake.)

The remains of the extinct mammiferous animals, from the two gravel beds have been described by Professor Owen in the "Zoology of the Voyage of the 'Beagle':" they consist of, 1st, one nearly perfect head and three fragments of heads of the Megatherium Cuvierii; 2nd, a lower jaw of Megalonyx Jeffersonii; 3rd, lower jaw of Mylodon Darwinii; 4th, fragments of a head of some gigantic Edental quadruped; 5th, an almost entire skeleton of the great Scelidotherium leptocephalum, with most of the bones, including the head, vertebrae, ribs, some of the extremities to the claw-bone, and even, as remarked by Professor Owen, the knee-cap, all nearly in their proper relative positions; 6th, fragments of the jaw and a separate tooth of a Toxodon, belonging either to T. Platensis, or to a second species lately discovered near Buenos Ayres; 7th, a tooth of Equus curvidens; 8th, tooth of a Pachyderm, closely allied to Palaeotherium, of which parts of the head have been lately sent from Buenos Ayres to the British Museum; in all probability this pachyderm is identical with the Macrauchenia Patagonica from Port S. Julian, hereafter to be referred to. Lastly, and 9thly, in a cliff of the red clayey bed B, there was a double piece, about three feet long and two wide, of the bony armour of a large Dasypoid quadruped, with the two sides pressed nearly close together: as the cliff is now rapidly washing away, this fossil probably was lately much more perfect; from between its doubled-up sides, I extracted the middle and ungual phalanges, united together, of one of the feet, and likewise a separate phalanx: hence one or more of the limbs must have been attached to the dermal case, when it was embedded. Besides these several remains in a distinguishable condition, there were very many single bones: the greater number were embedded in a space 200 yards square. The preponderance of the Edental quadrupeds is remarkable; as is, in contrast with the beds of Monte Hermoso, the absence of Rodents. Most of the bones are now in a soft and friable condition, and, like the shells, do not emit when burnt an animal odour. The decayed state of the bones may be partly owing to their late exposure to the air and tidal-waves. Barnacles, Serpulae, and corallines are attached to many of the bones, but I neglected to observe whether these might not have grown on them since being exposed to the present tidal action (After having packed up my specimens at Bahia Blanca, this point occurred to me, and I noted it; but forgot it on my return, until the remains had been cleaned and oiled: my attention has been lately called to the subject by some remarks by M. d'Orbigny.); but I believe that some of the barnacles must have grown on the Scelidotherium, soon after being deposited, and before being WHOLLY covered up by the gravel. Besides the remains in the condition here described, I found one single fragment of bone very much rolled, and as black as jet, so as perfectly to resemble some of the remains from Monte Hermoso.

Very many of the bones had been broken, abraded, and rolled, before being embedded. Others, even some of those included in the coarsest parts of the the now hard conglomerate, still retain all their minutest prominences perfectly preserved; so that I conclude that they probably were protected by skin, flesh, or ligaments, whilst being covered up. In the case of the Scelidotherium, it is quite certain that the whole skeleton was held together by its ligaments, when deposited in the gravel in which I found it. Some cervical vertebrae and a humerus of corresponding size lay so close together, as did some ribs and the bones of a leg, that I thought that they must originally have belonged to two skeletons, and not have been washed in single; but as remains were here very numerous, I will not lay much stress on these two cases. We have just seen that the armour of the Dasypoid quadruped was certainly embedded together with some of the bones of the feet.

Professor Ehrenberg has examined for me specimens of the finer matter from in contact with these mammiferous remains: he finds in them two Polygastrica, decidedly marine forms; and six Phytolitharia, of which one is probably marine, and the others either of fresh-water or terrestrial origin. ("Monatsberichten der Akad. zu Berlin" April 1845. The list consists of:—

POLYGASTRICA.
Gallionella sulcata.
Stauroptera aspera? fragm.
PHYTOLITHARIA.
Lithasteriscus tuberculatus.
Lithostylidium Clepsammidium.
Lithostylidium quadratum.
Lithostylidium rude.
Lithostylidium unidentatum.
Spongolithis acicularis.)

Only one of these eight microscopical bodies is common to the nine from Monte Hermoso: but five of them are in common with those from the Pampean mud on the banks of the Parana. The presence of any fresh-water infusoria, considering the aridity of the surrounding country, is here remarkable: the most probable explanation appears to be, that these microscopical organisms were washed out of the adjoining great Pampean formation during its denudation, and afterwards redeposited.

We will now see what conclusions may be drawn from the facts above detailed. It is certain that the gravel-beds and intermediate red mud were deposited within the period, when existing species of Mollusca held to each other nearly the same relative proportions as they do on the present coast. These beds, from the number of littoral species, must have been accumulated in shallow water; but not, judging from the stratification of the gravel and the layers of marl, on a beach. From the manner in which the red clay fills up furrows in the underlying gravel, and is in some parts itself furrowed by the overlying gravel, whilst in other parts it either insensibly passes into, or alternates with, this upper gravel, we may infer several local changes in the currents, perhaps caused by slight changes, up or down, in the level of the land. By the elevation of these beds, to which period the alluvial mantle with pumice-pebbles, land and sea-shells belongs, the plain of Punta Alta, from twenty to thirty feet in height, was formed. In this neighbourhood there are other and higher sea-formed plains and lines of cliffs in the Pampean formation worn by the denuding action of the waves at different levels. Hence we can easily understand the presence of rounded masses of tosca-rock in this lowest plain; and likewise, as the cliffs at Monte Hermoso with their mammiferous remains stand at a higher level, the presence of the one much-rolled fragment of bone which was as black as jet: possibly some few of the other much-rolled bones may have been similarly derived, though I saw only the one fragment, in the same condition with those from Monte Hermoso. M. d'Orbigny has suggested that all these mammiferous remains may have been washed out of the Pampean formation, and afterwards redeposited together with the recent shells. ("Voyage" Part. Geolog.) Undoubtedly it is a marvellous fact that these numerous gigantic quadrupeds, belonging, with the exception of the Equus curvidens, to seven extinct genera, and one, namely, the Toxodon, not falling into any existing family, should have co-existed with Mollusca, all of which are still living species; but analogous facts have been observed in North America and in Europe. In the first place, it should not be overlooked, that most of the co-embedded shells have a more ancient and altered appearance than the bones. In the second place, is it probable that numerous bones not hardened by silex or any other mineral, could have retained their delicate prominences and surfaces perfect if they had been washed out of one deposit, and re-embedded in another:—this later deposit being formed of large, hard pebbles, arranged by the action of currents or breakers in shallow water into variously curved and inclined layers? The bones which are now in so perfect a state of preservation, must, I conceive, have been fresh and sound when embedded, and probably were protected by skin, flesh, or ligaments. The skeleton of the Scelidotherium indisputably was deposited entire: shall we say that when held together by its matrix it was washed out of an old gravel-bed (totally unlike in character to the Pampean formation), and re-embedded in another gravel-bed, composed (I speak after careful comparison) of exactly the same kind of pebbles, in the same kind of cement? I will lay no stress on the two cases of several ribs and bones of the extremities having APPARENTLY been embedded in their proper relative position: but will any one be so bold as to affirm that it is possible, that a piece of the thin tessellated armour of a Dasypoid quadruped, at least three feet long and two in width, and now so tender that I was unable with the utmost care to extract a fragment more than two or three inches square, could have been washed out of one bed, and re-embedded in another, together with some of the small bones of the feet, without having been dashed into atoms? We must then wholly reject M. d'Orbigny's supposition, and admit as certain, that the Scelidotherium and the large Dasypoid quadruped, and as highly probable, that the Toxodon, Megatherium, etc., some of the bones of which are perfectly preserved, were embedded for the first time, and in a fresh condition, in the strata in which they were found entombed. These gigantic quadrupeds, therefore, though belonging to extinct genera and families, coexisted with the twenty above-enumerated Mollusca, the barnacle and two corals, still living on this coast. From the rolled fragment of black bone, and from the plain of Punta Alta being lower than that of Monte Hermoso, I conclude that the coarse sub-littoral deposits of Punta Alta, are of subsequent origin to the Pampean mud of Monte Hermoso; and the beds at this latter place, as we have seen, are probably of subsequent origin to the high tosca-plain round the Sierra Ventana: we shall, however, return, at the end of this chapter, to the consideration of these several stages in the great Pampean formation.

BUENOS AYRES TO ST. FE BAJADA, IN ENTRE RIOS.

For some distance northward of Buenos Ayres, the escarpment of the Pampean formation does not approach very near to the Plata, and it is concealed by vegetation: but in sections on the banks of the Rios Luxan, Areco, and Arrecifes, I observed both pale and dark reddish Pampean mud, with small, whitish concretions of tosca; at all these places mammiferous remains have been found. In the cliffs on the Parana, at San Nicolas, the Pampean mud contains but little tosca; here M. d'Orbigny found the remains of two rodents (Ctenomys Bonariensis and Kerodon antiquus) and the jaw of a Canis: when on the river I could clearly distinguish in this fine line of cliffs, "horizontal lines of variation both in tint and compactness." (I quote these words from my note-book, as written down on the spot, on account of the general absence of stratification in the Pampean formation having been insisted on by M. d'Orbigny as a proof of the diluvial origin of this great deposit.) The plain northward of this point is very level, but with some depressions and lakes; I estimated its height at from forty to sixty feet above the Parana. At the A. Medio the bright red Pampean mud contains scarcely any tosca-rock; whilst at a short distance the stream of the Pabon, forms a cascade, about twenty feet in height, over a cavernous mass of two varieties of tosca-rock; of which one is very compact and semi- crystalline, with seams of crystallised carbonate of lime: similar compact varieties are met with on the Salidillo and Seco. The absolute identity (I speak after a comparison of my specimens) between some of these varieties, and those from Tapalguen, and from the ridge south of Bahia Blanca, a distance of

400 miles of latitude, is very striking.

At Rosario there is but little tosca-rock: near this place I first noticed at the edge of the river traces of an underlying formation, which, twenty- five miles higher up in the estancia of Gorodona, consists of a pale yellowish clay, abounding with concretionary cylinders of a ferruginous sandstone. This bed, which is probably the equivalent of the older tertiary marine strata, immediately to be described in Entre Rios, only just rises above the level of the Parana when low. The rest of the cliff at Gorodona, is formed of red Pampean mud, with, in the lower part, many concretions of tosca, some stalacti-formed, and with only a few in the upper part: at the height of six feet above the river, two gigantic skeletons of the Mastodon Andium were here embedded; their bones were scattered a few feet apart, but many of them still held their proper relative positions: they were much decayed and as soft as cheese, so that even one of the great molar teeth fell into pieces in my hand. We here see that the Pampean deposit contains mammiferous remains close to its base. On the banks of the Carcarana, a few miles distant, the lowest bed visible was pale Pampean mud, with masses of tosca-rock, in one of which I found a much decayed tooth of the Mastodon: above this bed, there was a thin layer almost composed of small concretions of white tosca, out of which I extracted a well preserved, but slightly broken tooth of Toxodon Platensis: above this there was an unusual bed of very soft impure sandstone. In this neighbourhood I noticed many single embedded bones, and I heard of others having been found in so perfect a state that they were long used as gate-posts: the Jesuit Falkner found here the dermal armour of some gigantic Edental quadruped.

In some of the red mud scraped from a tooth of one of the Mastodons at Gorodona, Professor Ehrenberg finds seven Polygastrica and thirteen Phytolitharia, all of them, I believe, with two exceptions, already known species. ("Monatsberichten der konig. Akad. zu Berlin" April 1845. The list consists of:—

POLYGASTRICA.
Campylodiscus clypeus.
Coscinodiscus subtilis.
Coscinodiscus al. sp.
Eunotia.
Gallionella granulata.
Himantidium gracile.
Pinnularia borealis.)

Of these twenty, the preponderating number are of fresh-water origin; only two species of Coscinodiscus and a Spongolithis show the direct influence of the sea; therefore Professor Ehrenberg arrives at the important conclusion that the deposit must have been of brackish-water origin. Of the thirteen Phytolitharia, nine are met with in the two deposits in Bahia Blanca, where there is evidence from two other species of Polygastrica that the beds were accumulated in brackish water. The traces of coral, sponges, and Polythalamia, found by Dr. Carpenter in the tosca-rock (of which I must observe the greater number of specimens were from the upper beds in the southern parts of the formation), apparently show a more purely marine origin.

At ST. FE BAJADA, in Entre Rios, the cliffs, estimated at between sixty and seventy feet in height, expose an interesting section: the lower half consists of tertiary strata with marine shells, and the upper half of the Pampean formation. The lowest bed is an obliquely laminated, blackish, indurated mud, with distinct traces of vegetable remains. (M. d'Orbigny "Voyage" Part. Geolog, has given a detailed description of this section, but as he does not mention this lowest bed, it may have been concealed when he was there by the river. There is a considerable discrepancy between his description and mine, which I can only account for by the beds themselves varying considerably in short distances.) Above this there is a thick bed of yellowish sandy clay, with much crystallised gypsum and many shells of Ostreae, Pectens, and Arcae: above this there generally comes an arenaceous crystalline limestone, but there is sometimes interposed a bed, about twelve feet thick, of dark green, soapy clay, weathering into small angular fragments. The limestone, where purest, is white, highly crystalline, and full of cavities: it includes small pebbles of quartz, broken shells, teeth of sharks, and sometimes, as I was informed, large bones: it often contains so much sand as to pass into a calcareous sandstone, and in such parts the great Ostrea Patagonica chiefly abounds. (Captain Sulivan, R.N., has given me a specimen of this shell, which he found in the cliffs at Point Cerrito, between twenty and thirty miles above the Bajada.) In the upper part, the limestone alternates with layers of fine white sand. The shells included in these beds have been named for me by M. d'Orbigny: they consist of:—

1. Ostrea Patagonica, d'Orbigny, "Voyage" Part. Pal. 2. Ostrea Alvarezii, d'Orbigny, "Voyage" Part. Pal. 3. Pecten Paranensis, d'Orbigny, "Voyage" Part. Pal. 4. Pecten Darwinianus, d'Orbigny, "Voyage" Part. Pal. 5. Venus Munsterii, d'Orbigny, "Voyage" Pal. 6. Arca Bonplandiana, d'Orbigny, "Voyage" Pal. 7. Cardium Platense, d'Orbigny, "Voyage" Pal. 8. Tellina, probably nov. species, but too imperfect for description.

PHYTOLITHARIA.
Lithasteriscus tuberculatus.
Lithodontium bursa.
Lithodontium furcatum.
Lithodontium rostratum.
Lithostylidium Amphiodon.
Lithostylidium Clepsammidium.
Lithostylidium Hamus.
Lithostylidium polyedrum.
Lithostylidium quadratum.
Lithostylidium rude.
Lithostylidium Serra.
Lithostylidium unidentatum.
Spongolithis Fustis.

These species are all extinct: the six first were found by M. d'Orbigny and myself in the formations of the Rio Negro, S. Josef, and other parts of Patagonia; and therefore, as first observed by M. d'Orbigny, these beds certainly belong to the great Patagonian formation, which will be described in the ensuing chapter, and which we shall see must be considered as a very ancient tertiary one. North of the Bajada, M. d'Orbigny found, in beds which he considers as lying beneath the strata here described, remains of a Toxodon, which he has named as a distinct species from the T. Platensis of the Pampean formation. Much silicified wood is found on the banks of the Parana (and likewise on the Uruguay), and I was informed that they come out of these lower beds; four specimens collected by myself are dicotyledonous.

The upper half of the cliff, to a thickness of about thirty feet, consists of Pampean mud, of which the lower part is pale-coloured, and the upper part of a brighter red, with some irregular layers of an arenaceous variety of tosca, and a few small concretions of the ordinary kind. Close above the marine limestone, there is a thin stratum with a concretionary outline of white hard tosca-rock or marl, which may be considered either as the uppermost bed of the inferior deposits, or the lowest of the Pampean formation; at one time I considered this bed as marking a passage between the two formations: but I have since become convinced that I was deceived on this point. In the section on the Parana, I did not find any mammiferous remains; but at two miles distance on the A. Tapas (a tributary of the Conchitas), they were extremely numerous in a low cliff of red Pampean mud with small concretions, precisely like the upper bed on the Parana. Most of the bones were solitary and much decayed; but I saw the dermal armour of a gigantic Edental quadruped, forming a caldron-like hollow, four or five feet in diameter, out of which, as I was informed, the almost entire skeleton had been lately removed. I found single teeth of the Mastodon Andium, Toxodon Platensis, and Equus curvidens, near to each other. As this latter tooth approaches closely to that of the common horse, I paid particular attention to its true embedment, for I did not at that time know that there was a similar tooth hidden in the matrix with the other mammiferous remains from Punta Alta. It is an interesting circumstance, that Professor Owen finds that the teeth of this horse approach more closely in their peculiar curvature to a fossil specimen brought by Mr. Lyell from North America, than to those of any other species of Equus. (Lyell "Travels in North America" volume 1 and "Proceedings of Geological Society" volume 4)

The underlying marine tertiary strata extend over a wide area: I was assured that they can be traced in ravines in an east and west line across Entre Rios to the Uruguay, a distance of about 135 miles. In a S.E. direction I heard of their existence at the head of the R. Nankay; and at P. Gorda in Banda Oriental, a distance of 170 miles, I found the same limestone, containing the same fossil shells, lying at about the same level above the river as at St. Fe. In a southerly direction, these beds sink in height, for at another P. Gorda in Entre Rios, the limestone is seen at a much less height; and there can be little doubt that the yellowish sandy clay, on a level with the river, between the Carcarana and S. Nicholas, belongs to this same formation; as perhaps do the beds of sand at Buenos Ayres, which lie at the bottom of the Pampean formation, about sixty feet beneath the surface of the Plata. The southerly declination of these beds may perhaps be due, not to unequal elevation, but to the original form of the bottom of the sea, sloping from land situated to the north; for that land existed at no great distance, we have evidence in the vegetable remains in the lowest bed at St. Fe; and in the silicified wood and in the bones of Toxodon Paranensis, found (according to M. d'Orbigny) in still lower strata.

BANDA ORIENTAL.

This province lies on the northern side of the Plata, and eastward of the Uruguay: it has a gentle undulatory surface, with a basis of primary rocks; and is in most parts covered up with an unstratified mass, of no great thickness, of reddish Pampean mud. In the eastern half, near Maldonado, this deposit is more arenaceous than in the Pampas, it contains many though small concretions of marl or tosca-rock, and others of highly ferruginous sandstone; in one section, only a few yards in depth, it rested on stratified sand. Near Monte Video this deposit in some spots appears to be of greater thickness; and the remains of the Glyptodon and other extinct mammifers have been found in it. In the long line of cliffs, between fifty and sixty feet in height, called the Barrancas de S. Gregorio, which extend westward of the Rio S. Lucia, the lower half is formed of coarse sand of quartz and feldspar without mica, like that now cast up on the beach near Maldonado; and the upper half of Pampean mud, varying in colour and containing honeycombed veins of soft calcareous matter and small concretions of tosca-rock arranged in lines, and likewise a few pebbles of quartz. This deposit fills up hollows and furrows in the underlying sand; appearing as if water charged with mud had invaded a sandy beach. These cliffs extend far westward, and at a distance of sixty miles, near Colonia del Sacramiento, I found the Pampean deposit resting in some places on this sand, and in others on the primary rocks: between the sand and the reddish mud, there appeared to be interposed, but the section was not a very good one, a thin bed of shells of an existing Mytilus, still partially retaining their colour. The Pampean formation in Banda Oriental might readily be mistaken for an alluvial deposit: compared with that of the Pampas, it is often more sandy, and contains small fragments of quartz; the concretions are much smaller, and there are no extensive masses of tosca-rock.

In the extreme western parts of this province, between the Uruguay and a line drawn from Colonia to the R. Perdido (a tributary of the R. Negro), the formations are far more complicated. Besides primary rocks, we meet with extensive tracts and many flat-topped, horizontally stratified, cliff- bounded, isolated hills of tertiary strata, varying extraordinarily in mineralogical nature, some identical with the old marine beds of St. Fe Bajada, and some with those of the much more recent Pampean formation. There are, also, extensive LOW tracts of country covered with a deposit containing mammiferous remains, precisely like that just described in the more eastern parts of the province. Although from the smooth and unbroken state of the country, I never obtained a section of this latter deposit close to the foot of the higher tertiary hills, yet I have not the least doubt that it is of quite subsequent origin; having been deposited after the sea had worn the tertiary strata into the cliff-bounded hills. This later formation, which is certainly the equivalent of that of the Pampas, is well seen in the valleys in the estancia of Berquelo, near Mercedes; it here consists of reddish earth, full of rounded grains of quartz, and with some small concretions of tosca-rock arranged in horizontal lines, so as perfectly to resemble, except in containing a little calcareous matter, the formation in the eastern parts of Banda Oriental, in Entre Rios, and at other places: in this estancia the skeleton of a great Edental quadruped was found. In the valley of the Sarandis, at the distance of only a few miles, this deposit has a somewhat different character, being whiter, softer, finer-grained, and full of little cavities, and consequently of little specific gravity; nor does it contain any concretions or calcareous matter: I here procured a head, which when first discovered must have been quite perfect, of the Toxodon Platensis, another of a Mylodon (This head was at first considered by Professor Owen (in the "Zoology of the 'Beagle's' Voyage") as belonging to a distinct genus, namely, Glossotherium.), perhaps M. Darwinii, and a large piece of dermal armour, differing from that of the Glyptodon clavipes. These bones are remarkable from their extraordinarily fresh appearance; when held over a lamp of spirits of wine, they give out a strong odour and burn with a small flame; Mr. T. Reeks has been so kind as to analyse some of the fragments, and he finds that they contain about 7 per cent of animal matter, and 8 per cent of water. (Liebig "Chemistry of Agriculture" states that fresh dry bones contain from 32 to 33 per cent of dry gelatine. See also Dr. Daubeny, in "Edinburgh New Philosophical Journal" volume 37)

The older tertiary strata, forming the higher isolated hills and extensive tracts of country, vary, as I have said, extraordinarily in composition: within the distance of a few miles, I sometimes passed over crystalline limestone with agate, calcareous tuffs, and marly rocks, all passing into each other,—red and pale mud with concretions of tosca-rock, quite like the Pampean formation,—calcareous conglomerates and sandstones,—bright red sandstones passing either into red conglomerate, or into white sandstone,—hard siliceous sandstones, jaspery and chalcedonic rocks, and numerous other subordinate varieties. I was unable to mark out the relations of all these strata, and will describe only a few distinct sections:—in the cliffs between P. Gorda on the Uruguay and the A. de Vivoras, the upper bed is crystalline cellular limestone often passing into calcareous sandstone, with impressions of some of the same shells as at St. Fe Bajada; at P. Gorda, this limestone is interstratified with and rests on, white sand, which covers a bed about thirty feet thick of pale-coloured clay, with many shells of the great Ostrea Patagonica (In my "Journal" 1st edition, I have hastily and inaccurately stated that the Pampean mud, which is found over the eastern part of B. Oriental, lies OVER the limestone at P. Gorda; I should have said that there was reason to infer that it was a subsequent or superior deposit.): beneath this, in the vertical cliff, nearly on a level with the river, there is a bed of red mud absolutely like the Pampean deposit, with numerous often large concretions of perfectly characterised white, compact tosca-rock. At the mouth of the Vivoras, the river flows over a pale cavernous tosca-rock, quite like that in the Pampas, and this APPEARED to underlie the crystalline limestone; but the section was not unequivocal like that at P. Gorda. These beds now form only a narrow and much denuded strip of land; but they must once have extended much further; for on the next stream, south of the S. Juan, Captain Sulivan, R.N., found a little cliff, only just above the surface of the river, with numerous shells of the Venus Munsterii, D'Orbigny,—one of the species occurring at St. Fe, and of which there are casts at P. Gorda: the line of cliffs of the subsequently deposited true Pampean mud, extend from Colonia to within half a mile of this spot, and no doubt once covered up this denuded marine stratum. Again at Colonia, a Frenchman found, in digging the foundations of a house, a great mass of the Ostrea Patagonica (of which I saw many fragments), packed together just beneath the surface, and directly superimposed on the gneiss. These sections are important: M. d'Orbigny is unwilling to believe that beds of the same nature with the Pampean formation ever underlie the ancient marine tertiary strata; and I was as much surprised at it as he could have been; but the vertical cliff at P. Gorda allowed of no mistake, and I must be permitted to affirm, that after having examined the country from the Colorado to St. Fe Bajada, I could not be deceived in the mineralogical character of the Pampean deposit.

Moreover, in a precipitous part of the ravine of Las Bocas, a red sandstone is distinctly seen to overlie a thick bed of pale mud, also quite like the Pampean formation, abounding with concretions of true tosca-rock. This sandstone extends over many miles of country: it is as red as the brightest volcanic scoriae; it sometimes passes into a coarse red conglomerate composed of the underlying primary rocks; and often passes into a soft white sandstone with red streaks. At the Calera de los Huerfanos, only a quarter of a mile south of where I first met with the red sandstone, the crystalline white limestone is quarried: as this bed is the uppermost, and as it often passes into calcareous sandstone, interstratified with pure sand; and as the red sandstone likewise passes into soft white sandstone, and is also the uppermost bed, I believe that these two beds, though so different, are equivalents. A few leagues southward of these two places, on each side of the low primary range of S. Juan, there are some flat-topped, cliff-bounded, separate little hills, very similar to those fringing the primary ranges in the great plain south of Buenos Ayres: they are composed- -1st, of calcareous tuff with many particles of quartz, sometimes passing into a coarse conglomerate; 2nd, of a stone undistinguishable on the closest inspection from the compacter varieties of tosca-rock; and 3rd, of semi-crystalline limestone, including nodules of agate: these three varieties pass insensibly into each other, and as they form the uppermost stratum in this district, I believe that they, also, are the equivalents of the pure crystalline limestone, and of the red and white sandstones and conglomerates.

Between these points and Mercedes on the Rio Negro, there are scarcely any good sections, the road passing over limestone, tosca-rock, calcareous and bright red sandstones, and near the source of the San Salvador over a wide extent of jaspery rocks, with much milky agate, like that in the limestone near San Juan. In the estancia of Berquelo, the separate, flat-topped, cliff-bounded hills are rather higher than in the other parts of the country; they range in a N.E. and S.W. direction; their uppermost beds consist of the same bright red sandstone, passing sometimes into a conglomerate, and in the lower part into soft white sandstone, and even into loose sand: beneath this sandstone, I saw in two places layers of calcareous and marly rocks, and in one place red Pampean-like earth; at the base of these sections, there was a hard, stratified, white sandstone, with chalcedonic layers. Near Mercedes, beds of the same nature and apparently of the same age, are associated with compact, white, crystalline limestone, including much botryoidal agate, and singular masses, like porcelain, but really composed of a calcareo-siliceous paste. In sinking wells in this district the chalcedonic strata seem to be the lowest. Beds, such as there described, occur over the whole of this neighbourhood; but twenty miles further up the R. Negro, in the cliffs of Perika, which are about fifty feet in height, the upper bed is a prettily variegated chalcedony, mingled with a pure white tallowy limestone; beneath this there is a conglomerate of quartz and granite; beneath this many sandstones, some highly calcareous; and the whole lower two-thirds of the cliff consists of earthy calcareous beds of various degrees of purity, with one layer of reddish Pampean-like mud.

When examining the agates, the chalcedonic and jaspery rocks, some of the limestones, and even the bright red sandstones, I was forcibly struck with their resemblance to deposits formed in the neighbourhood of volcanic action. I now find that M. Isabelle, in his "Voyage a Buenos Ayres," has described closely similar beds on Itaquy and Ibicuy (which enter the Uruguay some way north of the R. Negro) and these beds include fragments of red decomposed true scoriae hardened by zeolite, and of black retinite: we have then here good evidence of volcanic action during our tertiary period. Still further north, near S. Anna, where the Parana makes a remarkable bend, M. Bonpland found some singular amygdaloidal rocks, which perhaps may belong to this same epoch. (M. d'Orbigny "Voyage" Part. Geolog.) I may remark that, judging from the size and well-rounded condition of the blocks of rock in the above-described conglomerates, masses of primary formation probably existed at this tertiary period above water: there is, also, according to M. Isabelle, much conglomerate further north, at Salto.

From whatever source and through whatever means the great Pampean formation originated, we here have, I must repeat, unequivocal evidence of a similar action at a period before that of the deposition of the marine tertiary strata with extinct shells, at Santa Fe and P. Gorda. During also the deposition of these strata, we have in the intercalated layers of red Pampean-like mud and tosca-rock, and in the passage near S. Juan of the semi-crystalline limestones with agate into tosca undistinguishable from that of the Pampas, evidence of the same action, though continued only at intervals and in a feeble manner. We have further seen that in this district, at a period not only subsequent to the deposition of the tertiary strata, but to their upheavement and most extensive denudation, true Pampean mud with its usual characters and including mammiferous remains, was deposited round and between the hills or islets formed of these tertiary strata, and over the whole eastern and low primary districts of Banda Oriental.

EARTHY MASS, WITH EXTINCT MAMMIFEROUS REMAINS, OVER THE PORPHYRITIC GRAVEL AT S. JULIAN, LATITUDE 49 DEGREES 14' S., IN PATAGONIA.

(FIGURE 16. SECTION OF THE LOWEST PLAIN AT PORT S. JULIAN.
(Section through beds from top to bottom: A, B, C, D, E, F.)

AA. Superficial bed of reddish earth, with the remains of the Macrauchenia, and with recent sea-shells on the surface.

B. Gravel of porphyritic rocks.

C. and D. Pumiceous mudstone.—Ancient tertiary formation.

E. and F. Sandstone and argillaceous beds.—Ancient tertiary formation.)

This case, though not coming strictly under the Pampean formation, may be conveniently given here. On the south side of the harbour, there is a nearly level plain (mentioned in the First Chapter) about seven miles long, and three or four miles wide, estimated at ninety feet in height, and bordered by perpendicular cliffs, of which a section is represented in Figure 16.

The lower old tertiary strata (to be described in the next chapter) are covered by the usual gravel bed; and this by an irregular earthy, sometimes sandy mass, seldom more than two or three feet in thickness, except where it fills up furrows or gullies worn not only through the underlying gravel, but even through the upper tertiary beds. This earthy mass is of a pale reddish colour, like the less pure varieties of Pampean mud in Banda Oriental; it includes small calcareous concretions, like those of tosca- rock but more arenaceous, and other concretions of a greenish, indurated argillaceous substance: a few pebbles, also, from the underlying gravel-bed are also included in it, and these being occasionally arranged in horizontal lines, show that the mass is of sub-aqueous origin. On the surface and embedded in the superficial parts, there are numerous shells, partially retaining their colours, of three or four of the now commonest littoral species. Near the bottom of one deep furrow (represented in Figure 16), filled up with this earthy deposit, I found a large part of the skeleton of the Macrauchenia Patachonica—a gigantic and most extraordinary pachyderm, allied, according to Professor Owen, to the Palaeotherium, but with affinities to the Ruminants, especially to the American division of the Camelidae. Several of the vertebrae in a chain, and nearly all the bones of one of the limbs, even to the smallest bones of the foot, were embedded in their proper relative positions: hence the skeleton was certainly united by its flesh or ligaments, when enveloped in the mud. This earthy mass, with its concretions and mammiferous remains, filling up furrows in the underlying gravel, certainly presents a very striking resemblance to some of the sections (for instance, at P. Alta in B. Blanca, or at the

Barrancas de S. Gregorio) in the Pampean formation; but I must believe that this resemblance is only accidental. I suspect that the mud which at the present day is accumulating in deep and narrow gullies at the head of the harbour, would, after elevation, present a very similar appearance. The southernmost part of the true Pampean formation, namely, on the Colorado, lies 560 miles of latitude north of this point. (In the succeeding chapter I shall have to refer to a great deposit of extinct mammiferous remains, lately discovered by Captain Sulivan, R.N., at a point still further south, namely, at the R. Gallegos; their age must at present remain doubtful.)

With respect to the age of the Macrauchenia, the shells on the surface prove that the mass in which the skeleton was enveloped has been elevated above the sea within the recent period: I did not see any of the shells embedded at a sufficient depth to assure me (though it be highly probable) that the whole thickness of the mass was contemporaneous with these INDIVIDUAL SPECIMENS. That the Macrauchenia lived subsequently to the spreading out of the gravel on this plain is certain; and that this gravel, at the height of ninety feet, was spread out long after the existence of recent shells, is scarcely less certain. For, it was shown in the First Chapter, that this line of coast has been upheaved with remarkable equability, and that over a vast space both north and south of S. Julian, recent species of shells are strewed on (or embedded in) the surface of the 250 feet plain, and of the 350 feet plain up to a height of 400 feet. These wide step-formed plains have been formed by the denuding action of the coast-waves on the old tertiary strata; and therefore, when the surface of the 350 feet plain, with the shells on it, first rose above the level of the sea, the 250 feet plain did not exist, and its formation, as well as the spreading out of the gravel on its summit, must have taken place subsequently. So also the denudation and the gravel-covering of the 90 feet plain must have taken place subsequently to the elevation of the 250 feet plain, on which recent shells are also strewed. Hence there cannot be any doubt that the Macrauchenia, which certainly was entombed in a fresh state, and which must have been alive after the spreading out of the gravel on the 90 feet plain, existed, not only subsequently to the upraised shells on the surface of the 250 feet plain, but also to those on the 350 to 400 feet plain: these shells, eight in number (namely, three species of Mytilus, two of Patella, one Fusus, Voluta, and Balanus), are undoubtedly recent species, and are the commonest kinds now living on this coast. At Punta Alta in B. Blanca, I remarked how marvellous it was, that the Toxodon, a mammifer so unlike to all known genera, should have co-existed with twenty- three still living marine animals; and now we find that the Macrauchenia, a quadruped only a little less anomalous than the Toxodon, also co-existed with eight other still existing Mollusca: it should, moreover, be borne in mind, that a tooth of a pachydermatous animal was found with the other remains at Punta Alta, which Professor Owen thinks almost certainly belonged to the Macrauchenia.

Mr. Lyell has arrived at a highly important conclusion with respect to the age of the North American extinct mammifers (many of which are closely allied to, and even identical with, those of the Pampean formation), namely, that they lived subsequently to the period when erratic boulders were transported by the agency of floating ice in temperate latitudes. ("Geological Proceedings" volume 4) Now in the valley of the Santa Cruz, only fifty miles of latitude south of the spot where the Macrauchenia was entombed, vast numbers of gigantic, angular boulders, which must have been transported from the Cordillera on icebergs, lie strewed on the plain, at the height of 1,400 feet above the level of the sea. In ascending to this level, several step-formed plains must be crossed, all of which have necessarily required long time for their formation; hence the lowest or ninety feet plain, with its superficial bed containing the remains of the Macrauchenia, must have been formed very long subsequently to the period when the 1,400 feet plain was beneath the sea, and boulders were dropped on it from floating masses of ice. (It must not be inferred from these remarks, that the ice-action ceased in South America at this comparatively ancient period; for in Tierra del Fuego boulders were probably transported contemporaneously with, if not subsequently to, the formation of the ninety feet plain at S. Julian, and at other parts of the coast of Patagonia.)

Mr. Lyell's conclusion, therefore, is thus far confirmed in the southern hemisphere; and it is the more important, as one is naturally tempted to admit so simple an explanation, that it was the ice-period that caused the extinction of the numerous great mammifers which so lately swarmed over the two Americas.

SUMMARY AND CONCLUDING REMARKS ON THE PAMPEAN FORMATION.

One of its most striking features is its great extent; I passed continuously over it from the Colorado to St. Fe Bajada, a distance of 500 geographical miles; and M. d'Orbigny traced it for 250 miles further north. In the latitude of the Plata, I examined this formation at intervals over an east and west line of 300 miles from Maldonado to the R. Carcarana; and M. d'Orbigny believes it extends 100 miles further inland: from Mr. Caldcleugh's travels, however, I should have thought that it had extended, south of the Cordovese range, to near Mendoza, and I may add that I heard of great bones having been found high up the R. Quinto. Hence the area of the Pampean formation, as remarked by M. d'Orbigny, is probably at least equal to that of France, and perhaps twice or thrice as great. In a basin, surrounded by gravel-cliff (at a height of nearly three thousand feet), south of Mendoza, there is, as described in the Third Chapter, a deposit very like the Pampean, interstratified with other matter; and again at S. Julian's, in Patagonia, 560 miles south of the Colorado, a small irregular bed of a nearly similar nature contains, as we have just seen, mammiferous remains. In the provinces of Moxos and Chiquitos (1,000 miles northward of the Pampas), and in Bolivia, at a height of 4,000 metres, M. d'Orbigny has described similar deposits, which he believes to have been formed by the same agency contemporaneously with the Pampean formation. Considering the immense distances between these several points, and their different heights, it appears to me infinitely more probable, that this similarity has resulted not from contemporaneousness of origin, but from the similarity of the rocky framework of the continent: it is known that in Brazil an immense area consists of gneissic rocks, and we shall hereafter see, over how great a length the plutonic rocks of the Cordillera, the overlying purple porphyries, and the trachytic ejections, are almost identical in nature.

Three theories on the origin of the Pampean formation have been propounded:—First, that of a great debacle by M. d'Orbigny; this seems founded chiefly on the absence of stratification, and on the number of embedded remains of terrestrial quadrupeds. Although the Pampean formation (like so many argillaceous deposits) is not divided into distinct and separate strata, yet we have seen that in one good section it was striped with horizontal zones of colour, and that in several specified places the upper and lower parts differed, not only considerably in colour, but greatly in constitution. In the southern part of the Pampas the upper mass (to a certain extent stratified) generally consists of hard tosca-rock, and the lower part of red Pampean mud, often itself divided into two or more masses, varying in colour and in the quantity of included calcareous matter. In Western Banda Oriental, beds of a similar nature, but of a greater age, conformably underlie and are intercalated with the regularly stratified tertiary formation. As a general rule, the marly concretions are arranged in horizontal lines, sometimes united into irregular strata: surely, if the mud had been tumultuously deposited in mass, the included calcareous matter would have segregated itself irregularly, and not into nodules arranged in horizontal lines, one above the other and often far apart: this arrangement appears to me to prove that mud, differing slightly in composition, was successively and quietly deposited. On the theory of a debacle, a prodigious amount of mud, without a single pebble, is supposed to have been borne over the wide surface of the Pampas, when under water: on the other hand, over the whole of Patagonia, the same or another debacle is supposed to have borne nothing but gravel,—the gravel and the fine mud in the neighbourhood of the Rios Negro and Colorado having been borne to an equal distance from the Cordillera, or imagined line of disturbance: assuredly directly opposite effects ought not to be attributed to the same agency. Where, again, could a mass of fine sediment, charged with calcareous matter in a fit state for chemical segregation, and in quantity sufficient to cover an area at least 750 miles long, and 400 miles broad, to a depth of from twenty to thirty feet to a hundred feet, have been accumulated, ready to be transported by the supposed debacle? To my mind it is little short of demonstration, that a great lapse of time was necessary for the production and deposition of the enormous amount of mudlike matter forming the Pampas; nor should I have noticed the theory of a debacle, had it not been adduced by a naturalist so eminent as M. d'Orbigny.

A second theory, first suggested, I believe, by Sir W. Parish, is that the Pampean formation was thrown down on low and marshy plains by the rivers of this country before they assumed their present courses. The appearance and composition of the deposit, the manner in which it slopes up and round the primary ranges, the nature of the underlying marine beds, the estuary and sea-shells on the surface, the overlying sandstone beds at M. Hermoso, are all quite opposed to this view. Nor do I believe that there is a single instance of a skeleton of one of the extinct mammifers having been found in an upright position, as if it had been mired.

The third theory, of the truth of which I cannot entertain the smallest doubt, is that the Pampean formation was slowly accumulated at the mouth of the former estuary of the Plata and in the sea adjoining it. I have come to this conclusion from the reasons assigned against the two foregoing theories, and from simple geographical considerations. From the numerous shells of the Azara labiata lying loose on the surface of the plains, and near Buenos Ayres embedded in the tosca-rock, we know that this formation not only was formerly covered by, but that the uppermost parts were deposited in, the brackish water of the ancient La Plata. Southward and seaward of Buenos Ayres, the plains were upheaved from under water inhabited by true marine shells. We further know from Professor Ehrenberg's examination of the twenty microscopical organisms in the mud round the tooth of the Mastodon high up the course of the Parana, that the bottom-most part of this formation was of brackish-water origin. A similar conclusion must be extended to the beds of like composition, at the level of the sea and under it, at M. Hermoso in Bahia Blanca. Dr. Carpenter finds that the harder varieties of tosca-rock, collected chiefly to the south, contain marine spongoid bodies, minute fragments of shells, corals, and Polythalamia; these perhaps may have been drifted inwards by the tides, from the more open parts of the sea. The absence of shells, throughout this deposit, with the exception of the uppermost layers near Buenos Ayres, is a remarkable fact: can it be explained by the brackish condition of the water, or by the deep mud at the bottom? I have stated that both the reddish mud and the concretions of tosca-rock are often penetrated by minute, linear cavities, such as frequently may be observed in fresh-water calcareous deposits:—were they produced by the burrowing of small worms? Only on this view of the Pampean formation having been of estuary origin, can the extraordinary numbers (presently to be alluded to) of the embedded mammiferous remains be explained. (It is almost superfluous to give the numerous cases (for instance, in Sumatra; Lyell "Principles" volume 3 sixth edition, of the carcasses of animals having been washed out to sea by swollen rivers; but I may refer to a recent account by Mr. Bettington "Asiatic Society" 1845 June 21st, of oxen, deer, and bears being carried into the Gulf of Cambray; see also the account in my "Journal" 2nd edition, of the numbers of animals drowned in the Plata during the great, often recurrent, droughts.)

With respect to the first origin of the reddish mud, I will only remark, that the enormous area of Brazil consists in chief part of gneissic and other granitic rocks, which have suffered decomposition, and been converted into a red, gritty, argillaceous mass, to a greater depth than in any other country which I have seen. The mixture of rounded grains, and even of small fragments and pebbles of quartz, in the Pampean mud of Banda Oriental, is evidently due to the neighbouring and underlying primary rocks. The estuary mud was drifted during the Pampean period in a much more southerly course, owing probably to the east and west primary ridges south of the Plata not having been then elevated, than the mud of the Plata at present is; for it was formerly deposited as far south as the Colorado. The quantity of calcareous matter in this formation, especially in those large districts where the whole mass passes into tosca-rock, is very great: I have already remarked on the close resemblance in external and microscopical appearance, between this tosca-rock and the strata at Coquimbo, which have certainly resulted from the decay and attrition of recent shells: I dare not, however, extend this conclusion to the calcareous rocks of the Pampas, more especially as the underlying tertiary strata in western Banda Oriental show that at that period there was a copious emission of carbonate of lime, in connection with volcanic action. (I may add, that there are nearly similar superficial calcareous beds at King George's Sound in Australia; and these undoubtedly have been formed by the disintegration of marine remains see "Volcanic Islands" etc.. There is, however, something very remarkable in the frequency of superficial, thin beds of earthy calcareous matter, in districts where the surrounding rocks are not calcareous. Major Charters, in a Paper read before the Geographical Society April 13, 1840 and abstracted in the "Athenaeum", states that this is the case in parts of Mexico, and that he has observed similar appearances in many parts of South Africa. The circumstance of the uppermost stratum round the ragged Sierra Ventana, consisting of calcareous or marly matter, without any covering of alluvial matter, strikes me as very singular, in whatever manner we view the deposition and elevation of the Pampean formation.)

The Pampean formation, judging from its similar composition, and from the apparent absolute specific identity of some of its mammiferous remains, and from the generic resemblance of others, belongs over its vast area— throughout Banda Oriental, Entre Rios, and the wide extent of the Pampas as far south as the Colorado,—to the same geological epoch. The mammiferous remains occur at all depths from the top to the bottom of the deposit; and I may add that nowhere in the Pampas is there any appearance of much superficial denudation: some bones which I found near the Guardia del Monte were embedded close to the surface; and this appears to have been the case with many of those discovered in Banda Oriental: on the Matanzas, twenty miles south of Buenos Ayres, a Glyptodon was embedded five feet beneath the surface; numerous remains were found by S. Muniz, near Luxan, at an average depth of eighteen feet; in Buenos Ayres a skeleton was disinterred at sixty feet depth, and on the Parana I have described two skeletons of the Mastodon only five or six feet above the very base of the deposit. With respect to the age of this formation, as judged of by the ordinary standard of the existence of Mollusca, the only evidence within the limits of the true Pampas which is at all trustworthy, is afforded by the still living Azara labiata being embedded in tosca-rock near Buenos Ayres. At Punta Alta, however, we have seen that several of the extinct mammifers, most characteristic of the Pampean formation, co-existed with twenty species of Mollusca, a barnacle and two corals, all still living on this same coast;— for when we remember that the shells have a more ancient appearance than the bones; that many of the bones, though embedded in a coarse conglomerate, are perfectly preserved; that almost all the parts of the skeleton of the Scelidotherium, even to the knee-cap, were lying in their proper relative positions; and that a large piece of the fragile dermal armour of a Dasypoid quadruped, connected with some of the bones of the foot, had been entombed in a condition allowing the two sides to be doubled together, it must assuredly be admitted that these mammiferous remains were embedded in a fresh state, and therefore that the living animals co-existed with the co-embedded shells. Moreover, the Macrauchenia Patachonica (of which, according to Professor Owen, remains also occur in the Pampas of Buenos Ayres, and at Punta Alta) has been shown by satisfactory evidence of another kind, to have lived on the plains of Patagonia long after the period when the adjoining sea was first tenanted by its present commonest molluscous animals. We must, therefore, conclude that the Pampean formation belongs, in the ordinary geological sense of the word, to the Recent Period. (M. d'Orbigny believes "Voyage" Part. Geolog, that this formation, though "tres voisine de la notre, est neanmoins de beaucoup anterieure a notre creation.")

At St. Fe Bajada, the Pampean estuary formation, with its mammiferous remains, conformably overlies the marine tertiary strata, which (as first shown by M. d'Orbigny) are contemporaneous with those of Patagonia, and which, as we shall hereafter see, belong to a very ancient tertiary stage. When examining the junction between these two formations, I thought that the concretionary layer of marl marked a passage between the marine and estuary stages. M. d'Orbigny disputes this view (as given in my "Journal"), and I admit that it is erroneous, though in some degree excusable, from their conformability and from both abounding with calcareous matter. It would, indeed, have been a great anomaly if there had been a true passage between a deposit contemporaneous with existing species of mollusca, and one in which all the mollusca appear to be extinct. Northward of Santa Fe, M. d'Orbigny met with ferruginous sandstones, marly rocks, and other beds, which he considers as a distinct and lower formation; but the evidence that they are not parts of the same with an altered mineralogical character, does not appear to me quite satisfactory.

In Western Banda Oriental, while the marine tertiary strata were accumulating, there were volcanic eruptions, much silex and lime were precipitated from solution, coarse conglomerates were formed, being derived probably from adjoining land, and layers of red mud and marly rocks, like those of the Pampean formation, were occasionally deposited. The true Pampean deposit, with mammiferous remains, instead of as at Santa Fe overlying conformably the tertiary strata, is here seen at a lower level folding round and between the flat-topped, cliff-bounded hills, formed by a upheaval and denudation of these same tertiary strata. The upheaval, having occurred here earlier than at Santa Fe, may be naturally accounted for by the contemporaneous volcanic action. At the Barrancas de S. Gregorio, the Pampean deposit, as we have seen, overlies and fills up furrows in coarse sand, precisely like that now accumulating on the shores near the mouth of the Plata. I can hardly believe that this loose and coarse sand is contemporaneous with the old tertiary and often crystalline strata of the more western parts of the province; and am induced to suspect that it is of subsequent origin. If that section near Colonia could be implicitly trusted, in which, at a height of only fifteen feet above the Plata, a bed of fresh-looking mussels, of an existing littoral species, appeared to lie between the sand and the Pampean mud, I should conclude that Banda Oriental must have stood, when the coarse sand was accumulating, at only a little below its present level, and had then subsided, allowing the estuary Pampean mud to cover far and wide its surface up to a height of some hundred feet; and that after this subsidence the province had been uplifted to its present level.

In Western Banda Oriental, we know, from two unequivocal sections that there is a mass, absolutely undistinguishable from the true Pampean deposit, beneath the old tertiary strata. This inferior mass must be very much more ancient than the upper deposit with its mammiferous remains, for it lies beneath the tertiary strata in which all the shells are extinct. Nevertheless, the lower and upper masses, as well as some intermediate layers, are so similar in mineralogical character, that I cannot doubt that they are all of estuary origin, and have been derived from the same great source. At first it appeared to me extremely improbable, that mud of the same nature should have been deposited on nearly the same spot, during an immense lapse of time, namely, from a period equivalent perhaps to the Eocene of Europe to that of the Pampean formation. But as, at the very commencement of the Pampean period, if not at a still earlier period, the Sierra Ventana formed a boundary to the south,—the Cordillera or the plains in front of them to the west,—the whole province of Corrientes probably to the north, for, according to M. d'Orbigny, it is not covered by the Pampean deposit,—and Brazil, as known by the remains in the caves, to the north-east; and as again, during the older tertiary period, land already existed in Western Banda Oriental and near St. Fe Bajada, as may be inferred from the vegetable debris, from the quantities of silicified wood, and from the remains of a Toxodon found, according to M. d'Orbigny, in still lower strata, we may conclude, that at this ancient period a great expanse of water was surrounded by the same rocky framework which now bounds the plains of Pampean formation. This having been the case, the circumstance of sediment of the same nature having been deposited in the same area during an immense lapse of time, though highly remarkable, does not appear incredible.

The elevation of the Pampas, at least of the southern parts, has been slow and interrupted by several periods of rest, as may be inferred from the plains, cliffs, and lines of sand-dunes (with shells and pumice-pebbles) standing at different heights. I believe, also, that the Pampean mud continued to be deposited, after parts of this formation had already been elevated, in the same manner as mud would continue to be deposited in the estuary of the Plata, if the mud-banks on its shores were now uplifted and changed into plains: I believe in this from the improbability of so many skeletons and bones having been accumulated at one spot, where M. Hermoso now stands, at a depth of between eight hundred and one thousand feet, and at a vast distance from any land except small rocky islets,—as must have been the case, if the high tosca-plain round the Ventana and adjoining Sierras, had not been already uplifted and converted into land, supporting mammiferous animals. At Punta Alta we have good evidence that the gravel-strata, which certainly belong to the true Pampean period, were accumulated after the elevation in that neighbourhood of the main part of the Pampean deposit, whence the rounded masses of tosca-rock were derived, and that rolled fragment of black bone in the same peculiar condition with the remains at Monte Hermoso.

The number of the mammiferous remains embedded in the Pampas is, as I have remarked, wonderful: it should be borne in mind that they have almost exclusively been found in the cliffs and steep banks of rivers, and that, until lately, they excited no attention amongst the inhabitants: I am firmly convinced that a deep trench could not be cut in any line across the Pampas, without intersecting the remains of some quadruped. It is difficult to form an opinion in what part of the Pampas they are most numerous; in a limited spot they could not well have been more numerous than they were at P. Alta; the number, however, lately found by Senor F. Muniz, near Luxan, in a central spot in the Pampas, is extraordinarily great: at the end of this chapter I will give a list of all the localities at which I have heard of remains having been discovered. Very frequently the remains consist of almost perfect skeletons; but there are, also, numerous single bones, as for instance at St. Fe. Their state of preservation varies much, even when embedded near each other: I saw none others so perfectly preserved as the heads of the Toxodon and Mylodon from the white soft earthy bed on the Sarandis in Banda Oriental. It is remarkable that in two limited sections I found no less than five teeth separately embedded, and I heard of teeth having been similarly found in other parts: may we suppose that the skeletons or heads were for a long time gently drifted by currents over the soft muddy bottom, and that the teeth occasionally, here and there, dropped out?

It may be naturally asked, where did these numerous animals live? From the remarkable discoveries of MM. Lund and Clausen, it appears that some of the species found in the Pampas inhabited the highlands of Brazil: the Mastodon Andium is embedded at great heights in the Cordillera from north of the equator to at least as far south as Tarija (Humboldt states that the Mastodon has been discovered in New Granada: it has been found in Quito. When at Lima, I saw a tooth of a Mastodon in the possession of Don M. Rivero, found at Playa Chica on the Maranon, near the Guallaga. Every one has heard of the numerous remains of Mastodon in Bolivia.); and as there is no higher land, there can be little doubt that this Mastodon must have lived on the plains and valleys of that great range. These countries, however, appear too far distant for the habitation of the individuals entombed in the Pampas: we must probably look to nearer points, for instance to the province of Corrientes, which, as already remarked, is said not to be covered by the Pampean formation, and may therefore, at the period of its deposition, have existed as dry land. I have already given my reasons for believing that the animals embedded at M. Hermoso and at P. Alta in Bahia Blanca, lived on adjoining land, formed of parts of the already elevated Pampean deposit. With respect to the food of these many great extinct quadrupeds, I will not repeat the facts given in my "Journal" (second edition), showing that there is no correlation between the luxuriance of the vegetation of a country and the size of its mammiferous inhabitants. I do not doubt that large animals could now exist, as far as the amount, not kind, of vegetation is concerned, on the sterile plains of Bahia Blanca and of the R. Negro, as well as on the equally, if not more sterile plains of Southern Africa. The climate, however, may perhaps have somewhat deteriorated since the mammifers embedded at Bahia Blanca lived there; for we must not infer, from the continued existence of the same shells on the present coasts, that there has been no change in climate; for several of these shells now range northward along the shores of Brazil, where the most luxuriant vegetation flourishes under a tropical temperature. With respect to the extinction, which at first fills the mind with astonishment, of the many great and small mammifers of this period, I may also refer to the work above cited (second edition), in which I have endeavoured to show, that however unable we may be to explain the precise cause, we ought not properly to feel more surprised at a species becoming extinct than at one being rare; and yet we are accustomed to view the rarity of any particular species as an ordinary event, not requiring any extraordinary agency.

The several mammifers embedded in the Pampean formation, which mostly belong to extinct genera, and some even to extinct families or orders, and which differ nearly, if not quite, as much as do the Eocene mammifers of Europe from living quadrupeds having existed contemporaneously with mollusca, all still inhabiting the adjoining sea, is certainly a most striking fact. It is, however, far from being an isolated one; for, during the late tertiary deposits of Britain, an elephant, rhinoceros, and hippopotamus co-existed with many recent land and fresh-water shells; and in North America, we have the best evidence that a mastodon, elephant, megatherium, megalonyx, mylodon, an extinct horse and ox, likewise co- existed with numerous land, fresh-water, and marine recent shells. (Many original observations, and a summary on this subject, are given in Mr. Lyell's paper in the "Geological Proceedings" volume 4 and in his "Travels in North America" volume 1 and volume 2. For the European analogous cases see Mr. Lyell's "Principles of Geology" 6th edition volume 1.) The enumeration of these extinct North American animals naturally leads me to refer to the former closer relation of the mammiferous inhabitants of the two Americas, which I have discussed in my "Journal," and likewise to the vast extent of country over which some of them ranged: thus the same species of the Megatherium, Megalonyx, Equus (as far as the state of their remains permits of identification), extended from the Southern United States of North America to Bahia Blanca, in latitude 39 degrees S., on the coast of Patagonia. The fact of these animals having inhabited tropical and temperate regions, does not appear to me any great difficulty, seeing that at the Cape of Good Hope several quadrupeds, such as the elephant and hippopotamus, range from the equator to latitude 35 degrees south. The case of the Mastodon Andium is one of more difficulty, for it is found from latitude 36 degrees S., over, as I have reason to believe, nearly the whole of Brazil, and up the Cordillera to regions which, according to M. d'Orbigny, border on perpetual snow, and which are almost destitute of vegetation: undoubtedly the climate of the Cordillera must have been different when the mastodon inhabited it; but we should not forget the case of the Siberian mammoth and rhinoceros, as showing how severe a climate the larger pachydermata can endure; nor overlook the fact of the guanaco ranging at the present day over the hot low deserts of Peru, the lofty pinnacles of the Cordillera, and the damp forest-clad land of Southern Tierra del Fuego; the puma, also, is found from the equator to the Strait of Magellan, and I have seen its footsteps only a little below the limits of perpetual snow in the Cordillera of Chile.

At the period, so recent in a geological sense, when these extinct mammifers existed, the two Americas must have swarmed with quadrupeds, many of them of gigantic size; for, besides those more particularly referred to in this chapter, we must include in this same period those wonderfully numerous remains, some few of them specifically, and others generically related to those of the Pampas, discovered by MM. Lund and Clausen in the caves of Brazil. Finally, the facts here given show how cautious we ought to be in judging of the antiquity of a formation from even a great amount of difference between the extinct and living species in any one class of animals;—we ought even to be cautious in accepting the general proposition, that change in organic forms and lapse of time are at all, necessarily,

correlatives.
...

LOCALITIES WITHIN THE REGION OF THE PAMPAS WHERE GREAT BONES HAVE BEEN FOUND.

The following list, which includes every account which I have hitherto met with of the discovery of fossil mammiferous remains in the Pampas, may be hereafter useful to a geologist investigating this region, and it tends to show their extraordinary abundance. I heard of and saw many fossils, the original position of which I could not ascertain; and I received many statements too vague to be here inserted. Beginning to the south:—we have the two stations in Bahia Blanca, described in this chapter, where at P. Alta, the Megatherium, Megalonyx, Scelidotherium, Mylodon, Holophractus (or an allied genus), Toxodon, Macrauchenia, and an Equus were collected; and at M. Hermoso a Ctenomys, Hydrochaerus, some other rodents and the bones of a great megatheroid quadruped. Close north-east of the S. Tapalguen, we have the Rios 'Huesos' (i.e. "bones"), which probably takes its name from large fossil bones. Near Villa Nuevo, and at Las Averias, not far from the Salado, three nearly perfect skeletons, one of the Megatherium, one of the Glyptodon clavipes, and one of some great Dasypoid quadruped, were found by the agent of Sir W. Parish (see his work "Buenos Ayres" etc.). I have seen the tooth of a Mastodon from the Salado; a little northward of this river, on the borders of a lake near the G. del Monte, I saw many bones, and one large piece of dermal armour; higher up the Salado, there is a place called Monte "Huesos." On the Matanzas, about twenty miles south of Buenos Ayres, the skeleton (vide of "Buenos Ayres" etc. by Sir W. Parish) of a Glyptodon was found about five feet beneath the surface; here also (see Catalogue of Royal College of Surgeons) remains of Glyptodon clavipes, G. ornatus, and G. reticulatus were found. Signor Angelis, in a letter which I have seen, refers to some great remains found in Buenos Ayres, at a depth of twenty varas from the surface. Seven leagues north of this city the same author found the skeletons of Mylodon robustus and Glyptodon ornatus. From this neighbourhood he has lately sent to the British Museum the following fossils:—Remains of three or four individuals of Megatherium; of three species of Glyptodon; of three individuals of the Mastodon Andium; of Macrauchenia; of a second species of Toxodon, different from T. Platensis; and lastly, of the Machairodus, a wonderful large carnivorous animal. M. d'Orbigny has lately received from the Recolate "Voyage" Pal.), near Buenos Ayres, a tooth of Toxodon Platensis.

Proceeding northward, along the west bank of the Parana, we come to the Rio Luxan, where two skeletons of the Megatherium have been found; and lately, within eight leagues of the town of Luxan, Dr. F. X. Muniz has collected ("British Packet" Buenos Ayres September 25, 1841), from an average depth of eighteen feet, very numerous remains, of no less than, as he believes, nine distinct species of mammifers. At Areco, large bones have been found, which are believed, by the inhabitants, to have been changed from small bones, by the water of the river! At Arrecifes, the Glyptodon, sent to the College of Surgeons, was found; and I have seen two teeth of a Mastodon from this quarter. At S. Nicolas, M. d'Orbigny found remains of a Canis, Ctenomys, and Kerodon; and M. Isabelle ("Voyage") refers to a gigantic Armadillo found there. At S. Carlos, I heard of great bones. A little below the mouth of the Carcarana, the two skeletons of Mastodon were found; on the banks of this river, near S. Miguel, I found teeth of the Mastodon and Toxodon; and "Falkner" describes the osseous armour of some great animal; I heard of many other bones in this neighbourhood. I have seen, I may add, in the possession of Mr. Caldcleugh, the tooth of a Mastodon Andium, said to have been found in Paraguay; I may here also refer to a statement in this gentleman's travels (volume 1), of a great skeleton having been found in the province of Bolivia in Brazil, on the R. de las Contas. The furthest point westward in the Pampas, at which I have HEARD of fossil bones, was high up on the banks of R. Quinto.

In Entre Rios, besides the remains of the Mastodon, Toxodon, Equus, and a great Dasypoid quadruped near St. Fe Bajada, I received an account of bones having been found a little S.E. of P. Gorda (on the Parana), and of an entire skeleton at Matanzas, on the Arroyo del Animal.

In Banda Oriental, besides the remains of the Toxodon, Mylodon, and two skeletons of great animals with osseous armour (distinct from that of the Glyptodon), found on the Arroyos Sarandis and Berquelo, M. Isabelle ("Voyage") says, many bones have been found near the R. Negro, and on the R. Arapey, an affluent of the Paraguay, in latitude 30 degrees 40 minutes south. I heard of bones near the source of the A. Vivoras. I saw the remains of a Dasypoid quadruped from the Arroyo Seco, close to M. Video; and M. d'Orbigny refers ("Voyage" Geolog.), to another found on the Pedernal, an affluent of the St. Lucia; and Signor Angelis, in a letter, states that a third skeleton of this family has been found, near Canelones. I saw a tooth of the Mastodon from Talas, another affluent of the St. Lucia. The most eastern point at which I heard of great bones having been found, was at Solis Grande, between M. Video and Maldonado.

CHAPTER V. ON THE OLDER TERTIARY FORMATIONS OF PATAGONIA AND CHILE.

Rio Negro.
S. Josef.
Port Desire, white pumiceous mudstone with Infusoria.
Port S. Julian.
Santa Cruz, basaltic lava of.
P. Gallegos.
Eastern Tierra del Fuego; leaves of extinct beech-trees.
Summary on the Patagonian tertiary formations.
Tertiary formations of the Western Coast.
Chonos and Chiloe groups, volcanic rocks of.
Concepcion.
Navidad.
Coquimbo.
Summary.
Age of the tertiary formations.
Lines of elevation.
Silicified wood.
Comparative ranges of the extinct and living mollusca on the West Coast of S. America.
Climate of the tertiary period.
On the causes of the absence of recent conchiferous deposits on the coast of S. America.
On the contemporaneous deposition and preservation of sedimentary formations.

RIO NEGRO.

I can add little to the details given by M. d'Orbigny on the sandstone formation of this district. ("Voyage" Part Geolog.) The cliffs to the south of the river are about two hundred feet in height, and are composed of sandstone of various tints and degrees of hardness. One layer, which thinned out at both ends, consisted of earthy matter, of a pale reddish colour, with some gypsum, and very like (I speak after comparison of the specimens brought home) Pampean mud: above this was a layer of compact marly rock with dendritic manganese. Many blocks of a conglomerate of pumice-pebbles embedded in hard sandstone were strewed at the foot of the cliff, and had evidently fallen from above. A few miles N.E. of the town, I found, low down in the sandstone, a bed, a few inches in thickness, of a white, friable, harsh-feeling sediment, which adheres to the tongue, is of easy fusibility, and of little specific gravity; examined under the microscope, it is seen to be pumiceous tuff, formed of broken transparent crystals. In the cliffs south of the river, there is, also, a thin layer of nearly similar nature, but finer grained, and not so white; it might easily have been mistaken for a calcareous tuff, but it contains no lime: this substance precisely resembles a most widely extended and thick formation in Southern Patagonia, hereafter to be described, and which is remarkable for being partially formed of infusoria. These beds, conjointly with the conglomerate of pumice, are interesting, as showing the nature of the volcanic action in the Cordillera during this old tertiary period.

In a bed at the base of the southern cliffs, M. d'Orbigny found two extinct fresh-water shells, namely, a Unio and Chilina. This bed rested on one with bones of an extinct rodent, namely, the Megamys Patagoniensis; and this again on another with extinct marine shells. The species found by M. d'Orbigny in different parts of this formation consist of:—
1. Ostrea Patagonica, d'Orbigny, "Voyage, Pal." (also at St. Fe, and whole coast of Patagonia). 2. Ostrea Ferrarisi, d'Orbigny, "Voyage, Pal." 3. Ostrea Alvarezii, d'Orbigny, "Voyage, Pal." (also at St. Fe, and S. Josef). 4. Pecten Patagoniensis, d'Orbigny, "Voyage, Pal." 5. Venus Munsterii, d'Orbigny, "Voyage, Pal." (also at St. Fe). 6. Arca Bonplandiana, d'Orbigny, "Voyage, Pal." (also at St. Fe).

According to M. d'Orbigny, the sandstone extends westward along the coast as far as Port S. Antonio, and up the R. Negro far into the interior: northward I traced it to the southern side of the Rio Colorado, where it forms a low denuded plain. This formation, though contemporaneous with that of the rest of Patagonia, is quite different in mineralogical composition, being connected with it only by the one thin white layer: this difference may be reasonably attributed to the sediment brought down in ancient times by the Rio Negro; by which agency, also, we can understand the presence of the fresh-water shells, and of the bones of land animals. Judging from the identity of four of the above shells, this formation is contemporaneous (as remarked by M. d'Orbigny) with that under the Pampean deposit in Entre Rios and in Banda Oriental. The gravel capping the sandstone plain, with its calcareous cement and nodules of gypsum, is probably, from the reasons given in the First Chapter, contemporaneous with the uppermost beds of the Pampean formation on the upper plain north of the Colorado.

SAN JOSEF.

My examination here was very short: the cliffs are about a hundred feet high; the lower third consists of yellowish-brown, soft, slightly calcareous, muddy sandstone, parts of which when struck emit a fetid smell. In this bed the great Ostraea Patagonica, often marked with dendritic manganese and small coral-lines, were extraordinarily numerous. I found here the following shells:—
1. Ostrea Patagonica, d'Orbigny, "Voyage, Pal." (also at St. Fe and whole coast of Patagonia). 2. Ostrea Alvarezii, d'Orbigny, "Voyage, Pal." (also at St. Fe and R. Negro). 3. Pecten Paranensis, d'Orbigny, "Voyage, Pal." (also at St. Fe, S. Julian, and Port Desire). 4. Pecten Darwinianus, d'Orbigny, "Voyage, Pal." (also at St. Fe). 5. Pecten actinodes, G.B. Sowerby. 6. Terebratula Patagonica, G.B. Sowerby (also S. Julian). 7. Casts of a Turritella.

The four first of these species occur at St. Fe in Entre Rios, and the two first in the sandstone of the Rio Negro. Above this fossiliferous mass, there is a stratum of very fine-grained, pale brown mudstone, including numerous laminae of selenite. All the strata appear horizontal, but when followed by the eye for a long distance, they are seen to have a small easterly dip. On the surface we have the porphyritic gravel, and on it sand with recent shells.

NUEVO GULF.

From specimens and notes given me by Lieutenant Stokes, it appears that the lower bed consists of soft muddy sandstone, like that of S. Josef, with many imperfect shells, including the Pecten Paranensis, d'Orbigny, casts of a Turritella and Scutella. On this there are two strata of the pale brown mudstone, also like that of S. Josef, separated by a darker-coloured, more argillaceous variety, including the Ostrea Patagonica. Professor Ehrenberg has examined this mudstone for me: he finds in it three already known microscopic organisms, enveloped in a fine-grained pumiceous tuff, which I shall have immediately to describe in detail. Specimens brought to me from the uppermost bed, north of the Rio Chupat, consist of this same substance, but of a whiter colour.

Tertiary strata, such as here described, appear to extend along the whole coast between Rio Chupat and Port Desire, except where interrupted by the underlying claystone porphyry, and by some metamorphic rocks; these hard rocks, I may add, are found at intervals over a space of about five degrees of latitude, from Point Union to a point between Port S. Julian and S. Cruz, and will be described in the ensuing chapter. Many gigantic specimens of the Ostraea Patagonica were collected in the Gulf of St. George.

PORT DESIRE.
A good section of the lowest fossiliferous mass, about forty feet in thickness, resting on claystone porphyry, is exhibited a few miles south of the harbour. The shells sufficiently perfect to be recognised consist of:—
1. Ostrea Patagonica, d'Orbigny, (also at St. Fe, and whole coast of Patagonia). 2. Pecten Paranensis, d'Orbigny, "Voyage, Pal." (also at St. Fe, S. Josef, S. Julian). 3. Pecten centralis, G.B. Sowerby (also at S. Julian and S. Cruz). 4. Cucullaea alta, G.B. Sowerby (also at S. Cruz). 5. Nucula ornata, G.B. Sowerby. 6. Turritella Patagonica, G.B. Sowerby.

The fossiliferous strata, when not denuded, are conformably covered by a considerable thickness of the fine-grained pumiceous mudstone, divided into two masses: the lower half is very fine-grained, slightly unctuous, and so compact as to break with a semi-conchoidal fracture, though yielding to the nail; it includes laminae of selenite: the upper half precisely resembles the one layer at the Rio Negro, and with the exception of being whiter, the upper beds at San Josef and Nuevo Gulf. In neither mass is there any trace to the naked eye of organic forms. Taking the entire deposit, it is generally quite white, or yellowish, or feebly tinted with green; it is either almost friable under the finger, or as hard as chalk; it is of easy fusibility, of little specific gravity, is not harsh to the touch, adheres to the tongue, and when breathed on exhales a strong aluminous odour; it sometimes contains a very little calcareous matter, and traces (besides the included laminae) of gypsum. Under the microscope, according to Professor Ehrenberg, it consists of minute, triturated, cellular, glassy fragments of pumice, with some broken crystals. ("Monatsberichten de konig. Akad. zu Berlin" vom April 1845.) In the minute glassy fragments, Professor Ehrenberg recognises organic structures, which have been affected by volcanic heat: in the specimens from this place, and from Port S. Julian, he finds sixteen Polygastrica and twelve Phytolitharia. Of these organisms, seven are new forms, the others being previously known: all are of marine, and chiefly of oceanic, origin. This deposit to the naked eye resembles the crust which often appears on weathered surfaces of feldspathic rocks; it likewise resembles those beds of earthy feldspathic matter, sometimes interstratified with porphyritic rocks, as is the case in this very district with the underlying purple claystone porphyry. From examining specimens under a common microscope, and comparing them with other specimens undoubtedly of volcanic origin, I had come to the same conclusion with Professor Ehrenberg, namely, that this great deposit, in its first origin, is of volcanic nature.

PORT S. JULIAN.
(FIGURE 17. SECTION OF THE STRATA EXHIBITED IN THE CLIFFS OF THE NINETY FEET PLAIN AT PORT S. JULIAN.
(Section through beds from top to bottom: A, B, C, D, E, F.))
On the south side of the harbour, Figure 17 gives the nature of the beds seen in the cliffs of the ninety feet plain. Beginning at the top:—
1st, the earthy mass (AA), including the remains of the Macrauchenia, with recent shells on the surface.
Second, the porphyritic shingle (B), which in its lower part is interstratified (owing, I

believe, to redisposition during denudation) with the white pumiceous mudstone.

Third, this white mudstone, about twenty feet in thickness, and divided into two varieties (C and D), both closely resembling the lower, fine- grained, more unctuous and compact kind at Port Desire; and, as at that place, including much selenite.

Fourth, a fossiliferous mass, divided into three main beds, of which the uppermost is thin, and consists of ferruginous sandstone, with many shells of the great oyster and Pecten Paranensis; the middle bed (E) is a yellowish earthy sandstone abounding with Scutellae; and the lowest bed (F) is an indurated, greenish, sandy clay, including large concretions of calcareous sandstone, many shells of the great oyster, and in parts almost made up of fragments of Balanidae. Out of these three beds, I procured the following twelve species, of which the two first were exceedingly numerous in individuals, as were the Terebratulae and Turritellae in certain layers:—

1. Ostrea Patagonica, d'Orbigny, "Voyage, Pal." (also at St. Fe, and whole coast of Patagonia). 2. Pecten Paranensis, d'Orbigny, "Voyage, Pal." (St. Fe, S. Josef, Port Desire). 3. Pecten centralis, G.B. Sowerby (also at Port Desire and S. Cruz). 4. Pecten geminatus, G.B. Sowerby. 5. Terebratula Patagonica, G.B. Sowerby (also S. Josef). 6. Struthiolaria ornata, G.B. Sowerby (also S. Cruz). 7. Fusus Patagonicus, G.B. Sowerby. 8. Fusus Noachinus, G.B. Sowerby. 9. Scalaria rugulosa, G.B. Sowerby. 10. Turritella ambulacrum, G.B. Sowerby (also S. Cruz). 11. Pyrula, cast of, like P. ventricosa of Sowerby, Tank Cat. 12. Balanus varians, G.B. Sowerby. 13. Scutella, differing from the species from Nuevo Gulf.

At the head of the inner harbour of Port S. Julian, the fossiliferous mass is not displayed, and the sea-cliffs from the water's edge to a height of between one and two hundred feet are formed of the white pumiceous mudstone, which here includes innumerable, far-extended, sometimes horizontal, sometimes inclined or vertical laminae of transparent gypsum, often about an inch in thickness. Further inland, with the exception of the superficial gravel, the whole thickness of the truncated hills, which represent a formerly continuous plain 950 feet in height, appears to be formed of this white mudstone: here and there, however, at various heights, thin earthy layers, containing the great oyster, Pecten Paranensis and Turritella ambulacrum, are interstratified; thus showing that the whole mass belongs to the same epoch. I nowhere found even a fragment of a shell actually in the white deposit, and only a single cast of a Turritella. Out of the eighteen microscopic organisms discovered by Ehrenberg in the specimens from this place, ten are common to the same deposit at Port Desire. I may add that specimens of this white mudstone, with the same identical characters were brought me from two points,—one twenty miles north of S. Julian, where a wide gravel-capped plain, 350 feet in height, is thus composed; and the other forty miles south of S. Julian, where, on the old charts, the cliffs are marked as "Chalk Hills."

SANTA CRUZ.

The gravel-capped cliffs at the mouth of the river are 355 feet in height: the lower part, to a thickness of fifty or sixty feet, consists of a more or less hardened, darkish, muddy, or argillaceous sandstone (like the lowest bed of Port Desire), containing very many shells, some silicified and some converted into yellow calcareous spar. The great oyster is here numerous in layers; the Trigonocelia and Turritella are also very numerous: it is remarkable that the Pecten Paranensis, so common in all other parts of the coast, is here absent: the shells consist of:—

1. Ostrea Patagonica, d'Orbigny; "Voyage Pal." (also at St. Fe and whole coast of Patagonia). 2. Pecten centralis, G.B. Sowerby (also P. Desire and S. Julian). 3. Venus meridionalis of G.B. Sowerby. 4. Crassatella Lyellii, G.B. Sowerby. 5. Cardium puelchum, G.B. Sowerby. 6. Cardita Patagonica, G.B. Sowerby. 7. Mactra rugata, G.B. Sowerby. 8. Mactra Darwinii, G.B. Sowerby. 9. Cucullaea alta, G.B. Sowerby (also P. Desire). 10. Trigonocelia insolita, G.B. Sowerby. 11. Nucula (?) glabra, G.B. Sowerby. 12. Crepidula gregaria, G.B. Sowerby. 13. Voluta alta, G.B. Sowerby. 14. Trochus collaris, G.B. Sowerby. 15. Natica solida (?), G.B. Sowerby 16. Struthiolaria ornata, G.B. Sowerby (also P. Desire).

17. Turritella ambulacrum, G.B. Sowerby (also P. S. Julian). Imperfect fragments of the genera Byssoarca, Artemis, and Fusus.

The upper part of the cliff is generally divided into three great strata, differing slightly in composition, but essentially resembling the pumiceous mudstone of the places farther north; the deposit, however, here is more arenaceous, of greater specific gravity, and not so white: it is interlaced with numerous thin veins, partially or quite filled with transverse fibres of gypsum; these fibres were too short to reach across the vein, have their extremities curved or bent: in the same veins with the gypsum, and likewise in separate veins as well as in little nests, there is much powdery sulphate of magnesia (as ascertained by Mr. Reeks) in an uncompressed form: I believe that this salt has not heretofore been found in veins. Of the three beds, the central one is the most compact, and more like ordinary sandstone: it includes numerous flattened spherical concretions, often united like a necklace, composed of hard calcareous sandstone, containing a few shells: some of these concretions were four feet in diameter, and in a horizontal line nine feet apart, showing that the calcareous matter must have been drawn to the centres of attraction, from a distance of four feet and a half on both sides. In the upper and lower finer-grained strata, there were other concretions of a grey colour, containing calcareous matter, and so fine-grained and compact, as almost to resemble porcelain-rock: I have seen exactly similar concretions in a volcanic tufaceous bed in Chiloe. Although in this upper fine-grained strata, organic remains were very rare, yet I noticed a few of the great oyster; and in one included soft ferruginous layer, there were some specimens of the Cucullaea alta (found at Port Desire in the lower fossiliferous mass) and of the Mactra rugata, which latter shell has been partially converted into gypsum.

(FIGURE 18. SECTION OF THE PLAINS OF PATAGONIA, ON THE BANKS OF THE S. CRUZ.

(Section through strata (from top to bottom)): Surface of plain with erratic boulders; 1,146 feet above the sea. a. Gravel and boulders, 212 feet thick. b. Basaltic lava, 322 feet thick. c, d and e. Sedimentary layers, bed of small pebbles and talus respectively, total 592 feet thick. River of S. Cruz; here 280 feet above sea.)

In ascending the valley of the S. Cruz, the upper strata of the coast-cliffs are prolonged, with nearly the same characters, for fifty miles: at about this point, they begin in the most gradual and scarcely perceptible manner, to be banded with white lines; and after ascending ten miles farther, we meet with distinct thin layers of whitish, greenish, and yellowish fine-grained, fusible sediments. At eighty miles from the coast, in a cliff thus composed, there were a few layers of ferruginous sandstone, and of an argillaceous sandstone with concretions of marl like those in the Pampas. (At this spot, for a space of three-quarters of a mile along the north side of the river, and for a width of half a mile, there has been a great slip, which has formed hills between sixty and seventy feet in height, and has tilted the strata into highly inclined and even vertical positions. The strata generally dipped at an angle of 45 degrees towards the cliff from which they had slided. I have observed in slips, both on a small and large scale, that this inward dip is very general. Is it due to the hydrostatic pressure of water percolating with difficulty through the strata acting with greater force at the base of the mass than against the upper part?) At one hundred miles from the coast, that is at a central point between the Atlantic and the Cordillera, we have the section in Figure 18.

The upper half of the sedimentary mass, under the basaltic lava, consists of innumerable zones of perfectly white bright green, yellowish and brownish, fine-grained, sometimes incoherent, sedimentary matter. The white, pumiceous, trachytic tuff-like varieties are of rather greater specific gravity than the pumiceous mudstone on the coast to the north; some of the layers, especially the browner ones, are coarser, so that the broken crystals are distinguishable with a weak lens. The layers vary in character in short distances. With the exception of a few of the Ostrea Patagonica, which appeared to have rolled down from the cliff above, no organic remains were found. The chief difference between these layers taken as a whole, and the upper beds both at the mouth of the river and on the coast northward, seems to lie in the occasional presence of more colouring matter, and in the supply having been intermittent; these characters, as we have seen, very gradually disappear in descending the valley, and this fact may perhaps be accounted for by the currents of a more open sea having blended together the sediment from a distant and intermittent source.

The coloured layers in the foregoing section rest on a mass, apparently of great thickness (but much hidden by the talus), of soft sandstone, almost composed of minute pebbles, from one-tenth to two-tenths of an inch in diameter, of the rocks (with the entire exception of the basaltic lava) composing the great boulders on the surface of the plain, and probably composing the neighbouring Cordillera. Five miles higher up the valley, and again thirty miles higher up (that is twenty miles from the nearest range of the Cordillera), the lower plain included within the upper escarpments, is formed, as seen on the banks of the river, of a nearly similar but finer-grained, more earthy, laminated sandstone, alternating with argillaceous beds, and containing numerous moderately sized pebbles of the same rocks, and some shells of the great Ostrea Patagonica. (I found at both places, but not in situ, quantities of coniferous and ordinary dicotyledonous silicified wood, which was examined for me by Mr. R. Brown.) As most of these shells had been rolled before being here embedded, their presence does not prove that the sandstone belongs to the great Patagonian tertiary formation, for they might have been redeposited in it, when the valley existed as a sea-strait; but as amongst the pebbles there were none of basalt, although the cliffs on both sides of the valley are composed of this rock, I believe that the sandstone does belong to this formation. At the highest point to which we ascended, twenty miles distant from the nearest slope of the Cordillera, I could see the horizontally zoned white beds, stretching under the black basaltic lava, close up to the mountains; so that the valley of the S. Cruz gives a fair idea of the constitution of the whole width of Patagonia.

BASALTIC LAVA OF THE S. CRUZ.

This formation is first met with sixty-seven miles from the mouth of the river; thence it extends uninterruptedly, generally but not exclusively on the northern side of the valley, close up to the Cordillera. The basalt is generally black and fine-grained, but sometimes grey and laminated; it contains some olivine, and high up the valley much glassy feldspar, where, also, it is often amygdaloidal; it is never highly vesicular, except on the sides of rents and on the upper and lower, spherically laminated surfaces. It is often columnar; and in one place I saw magnificent columns, each face twelve feet in width, with their interstices filled up with calcareous tuff. The streams rest conformably on the white sedimentary beds, but I nowhere saw the actual junction; nor did I anywhere see the white beds actually superimposed on the lava; but some way up the valley at the foot of the uppermost escarpments, they must be thus superimposed. Moreover, at the lowest point down the valley, where the streams thin out and terminate in irregular projections, the spaces or intervals between these projections are filled up to the level of the now denuded and gravel-capped surfaces of the plains, with the white-zoned sedimentary beds; proving that this matter continued to be deposited after the streams had flowed. Hence we may conclude that the basalt is contemporaneous with the upper parts of the great tertiary formation.

The lava where first met with is 130 feet in thickness: it there consists of two, three, or perhaps more streams, divided from each other by vesicular spheroids like those on the surface. From the streams having, as it appears, extended to different distances, the terminal points are of unequal heights. Generally the surface of the basalt is smooth them in one part high up the valley, it was so uneven and hummocky, that until I afterwards saw the streams extending continuously on both sides of the valley up to a height of about three thousand feet close to the Cordillera, I thought that the craters of eruption were probably close at hand. This hummocky surface I believe to have been caused by the crossing and heaping up of different streams. In one place, there were several rounded ridges about twenty feet in height, some of them as broad as high, and some broader, which certainly had been formed whilst the lava was fluid, for in transverse sections each ridge was seen to be concentrically laminated, and to be composed of imperfect columns radiating from common centres, like the spokes of wheels.

The basaltic mass where first met with is, as I have said, 130 feet in thickness, and, thirty-five miles higher up the valley, it increases to 322 feet. In the first fourteen and a half miles of this distance, the upper surface of the lava, judging from three measurements taken above the level of the river (of which the apparently very uniform inclination has been calculated from its total height at a point 135 miles from the mouth), slopes towards the Atlantic at an angle of only 0 degrees 7 minutes twenty seconds: this must be considered only as an approximate measurement, but it cannot be far wrong. Taking the whole thirty-five miles, the upper surface slopes at an angle of 0 degrees 10 minutes 53 seconds; but this result is of no value in showing the inclination of any one stream, for halfway between the two points of measurement, the surface suddenly rises between one hundred and two hundred feet, apparently caused by some of the uppermost streams having extended thus far and no farther. From the measurement made at these two points, thirty-five miles apart, the mean inclination of the sedimentary beds, over which the lava has flowed, is NOW (after elevation from under the sea) only 0 degrees 7 minutes 52 seconds: for the sake of comparison, it may be mentioned that the bottom of the present sea in a line from the mouth of the S. Cruz to the Falkland Islands, from a depth of seventeen fathoms to a depth of eighty-five fathoms, declines at an angle of 0 degrees 1 minute 22 seconds; between the beach and the depth of seventeen fathoms, the slope is greater. From a point about half-way up the valley, the basaltic mass rises more abruptly towards the foot of the Cordillera, namely, from a height of 1,204 feet, to about 3,000 feet above the sea.

This great deluge of lava is worthy, in its dimensions, of the great continent to which it belongs. The aggregate streams have flowed from the Cordillera to a distance (unparalleled, I believe, in any case yet known) of about one hundred geographical miles. Near their furthest extremity their total thickness is 130 feet, which increase thirty-five miles farther inland, as we have just seen, to 322 feet. The least inclination given by M. E. de Beaumont of the upper surface of a lava-stream, namely 0 degrees 30 minutes, is that of the great subaerial eruption in 1783 from Skaptar Jukul in Iceland; and M. E. de Beaumont shows that it must have flowed down a mean inclination of less than 0 degrees 20 minutes. ("Memoires pour servir" etc..) But we now see that under the pressure of the sea, successive streams have flowed over a smooth bottom with a mean inclination of not more than 0 degrees 7 minutes 52 seconds; and that the upper surface of the terminal portion (over a space of fourteen and a half miles) has an inclination of not more than 0 degrees 7 minutes 20 seconds. If the elevation of Patagonia has been greater nearer the Cordillera than near the Atlantic (as is probable), then these angles are now all too large. I must repeat, that although the foregoing measurements, which were all carefully taken with the barometer, may not be absolutely correct, they cannot be widely erroneous.

Southward of the S. Cruz, the cliffs of the 840 feet plain extend to Coy Inlet, and owing to the naked patches of the white sediment, they are said on the charts to be "like the coast of Kent." At Coy Inlet the high plain trends inland, leaving flat-topped outliers. At Port Gallegos (latitude 51 degrees 35 minutes, and ninety miles south of S. Cruz), I am informed by Captain Sulivan, R.N., that there is a gravel-capped plain from two to three hundred feet in height, formed of numerous strata, some fine-grained and pale-coloured, like the upper beds at the mouth of the S. Cruz, others rather dark and coarser, so as to resemble gritstones or tuffs; these latter include rather large fragments of apparently decomposed volcanic rocks; there are, also, included layers of gravel. This formation is highly remarkable, from abounding with mammiferous remains, which have not as yet been examined by Professor Owen, but which include some large, but mostly small, species of Pachydermata, Edentata, and Rodentia. From the appearance of the pale-coloured, fine-grained beds, I was inclined to believe that they corresponded with the upper beds of the S. Cruz; but Professor Ehrenberg, who has examined some of the specimens, informs me that the included microscopical organisms are wholly different, being fresh and brackish-water forms. Hence the two to three hundred feet plain at Port Gallegos is of unknown age, but probably of subsequent origin to the great Patagonian tertiary formation.

EASTERN TIERRA DEL FUEGO.

Judging from the height, the general appearance, and the white colour of the patches visible on the hill sides, the uppermost plain, both on the north and western side of the Strait of Magellan, and along the eastern coast of Tierra del Fuego as far south as near Port St. Polycarp, probably belongs to the great Patagonian tertiary formation, These higher table-ranges are fringed by low, irregular, extensive plains, belonging to the boulder formation (Described in the "Geological Transactions" volume 6.), and composed of coarse unstratified masses, sometimes associated (as north of C. Virgin's) with fine, laminated, muddy sandstones. The cliffs in Sebastian Bay are 200 feet in height, and are composed of fine sandstones, often in curvilinear layers, including hard concretions of calcareous sandstone, and layers of gravel. In these beds there are fragments of wood, legs of crabs, barnacles encrusted with corallines still partially retaining their colour, imperfect fragments of a Pholas distinct from any known species, and of a Venus, approaching very closely to, but slightly different in form from, the V. lenticularis, a species living on the coast of Chile. Leaves of trees are numerous between the laminae of the muddy sandstone; they belong, as I am informed by Dr. J.D. Hooker, to three species of deciduous beech, different from the two species which compose the great proportion of trees in this forest-clad land. ("Botany of the Antarctic Voyage") From these facts it is difficult to conjecture, whether we here see the basal part of the great Patagonian formation, or some later deposit.

SUMMARY ON THE PATAGONIAN TERTIARY FORMATION.

Four out of the seven fossil shells, from St. Fe in Entre Rios, were found by M. d'Orbigny in the sandstone of the Rio Negro, and by me at San Josef. Three out of the six from San Josef are identical with those from Port Desire and S. Julian, which two places have together fifteen species, out of which three are common to both. Santa Cruz has seventeen species, out of which five are common to Port Desire and S. Julian. Considering the difference in latitude between these several places, and the small number of species altogether collected, namely thirty-six, I conceive the above proportional number of species in common, is sufficient to show that the lower fossiliferous mass belongs nearly, I do not say absolutely, to the same epoch. What this epoch may be, compared with the European tertiary stages, M. d'Orbigny will not pretend to determine. The thirty-six species (including those collected by myself and by M. d'Orbigny) are all extinct, or at least unknown; but it should be borne in mind, that the present coast consists of shingle, and that no one, I believe, has dredged here for shells; hence it is not improbable that some of the species may hereafter be found living. Some few of the species are closely related with existing ones; this is especially the case, according to M. d'Orbigny and Mr. Sowerby, with the Fusus Patagonicus; and, according to Mr. Sowerby, with the Pyrula, the Venus meridionalis, the Crepidula gregaria, and the Turritella ambulacrum, and T. Patagonica. At least three of the genera, namely, Cucullaea, Crassatella, and (as determined by Mr. Sowerby) Struthiolaria, are not found in this quarter of the world; and Trigonocelia is extinct. The evidence taken altogether indicates that this great tertiary formation is of considerable antiquity; but when treating of the Chilean beds, I shall have to refer again to this subject.

The white pumiceous mudstone, with its abundant gypsum, belongs to the same general epoch with the underlying fossiliferous mass, as may be inferred from the shells included in the intercalated layers at Nuevo Gulf, S. Julian, and S. Cruz. Out of the twenty-seven marine microscopic structures found by Professor Ehrenberg in the specimens from S. Julian and Port Desire, ten are common to these two places: the three found at Nuevo Gulf are distinct. I have minutely described this deposit, from its remarkable characters and its wide extension. From Coy Inlet to Port Desire, a distance of 230 miles, it is certainly continuous; and I have reason to believe that it likewise extends to the Rio Chupat, Nuevo Gulf, and San Josef, a distance of 570 miles: we have, also, seen that a single layer occurs at the Rio Negro.

At Port S. Julian it is from eight to nine hundred feet in thickness; and at S. Cruz it extends, with a slightly altered character, up to the Cordillera. From its microscopic structure, and from its analogy with other formations in volcanic districts, it must be considered as originally of volcanic origin: it may have been formed by the long-continued attrition of vast quantities of pumice, or judging from the manner in which the mass becomes, in ascending the valley of S. Cruz, divided into variously coloured layers, from the long-continued eruption of clouds of fine ashes. In either case, we must conclude, that the southern volcanic orifices of the Cordillera, now in a dormant state, were at about this period over a wide space, and for a great length of time, in action. We have evidence of this fact, in the latitude of the Rio Negro, in the sandstone-conglomerate with pumice, and demonstrative proof of it, at S. Cruz, in the vast deluges of basaltic lava: at this same tertiary period, also, there is distinct evidence of volcanic action in Western Banda Oriental.

The Patagonian tertiary formation extends continuously, judging from fossils alone, from S. Cruz to near the Rio Colorado, a distance of above six hundred miles, and reappears over a wide area in Entre Rios and Banda Oriental, making a total distance of 1,100 miles; but this formation undoubtedly extends (though no fossils were collected) far south of the S. Cruz, and, according to M. d'Orbigny, 120 miles north of St. Fe. At S. Cruz we have seen that it extends across the continent; being on the coast about eight hundred feet in thickness (and rather more at S. Julian), and rising with the contemporaneous lava-streams to a height of about three thousand feet at the base of the Cordillera. It rests, wherever any underlying formation can be seen, on plutonic and metamorphic rocks. Including the newer Pampean deposit, and those strata in Eastern Tierra del Fuego of doubtful age, as well as the boulder formation, we have a line of more than twenty-seven degrees of latitude, equal to that from the Straits of Gibraltar to the south of Iceland, continuously composed of tertiary formations. Throughout this great space the land has been upraised, without the strata having been in a single instance, as far as my means of observation went, unequally tilted or dislocated by a fault.

TERTIARY FORMATIONS ON THE WEST COAST.
CHONOS ARCHIPELAGO.

The numerous islands of this group, with the exception of Lemus, Ypun, consist of metamorphic schists; these two islands are formed of softish grey and brown, fusible, often laminated sandstones, containing a few pebbles, fragments of black lignite, and numerous mammillated concretions of hard calcareous sandstone. Out of these concretions at Ypun (latitude 40 degrees 30 minutes S.), I extracted the four following extinct species of shells:—

1. Turritella suturalis, G.B. Sowerby (also Navidad). 2. Sigaretus subglobosus, G.B. Sowerby (also Navidad). 3. Cytheraea (?) sulculosa (?), G.B. Sowerby (also Chiloe and Huafo?). 4. Voluta, fragments of.

In the northern parts of this group there are some cliffs of gravel and of the boulder formation. In the southern part (at P. Andres in Tres Montes), there is a volcanic formation, probably of tertiary origin. The lavas attain a thickness of from two to three hundred feet; they are extremely variable in colour and nature, being compact, or brecciated, or cellular, or amygdaloidal with zeolite, agate and bole, or porphyritic with glassy albitic feldspar. There is also much imperfect rubbly pitchstone, with the interstices charged with powdery carbonate of lime apparently of contemporaneous origin. These lavas are conformably associated with strata of breccia and of brown tuff containing lignite. The whole mass has been broken up and tilted at an angle of 45 degrees, by a series of great volcanic dikes, one of which was thirty yards in breadth. This volcanic formation resembles one, presently to be described, in Chiloe.

HUAFO.

This island lies between the Chonos and Chiloe groups: it is about eight hundred feet high, and perhaps has a nucleus of metamorphic rocks. The strata which I examined consisted of fine-grained muddy sandstones, with fragments of lignite and concretions of calcareous sandstone. I collected the following extinct shells, of which the Turritella was in great numbers:—

1. Bulla cosmophila, G.B. Sowerby. 2. Pleurotoma subaequalis, G.B. Sowerby. 3. Fusus cleryanus, d'Orbigny, "Voyage Pal." (also at Coquimbo). 4. Triton leucostomoides, G.B. Sowerby. 5. Turritella Chilensis, G.B. Sowerby (also Mocha). 6. Venus, probably a distinct species, but very imperfect. 7. Cytheraea (?) sulculosa (?), probably a distinct species, but very imperfect. 8. Dentalium majus, G.B. Sowerby.

CHILOE.

This fine island is about one hundred miles in length. The entire southern part, and the whole western coast, consists of mica-schist, which likewise is seen in the ravines of the interior. The central mountains rise to a height of 3,000 feet, and are said to be partly formed of granite and greenstone: there are two small volcanic districts. The eastern coast, and large parts of the northern extremity of the island are composed of gravel, the boulder formation, and underlying horizontal strata. The latter are well displayed for twenty miles north and south of Castro; they vary in character from common sandstone to fine-grained, laminated mudstones: all the specimens which I examined are easily fusible, and some of the beds might be called volcanic grit-stones. These latter strata are perhaps related to a mass of columnar trachyte which occurs behind Castro. The sandstone occasionally includes pebbles, and many fragments and layers of lignite; of the latter, some are apparently formed of wood and others of leaves: one layer on the N.W. side of Lemuy is nearly two feet in thickness. There is also much silicified wood, both common dicotyledonous and coniferous: a section of one specimen in the direction of the medullary rays has, as I am informed by Mr. R. Brown, the discs in a double row placed alternately, and not opposite as in the true Araucaria. I found marine remains only in one spot, in some concretions of hard calcareous sandstone: in several other districts I have observed that organic remains were exclusively confined to such concretions; are we to account for this fact, by the supposition that the shells lived only at these points, or is it not more probable that their remains were preserved only where concretions were formed? The shells here are in a bad state, they consist of:—

1. Tellinides (?) oblonga, G.B. Sowerby (a solenella in M. d'Orbigny's opinion). 2. Natica striolata, G.B. Sowerby. 3. Natica (?) pumila, G.B. Sowerby. 4. Cytheraea (?) sulculosa, G.B. Sowerby (also Ypun and Huafo?).

At the northern extremity of the island, near S. Carlos, there is a large volcanic formation, between five and seven hundred feet in thickness. The commonest lava is blackish-grey or brown, either vesicular, or amygdaloidal with calcareous spar and bole: most even of the darkest varieties fuse into a pale-coloured glass. The next commonest variety is a rubbly, rarely well characterised pitchstone (fusing into a white glass) which passes in the most irregular manner into stony grey lavas. This pitchstone, as well as some purple claystone porphyry, certainly flowed in the form of streams. These various lavas often pass, at a considerable depth from the surface, in the most abrupt and singular manner into wacke. Great masses of the solid rock are brecciated, and it was generally impossible to discover whether the recementing process had been an igneous or aqueous action. (In a cliff of the hardest fragmentary mass, I found several tortuous, vertical veins, varying in thickness from a few tenths of an inch to one inch and a half, of a substance which I have not seen described. It is glossy, and of a brown colour; it is thinly laminated, with the laminae transparent and elastic; it is a little harder than calcareous spar; it is infusible under the blowpipe, sometimes decrepitates, gives out water, curls up, blackens, and becomes magnetic. Borax easily dissolves a considerable quantity of it, and gives a glass tinged with green. I have no idea what its true nature is. On first seeing it, I mistook it for lignite!) The beds are obscurely separated from each other; they are sometimes parted by seams of tuff and layers of pebbles. In one place they rested on, and in another place were capped by, tuffs and girt-stones, apparently of submarine origin.

The neighbouring peninsula of Lacuy is almost wholly formed of tufaceous deposits, connected probably in their origin with the volcanic hills just described. The tuffs are pale-coloured, alternating with laminated mudstones and sandstones (all easily fusible), and passing sometimes into fine-grained white beds strikingly resembling the great upper infusorial deposit of Patagonia, and sometimes into brecciolas with pieces of pumice in the last stage of decay; these again pass into ordinary coarse breccias and conglomerates of hard rocks. Within very short distances, some of the finer tuffs often passed into each other in a peculiar manner, namely, by irregular polygonal concretions of one variety increasing so much and so suddenly in size, that the second variety, instead of any longer forming the entire mass, was left merely in thin veins between the concretions. In a straight line of cliffs, at Point Tenuy, I examined the following remarkable section (Figure 19):—

(FIGURE 19.)

On the left hand, the lower part (AA) consists of regular, alternating strata of brown tuffs and greenish laminated mudstone, gently inclined to the right, and conformably covered by a mass (B left) of a white, tufaceous and brecciolated deposit. On the right hand, the whole cliff (BB right) consists of the same white tufaceous matter, which on this side presents scarcely a trace of stratification, but to the left becomes very gradually and rather indistinctly divided into strata quite conformable with the underlying beds (AA): moreover, a few hundred yards further to the left, where the surface has been less denuded, the tufaceous strata (B left) are conformably covered by another set of strata, like the underlying ones (AA) of this section. In the middle of the diagram, the beds (AA) are seen to be abruptly cut off, and to abut against the tufaceous non-stratified mass; but the line of junction has been accidentally not represented steep enough, for I particularly noticed that before the beds had been tilted to the right, this line must have been nearly vertical. It appears that a current of water cut for itself a deep and steep submarine channel, and at the same time or afterwards filled it up with the tufaceous and brecciolated matter, and spread the same over the surrounding submarine beds; the matter becoming stratified in these more distant and less troubled parts, and being moreover subsequently covered up by other strata (like AA) not shown in the diagram. It is singular that three of the beds (of AA) are prolonged in their proper direction, as represented, beyond the line of junction into the white tufaceous matter: the prolonged portions of two of the beds are rounded; in the third, the terminal fragment has been pushed upwards: how these beds could have been left thus prolonged, I will not pretend to explain. In another section on the opposite side of a promontory, there was at the foot of this same line of junction, that is at the bottom of the old submarine channel, a pile of fragments of the strata (AA), with their interstices filled up with white tufaceous matter: this is exactly what might have been anticipated under such circumstances.

(FIGURE 20. GROUND PLAN SHOWING THE RELATION BETWEEN VEINS AND CONCRETIONARY ZONES IN A MASS OF TUFF.)

The various tufaceous and other beds at this northern end of Chiloe probably belong to about the same age with those near Castro, and they contain, as there, many fragments of black lignite and of silicified and pyritous wood, often embedded close together. They also contain many and singular concretions: some are of hard calcareous sandstone, in which it would appear that broken volcanic crystals and scales of mica have been better preserved (as in the case of the organic remains near Castro) than in the surrounding mass. Other concretions in the white brecciola are of a hard, ferruginous, yet fusible, nature; they are as round as cannon-balls, and vary from two or three inches to two feet in diameter; their insides generally consist either of fine, scarcely coherent volcanic sand (The frequent tendency in iron to form hollow concretions or shell containing incoherent matter is singular; D'Aubuisson ("Traite de Geogn." tome 1) remarks on this circumstance.), or of an argillaceous tuff; in this latter case, the external crust was quite thin and hard. Some of these spherical balls were encircled in the line of their equators, by a necklace-like row of smaller concretions. Again there were other concretions, irregularly formed, and composed of a hard, compact, ash- coloured stone, with an almost porcelainous fracture, adhesive to the tongue, and without any calcareous matter. These beds are, also, interlaced by many veins, containing gypsum, ferruginous matter, calcareous spar, and agate. It was here seen with remarkable distinctness, how intimately concretionary action and the production of fissures and veins are related together. Figure 20 is an accurate representation of a horizontal space of tuff, about four feet long by two and a half in width: the double lines represent the fissures partially filled with oxide of iron and agate: the curvilinear lines show the course of the innumerable, concentric, concretionary zones of different shades of colour and of coarseness in the particles of tuff. The symmetry and complexity of the arrangement gave the surface an elegant appearance. It may be seen how obviously the fissures determine (or have been determined by) the shape, sometimes of the whole concretion, and sometimes only of its central parts. The fissures also determine the curvatures of the long undulating zones of concretionary action. From the varying composition of the veins and concretions, the amount of chemical action which the mass has undergone is surprisingly great; and it would likewise appear from the difference in size in the particles of the concretionary zones, that the mass, also, has been subjected to internal mechanical movements.

In the peninsula of Lacuy, the strata over a width of four miles have been upheaved by three distinct, and some other indistinct, lines of elevation, ranging within a point of north and south. One line, about two hundred feet in height, is regularly anticlinal, with the strata dipping away on both sides, at an angle of 15 degrees, from a central "valley of elevation," about three hundred yards in width. A second narrow steep ridge, only sixty feet high, is uniclinal, the strata throughout dipping westward; those on both flanks being inclined at an angle of from ten to fifteen degrees; whilst those on the ridge dip in the same direction at an angle of between thirty and forty degrees. This ridge, traced northwards, dies away; and the beds at its terminal point, instead of dipping westward, are inclined 12 degrees to the north. This case interested me, as being the first in which I found in South America, formations perhaps of tertiary origin, broken by lines of elevation.

VALDIVIA: ISLAND OF MOCHA.

The formations of Chiloe seem to extend with nearly the same character to Valdivia, and for some leagues northward of it: the underlying rocks are micaceous schists, and are covered up with sandstone and other sedimentary beds, including, as I was assured, in many places layers of lignite. I did not land on Mocha (latitude 38 degrees 20 minutes), but Mr. Stokes brought me specimens of the grey, fine-grained, slightly calcareous sandstone, precisely like that of Huafo, containing lignite and numerous Turritellae. The island is flat topped, 1,240 feet in height, and appears like an outlier of the sedimentary beds on the mainland. The few shells collected consist of:—

1. Turritella Chilensis, G.B. Sowerby (also at Huafo). 2. Fusus, very imperfect, somewhat resembling F. subreflexus of Navidad, but probably different. 3. Venus, fragments of.

CONCEPCION.

Sailing northward from Valdivia, the coast-cliffs are seen, first to assume near the R. Tolten, and thence for 150 miles northward, to be continued with the same mineralogical characters, immediately to be described at Concepcion. I heard in many places of beds of lignite, some of it fine and glossy, and likewise of silicified wood; near the Tolten the cliffs are low, but they soon rise in height; and the horizontal strata are prolonged, with a nearly level surface, until coming to a more lofty tract between points Rumena and Lavapie. Here the beds have been broken up by at least eight or nine parallel lines of elevation, ranging E. or E.N.E. and W. or W.S.W. These lines can be followed with the eye many miles into the interior; they are all uniclinal, the strata in each dipping to a point between S. and S.S.E. with an inclination in the central lines of about forty degrees, and in the outer ones of under twenty degrees. This band of symmetrically troubled country is about eight miles in width.

The island of Quiriquina, in the Bay of Concepcion, is formed of various soft and often ferruginous sandstones, with bands of pebbles, and with the lower strata sometimes passing into a conglomerate resting on the underlying metamorphic schists. These beds include subordinate layers of greenish impure clay, soft micaceous and calcareous sandstones, and reddish friable earthy matter with white specks like decomposed crystals of feldspar; they include, also, hard concretions, fragments of shells, lignite, and silicified wood. In the upper part they pass into white, soft sediments and brecciolas, very like those described at Chiloe; as indeed is the whole formation. At Lirguen and other places on the eastern side of the bay, there are good sections of the lower sandstones, which are generally ferruginous, but which vary in character, and even pass into an argillaceous nature; they contain hard concretions, fragments of lignite, silicified wood, and pebbles (of the same rocks with the pebbles in the sandstones of Quiriquina), and they alternate with numerous, often very thin layers of imperfect coal, generally of little specific gravity. The main bed here is three feet thick; and only the coal of this one bed has a glossy fracture. Another irregular, curvilinear bed of brown, compact lignite, is remarkable for being included in a mass of coarse gravel. These imperfect coals, when placed in a heap, ignite spontaneously. The cliffs on this side of the bay, as well as on the island of Quiriquina, are capped with red friable earth, which, as stated in the Second Chapter, is of recent formation. The stratification in this neighbourhood is generally horizontal; but near Lirguen the beds dip N.W. at an angle of 23 degrees; near Concepcion they are also inclined: at the northern end of Quiriquina they have been tilted at an angle of 30 degrees, and at the southern end at angles varying from 15 degrees to 40 degrees: these dislocations must have taken place under the sea.

A collection of shells, from the island of Quiriquina, has been described by M. d'Orbigny: they are all extinct, and from their generic character, M. d'Orbigny inferred that they were of tertiary origin: they consist of:—

1. Scalaria Chilensis, d'Orbigny, "Voyage, Part Pal."
2. Natica Araucana, d'Orbigny, "Voyage, Part Pal."
3. Natica australis, d'Orbigny, "Voyage, Part Pal."
4. Fusus difficilis, d'Orbigny, "Voyage, Part Pal."
5. Pyrula longirostra, d'Orbigny, "Voyage, Part Pal."
6. Pleurotoma Araucana, d'Orbigny, "Voyage, Part Pal."
7. Cardium auca, d'Orbigny, "Voyage, Part Pal."
8. Cardium acuticostatum, d'Orbigny, "Voyage, Part Pal."
9. Venus auca, d'Orbigny, "Voyage, Part Pal."
10. Mactra cecileana, d'Orbigny, "Voyage, Part Pal."
11. Mactra Araucana, d'Orbigny, "Voyage, Part Pal."
12. Arca Araucana, d'Orbigny, "Voyage, Part Pal."
13. Nucula Largillierti, d'Orbigny, "Voyage, Part Pal."
14. Trigonia Hanetiana, d'Orbigny, "Voyage, Part Pal."

During a second visit of the "Beagle" to Concepcion, Mr. Kent collected for me some silicified wood and shells out of the concretions in the sandstone from Tome, situated a short distance north of Lirguen. They consist of:—

1. Natica australis, d'Orbigny, "Voyage, Part Pal." 2. Mactra Araucana, d'Orbigny, "Voyage, Part Pal." 3. Trigonia Hanetiana, d'Orbigny, "Voyage, Part Pal." 4. Pecten, fragments of, probably two species, but too imperfect for description. 5. Baculites vagina, E. Forbes. 6. Nautilus d'Orbignyanus, E. Forbes.

Besides these shells, Captain Belcher found here an Ammonite, nearly three feet in diameter, and so heavy that he could not bring it away; fragments are deposited at Haslar Hospital: he also found the silicified vertebrae of some very large animal. ("Zoology of Captain Beechey's Voyage") From the identity in mineralogical nature of the rocks, and from Captain Belcher's minute description of the coast between Lirguen and Tome, the fossiliferous concretions at this latter place certainly belong to the same formation with the beds examined by myself at Lirguen; and these again are undoubtedly the same with the strata of Quiriquina; moreover; the three first of the shells from Tome, though associated in the same concretions with the Baculite, are identical with the species from Quiriquina. Hence all the sandstone and lignitiferous beds in this neighbourhood certainly belong to the same formation. Although the generic character of the Quiriquina fossils naturally led M.

d'Orbigny to conceive that they were of tertiary origin, yet as we now find them associated with the Baculites vagina and with an Ammonite, we must, in the opinion of M. d'Orbigny, and if we are guided by the analogy of the northern hemisphere, rank them in the Cretaceous system. Moreover, the Baculites vagina, which is in a tolerable state of preservation, appears to Professor E. Forbes certainly to be identical with a species, so named by him, from Pondicherry in India; where it is associated with numerous decidedly cretaceous species, which approach most nearly to Lower Greensand or Neocomian forms: this fact, considering the vast distance between Chile and India, is truly surprising. Again, the Nautilus d'Orbignyanus, as far as its imperfect state allows of comparison, resembles, as I am informed by Professor Forbes, both in its general form and in that of its chambers, two species from the Upper Greensand. It may be added that every one of the above-named genera from Quiriquina, which have an apparently tertiary character, are found in the Pondicherry strata. There are, however, some difficulties on this view of the formations at Concepcion being cretaceous, which I shall afterwards allude to; and I will here only state that the Cardium auca is found also at Coquimbo, the beds at which place, there can be no doubt, are tertiary.

NAVIDAD. (I was guided to this locality by the Report on M. Gay's "Geological Researches" in the "Annales des Scienc. Nat." 1st series tome 28.)

The Concepcion formation extends some distance northward, but how far I know not; for the next point at which I landed was at Navidad, 160 miles north of Concepcion, and 60 miles south of Valparaiso. The cliffs here are about eight hundred feet in height: they consist, wherever I could examine them, of fine-grained, yellowish, earthy sandstones, with ferruginous veins, and with concretions of hard calcareous sandstone. In one part, there were many pebbles of the common metamorphic porphyries of the Cordillera: and near the base of the cliff, I observed a single rounded boulder of greenstone, nearly a yard in diameter. I traced this sandstone formation beneath the superficial covering of gravel, for some distance inland: the strata are slightly inclined from the sea towards the Cordillera, which apparently has been caused by their having been accumulated against or round outlying masses of granite, of which some points project near the coast. The sandstone contains fragments of wood, either in the state of lignite or partially silicified, sharks' teeth, and shells in great abundance, both high up and low down the sea-cliffs. Pectunculus and Oliva were most numerous in individuals, and next to them Turritella and Fusus. I collected in a short time, though suffering from illness, the following thirty-one species, all of which are extinct, and several of the genera do not now range (as we shall hereafter show) nearly so far south:—

1. Gastridium cepa, G.B. Sowerby. 2. Monoceros, fragments of, considered by M. d'Orbigny as a new species. 3. Voluta alta, G.B. Sowerby (considered by M. d'Orbigny as distinct from the V. alta of Santa Cruz). 4. Voluta triplicata, G.B. Sowerby. 5. Oliva dimidiata, G.B. Sowerby. 6. Pleurotoma discors, G.B. Sowerby. 7. Pleurotoma turbinelloides, G.B. Sowerby. 8. Fusus subreflexus, G.B. Sowerby. 9. Fusus pyruliformis, G.B. Sowerby. 10. Fusus, allied to F. regularis (considered by M. d'Orbigny as a distinct species). 11. Turritella suturalis, G.B. Sowerby. 12. Turritella Patagonica, G.B. Sowerby (fragments of). 13. Trochus laevis, G.B. Sowerby. 14. Trochus collaris, G.B. Sowerby (considered by M. d'Orbigny as the young of the T. laevis). 15. Cassis monilifer, G.B. Sowerby. 16. Pyrula distans, G.B. Sowerby. 17. Triton verruculosus, G.B. Sowerby. 18. Sigaretus subglobosus, G.B. Sowerby. 19. Natica solida, G.B. Sowerby. (It is doubtful whether the Natica solida of S. Cruz is the same species with this.) 20. Terebra undulifera, G.B. Sowerby. 21. Terebra costellata, G.B. Sowerby. 22. Bulla (fragments of). 23. Dentalium giganteum, do. 24. Dentalium sulcosum, do. 25. Corbis (?) laevigata, do. 26. Cardium multiradiatum, do. 27. Venus meridionalis, do. 28. Pectunculus dispar, (?) Desh. (considered by M. d'Orbigny as a distinct species). 29, 30. Cytheraea and Mactra, fragments of (considered by M. d'Orbigny as new species). 31. Pecten, fragments of.

COQUIMBO.
(FIGURE 21. SECTION OF THE TERTIARY FORMATION AT COQUIMBO.
From Level of Sea to Surface of plain, 252 feet above sea, through levels F, E, D and C:
F.—Lower sandstone, with concretions and silicified bones, with fossil shells, all, or nearly all, extinct.
E.—Upper ferruginous sandstone, with numerous Balani, with fossil shells, all, or nearly all, extinct.
C and D.—Calcareous beds with recent shells.
A.—Stratified sand in a ravine, also with recent shells.)

For more than two hundred miles northward of Navidad, the coast consists of plutonic and metamorphic rocks, with the exception of some quite insignificant superficial beds of recent origin. At Tonguay, twenty-five miles south of Coquimbo, tertiary beds recommence. I have already minutely described in the Second Chapter, the step-formed plains of Coquimbo, and the upper calcareous beds (from twenty to thirty feet in thickness) containing shells of recent species, but in different proportions from those on the beach. There remains to be described only the underlying ancient tertiary beds, represented in Figure 21 by the letters F and E:—

I obtained good sections of bed F only in Herradura Bay: it consists of soft whitish sandstone, with ferruginous veins, some pebbles of granite, and concretionary layers of hard calcareous sandstone. These concretions are remarkable from the great number of large silicified bones, apparently of cetaceous animals, which they contain; and likewise of a shark's teeth, closely resembling those of the Carcharias megalodon. Shells of the following species, of which the gigantic Oyster and Perna are the most conspicuous, are numerously embedded in the concretions:—

1. Bulla ambigua, d'Orbigny "Voyage" Pal. 2. Monoceros Blainvillii, d'Orbigny "Voyage" Pal. 3. Cardium auca, d'Orbigny "Voyage" Pal. 4. Panopaea Coquimbensis, d'Orbigny "Voyage" Pal. 5. Perna Gaudichaudi, d'Orbigny "Voyage" Pal. 6. Artemis ponderosa; Mr. Sowerby can find no distinguishing character between this fossil and the recent A. ponderosa; it is certainly an Artemis, as shown by the pallial impression. 7. Ostrea Patagonica (?); Mr. Sowerby can point out no distinguishing character between this species and that so eminently characteristic of the great Patagonian formation; but he will not pretend to affirm that they are identical. 8. Fragments of a Venus and Natica.

The cliffs on one side of Herradura Bay are capped by a mass of stratified shingle, containing a little calcareous matter, and I did not doubt that it belonged to the same recent formation with the gravel on the surrounding plains, also cemented by calcareous matter, until to my surprise, I found in the midst of it, a single thin layer almost entirely composed of the above gigantic oyster.

At a little distance inland, I obtained several sections of the bed E, which, though different in appearance from the lower bed F, belongs to the same formation: it consists of a highly ferruginous sandy mass, almost composed, like the lowest bed at Port S. Julian, of fragments of Balanidae; it includes some pebbles, and layers of yellowish-brown mudstone. The embedded shells consist of:—

1. Monoceros Blainvillii, d'Orbigny "Voyage" Pal. 2. Monoceros ambiguus, G.B. Sowerby. 3. Anomia alternans, G.B. Sowerby. 4. Pecten rudis, G.B. Sowerby. 5. Perna Gaudichaudi, d'Orbigny "Voyage" Pal. 6. Ostrea Patagonica (?), d'Orbigny "Voyage" Pal. 7. Ostrea, small species, in imperfect state; it appeared to me like a small kind now living in, but very rare in the bay. 8. Mytilus Chiloensis; Mr. Sowerby can find no distinguishing character between this fossil, as far as its not very perfect condition allows of comparison, and the recent species. 9. Balanus Coquimbensis, G.B. Sowerby. 10. Balanus psittacus? King. This appears to Mr. Sowerby and myself identical with a very large and common species now living on the coast.

The uppermost layers of this ferrugino-sandy mass are conformably covered by, and impregnated to the depth of several inches with, the calcareous matter of the bed D called losa: hence I at one time imagined that there was a gradual passage between them; but as all the species are recent in the bed D, whilst the most characteristic shells of the uppermost layers of E are the extinct Perna, Pecten, and Monoceros, I agree with M. d'Orbigny, that this view is erroneous, and that there is only a mineralogical passage between them, and no gradual transition in the nature of their organic remains. Besides the fourteen species enumerated from these two lower beds, M. d'Orbigny has described ten other species given to him from this locality; namely:—

1. Fusus Cleryanus, d'Orbigny "Voyage" Pal.
2. Fusus petitianus, d'Orbigny "Voyage" Pal.
3. Venus hanetiana, d'Orbigny "Voyage" Pal.
4. Venus incerta (?) d'Orbigny "Voyage" Pal.
5. Venus Cleryana, d'Orbigny "Voyage" Pal.
6. Venus petitiana, d'Orbigny "Voyage" Pal.
7. Venus Chilensis, d'Orbigny "Voyage" Pal.
8. Solecurtus hanetianus, d'Orbigny "Voyage" Pal.
9. Mactra auca, d'Orbigny "Voyage" Pal.
10. Oliva serena, d'Orbigny "Voyage" Pal.

Of these twenty-four shells, all are extinct, except, according to Mr. Sowerby, the Artemis ponderosa, Mytilus Chiloensis, and probably the great Balanus.

COQUIMBO TO COPIAPO.

A few miles north of Coquimbo, I met with the ferruginous, balaniferous mass E with many silicified bones; I was informed that these silicified bones occur also at Tonguay, south of Coquimbo: their number is certainly remarkable, and they seem to take the place of the silicified wood, so common on the coast-formations of Southern Chile. In the valley of Chaneral, I again saw this same formation, capped with the recent calcareous beds. I here left the coast, and did not see any more of the tertiary formations, until descending to the sea at Copiapo: here in one place I found variously coloured layers of sand and soft sandstone, with seams of gypsum, and in another place, a comminuted shelly mass, with layers of rotten-stone and seams of gypsum, including many of the extinct gigantic oyster: beds with these oysters are said to occur at English Harbour, a few miles north of Copiapo.

COAST OF PERU.

With the exception of deposits containing recent shells and of quite insignificant dimensions, no tertiary formations have been observed on this coast, for a space of twenty-two degrees of latitude north of Copiapo, until coming to Payta, where there is said to be a considerable calcareous deposit: a few fossils have been described by M. d'Orbigny from this place, namely:—

1. Rostellaria Gaudichaudi, d'Orbigny "Voyage" Pal. 2. Pectunculus Paytensis, d'Orbigny "Voyage" Pal. 3. Venus petitiana, d'Orbigny "Voyage" Pal. 4. Ostrea Patagonica? This great oyster (of which specimens have been given me) cannot be distinguished by Mr. Sowerby from some of the varieties from Patagonia; though it would be hazardous to assert it is the same with that species, or with that from Coquimbo.

CONCLUDING REMARKS.

The formations described in this chapter, have, in the case of Chiloe and probably in that of Concepcion and Navidad, apparently been accumulated in troughs formed by submarine ridges extending parallel to the ancient shores of the continent; in the case of the islands of Mocha and Huafo it is highly probable, and in that of Ypun and Lemus almost certain, that they were accumulated round isolated rocky centres or nuclei, in the same manner as mud and sand are now collecting round the outlying islets and reefs in the West

Indian Archipelago. Hence, I may remark, it does not follow that the outlying tertiary masses of Mocha and Huafo were ever continuously united at the same level with the formations on the mainland, though they may have been of contemporaneous origin, and been subsequently upraised to the same height. In the more northern parts of Chile, the tertiary strata seem to have been separately accumulated in bays, now forming the mouths of valleys.

The relation between these several deposits on the shores of the Pacific, is not nearly so clear as in the case of the tertiary formations on the Atlantic. Judging from the form and height of the land (evidence which I feel sure is here much more trustworthy than it can ever be in such broken continents as that of Europe), from the identity of mineralogical composition, from the presence of fragments of lignite and of silicified wood, and from the intercalated layers of imperfect coal, I must believe that the coast-formations from Central Chiloe to Concepcion, a distance of 400 miles, are of the same age: from nearly similar reasons, I suspect that the beds of Mocha, Huafo, and Ypun, belong also to the same period. The commonest shell in Mocha and Huafo is the same species of Turritella; and I believe the same Cytheraea is found on the islands of Huafo, Chiloe, and Ypun; but with these trifling exceptions, the few organic remains found at these places are distinct. The numerous shells from Navidad, with the exception of two, namely, the Sigaretus and Turritella found at Ypun, are likewise distinct from those found in any other part of this coast. Coquimbo has Cardium auca in common with Concepcion, and Fusus Cleryanus with Huafo; I may add, that Coquimbo has Venus petitiana, and a gigantic oyster (said by M. d'Orbigny also to be found a little south of Concepcion) in common with Payta, though this latter place is situated twenty-two degrees northward of latitude 27 degrees, to which point the Coquimbo formation extends.

From these facts, and from the generic resemblance of the fossils from the different localities, I cannot avoid the suspicion that they all belong to nearly the same epoch, which epoch, as we shall immediately see, must be a very ancient tertiary one. But as the Baculite, especially considering its apparent identity with the Cretaceous Pondicherry species, and the presence of an Ammonite, and the resemblance of the Nautilus to two upper greensand species, together afford very strong evidence that the formation of Concepcion is a Secondary one; I will, in my remarks on the fossils from the other localities, put on one side those from Concepcion and from Eastern Chiloe, which, whatever their age may be, appear to me to belong to one group. I must, however, again call attention to the fact that the Cardium auca is found both at Concepcion and in the undoubtedly tertiary strata of Coquimbo: nor should the possibility be overlooked, that as Trigonia, though known in the northern hemisphere only as a Secondary genus, has living representatives in the Australian seas, so a Baculite, Ammonite, and Trigonia may have survived in this remote part of the southern ocean to a somewhat later period than to the north of the equator.

Before passing in review the fossils from the other localities, there are two points, with respect to the formations between Concepcion and Chiloe, which deserve some notice. First, that though the strata are generally horizontal, they have been upheaved in Chiloe in a set of parallel anticlinal and uniclinal lines ranging north and south,—in the district near P. Rumena by eight or nine far-extended, most symmetrical, uniclinal lines ranging nearly east and west,—and in the neighbourhood of Concepcion by less regular single lines, directed both N.E. and S.W., and N.W. and S.E. This fact is of some interest, as showing that within a period which cannot be considered as very ancient in relation to the history of the continent, the strata between the Cordillera and the Pacific have been broken up in the same variously directed manner as have the old plutonic and metamorphic rocks in this same district. The second point is, that the sandstone between Concepcion and Southern Chiloe is everywhere lignitiferous, and includes much silicified wood; whereas the formations in Northern Chile do not include beds of lignite or coal, and in place of the fragments of silicified wood there are silicified bones. Now, at the present day, from Cape Horn to near Concepcion, the land is entirely concealed by forests, which thin out at Concepcion, and in Central and Northern Chile entirely disappear. This coincidence in the distribution of the fossil wood and the living forests may be quite accidental; but I incline to take a different view of it; for, as the difference in climate, on which the presence of forests depends, is here obviously in chief part due to the form of the land, and as the Cordillera undoubtedly existed when the lignitiferous beds were accumulating, I conceive it is not improbable that the climate, during the lignitiferous period, varied on different parts of the coast in a somewhat similar manner as it now does. Looking to an earlier epoch, when the strata of the Cordillera were depositing, there were islands which even in the latitude of Northern Chile, where now all is irreclaimably desert, supported large coniferous forests.

TABLE 4.

Column 1. Genera, with living and tertiary species on the west coast of South America. (M. d'Orbigny states that the genus Natica is not found on the coast of Chile; but Mr. Cuming found it at Valparaiso. Scalaria was found at Valparaiso; Arca, at Iquique, in latitude 20, by Mr. Cuming; Arca, also, was found by Captain King, at Juan Fernandez, in latitude 33 degrees 30'S.)

Column 2. Latitudes, in which found fossil on the coasts of Chile and Peru. (In degrees and minutes.)

Column 3. Southernmost latitude, in which found living on the west coast of South America. (In degrees and minutes.)

Bulla : 30 to 43 30 : 12 near Lima.
Cassis : 34 : 1 37.
Pyrula : 34 (and 36 30 at Concepcion) : 5 Payta.
Fusus : 30 and 43 30 : 23 Mexillones; reappears at the St. of Magellan.
Pleurotoma : 34 to 43 30 : 2 18 St. Elena.
Terebra : 34 : 5 Payta.
Sigaretus : 34 to 44 30 : 12 Lima.
Anomia : 30 : 7 48.
Perna : 30 : 1 23 Xixappa.
Cardium : 30 to 34 (and 36 30 at Concepcion) : 5 Payta.
Artemis : 30 : 5 Payta.
Voluta : 34 to 44 30 : Mr. Cuming does not know of any species living on the west coast, between the equator and latitude 43 south; from this latitude a species is found as far south as Tierra del Fuego.

Seventy-nine species of fossil shells, in a tolerably recognisable condition, from the coast of Chile and Peru, are described in this volume, and in the Palaeontological part of M. d'Orbigny's "Voyage": if we put on one side the twenty species exclusively found at Concepcion and Chiloe, fifty-nine species from Navidad and the other specified localities remain. Of these fifty-nine species only an Artemis, a Mytilus and Balanus, all from Coquimbo, are (in the opinion of Mr. Sowerby, but not in that of M. d'Orbigny) identical with living shells; and it would certainly require a better series of specimens to render this conclusion certain. Only the Turritella Chilensis from Huafo and Mocha, the T. Patagonica and Venus meridionalis from Navidad, come very near to recent South American shells, namely, the two Turritellas to T. cingulata, and the Venus to V. exalbida: some few other species come rather less near; and some few resemble forms in the older European tertiary deposits: none of the species resemble secondary forms. Hence I conceive there can be no doubt that these formations are tertiary,—a point necessary to consider, after the case of Concepcion. The fifty-nine species belong to thirty-two genera; of these, Gastridium is extinct, and three or four of the genera (viz. Panopaea, Rostellaria, Corbis (?), and I believe Solecurtus) are not now found on the west coast of South America. Fifteen of the genera have on this coast living representatives in about the same latitudes with the fossil species; but twelve genera now range very differently to what they formerly did. The idea of Table 4, in which the difference between the extension in latitude of the fossil and existing species is shown, is taken from M. d'Orbigny's work; but the range of the living shells is given on the authority of Mr. Cuming, whose long-continued researches on the conchology of South

America are well-known.

When we consider that very few, if any, of the fifty-nine fossil shells are identical with, or make any close approach to, living species; when we consider that some of the genera do not now exist on the west coast of South America, and that no less than twelve genera out of the thirty-two formerly ranged very differently from the existing species of the same genera, we must admit that these deposits are of considerable antiquity, and that they probably verge on the commencement of the tertiary era. May we not venture to believe, that they are of nearly contemporaneous origin with the Eocene formations of the northern hemisphere?

Comparing the fossil remains from the coast of Chile (leaving out, as before, Concepcion and Chiloe) with those from Patagonia, we may conclude, from their generic resemblance, and from the small number of the species which from either coast approach closely to living forms, that the formations of both belong to nearly the same epoch; and this is the opinion of M. D'Orbigny. Had not a single fossil shell been common to the two coasts, it could not have been argued that the formations belonged to different ages; for Messrs. Cuming and Hinds have found, on the comparison of nearly two thousand living species from the opposite sides of South America, only one in common, namely, the Purpura lapillus from both sides of the Isthmus of Panama: even the shells collected by myself amongst the Chonos Islands and on the coast of Patagonia, are dissimilar, and we must descend to the apex of the continent, to Tierra del Fuego, to find these two great conchological provinces united into one. Hence it is remarkable that four or five of the fossil shells from Navidad, namely, Voluta alta, Turritella Patagonica, Trochus collaris, Venus meridionalis, perhaps Natica solida, and perhaps the large oyster from Coquimbo, are considered by Mr. Sowerby as identical with species from Santa Cruz and P. Desire. M. d'Orbigny, however, admits the perfect identity only of the Trochus.

ON THE TEMPERATURE OF THE TERTIARY PERIOD.

As the number of the fossil species and genera from the western and eastern coasts is considerable, it will be interesting to consider the probable nature of the climate under which they lived. We will first take the case of Navidad, in latitude 34 degrees, where thirty-one species were collected, and which, as we shall presently see, must have inhabited shallow water, and therefore will necessarily well exhibit the effects of temperature. Referring to Table 4 we find that the existing species of the genera Cassis, Pyrula, Pleurotoma, Terebra, and Sigaretus, which are generally (though by no means invariably) characteristic of warmer latitudes, do not at the present day range nearly so far south on this line of coast as the fossil species formerly did. Including Coquimbo, we have Perna in the same predicament. The first impression from this fact is, that the climate must formerly have been warmer than it now is; but we must be very cautious in admitting this, for Cardium, Bulla, and Fusus (and, if we include Coquimbo, Anomia and Artemis) likewise formerly ranged farther south than they now do; and as these genera are far from being characteristic of hot climates, their former greater southern range may well have been owing to causes quite distinct from climate: Voluta, again, though generally so tropical a genus, is at present confined on the west coast to colder or more southern latitudes than it was during the tertiary period. The Trochus collaris, moreover, and, as we have just seen according to Mr. Sowerby, two or three other species, formerly ranged from Navidad as far south as Santa Cruz in latitude 50 degrees. If, instead of comparing the fossils of Navidad, as we have hitherto done, with the shells now living on the west coast of South America, we compare them with those found in other parts of the world, under nearly similar latitudes; for instance, in the southern parts of the Mediterranean or of Australia, there is no evidence that the sea off Navidad was formerly hotter than what might have been expected from its latitude, even if it was somewhat warmer than it now is when cooled by the great southern polar current. Several of the most tropical genera have no representative fossils at Navidad; and there are only single species of Cassis, Pyrula, and Sigaretus, two of Pleurotoma and two of Terebra, but none of these species are of conspicuous size. In Patagonia, there is even still less evidence in the character of the fossils, of the climate having been formerly warmer. (It may be worth while to mention that the shells living at the present day on this eastern side of South America, in latitude 40 degrees, have perhaps a more tropical character than those in corresponding latitudes on the shores of Europe: for at Bahia Blanca and S. Blas, there are two fine species of Voluta and four of Oliva.) As from the various reasons already assigned, there can be little doubt that the formations of Patagonia and at least of Navidad and Coquimbo in Chile, are the equivalents of an ancient stage in the tertiary formations of the northern hemisphere, the conclusion that the climate of the southern seas at this period was not hotter than what might have been expected from the latitude of each place, appears to me highly important; for we must believe, in accordance with the views of Mr. Lyell, that the causes which gave to the older tertiary productions of the quite temperate zones of Europe a tropical character, WERE OF A LOCAL CHARACTER AND DID NOT AFFECT THE ENTIRE GLOBE. On the other hand, I have endeavoured to show, in the "Geological Transactions," that, at a much later period, Europe and North and South America were nearly contemporaneously subjected to ice- action, and consequently to a colder, or at least more equable, climate than that now characteristic of the same latitudes.

ON THE ABSENCE OF EXTENSIVE MODERN CONCHIFEROUS DEPOSITS IN SOUTH AMERICA; AND ON THE CONTEMPORANEOUSNESS OF THE OLDER TERTIARY DEPOSITS AT DISTANT POINTS BEING DUE TO CONTEMPORANEOUS MOVEMENTS OF SUBSIDENCE.

Knowing from the researches of Professor E. Forbes, that molluscous animals chiefly abound within a depth of 100 fathoms and under, and bearing in mind how many thousand miles of both coasts of South America have been upraised within the recent period by a slow, long-continued, intermittent movement,- -seeing the diversity in nature of the shores and the number of shells now living on them,—seeing also that the sea off Patagonia and off many parts of Chile, was during the tertiary period highly favourable to the accumulation of sediment,—the absence of extensive deposits including recent shells over these vast spaces of coast is highly remarkable. The conchiferous calcareous beds at Coquimbo, and at a few isolated points northward, offer the most marked exception to this statement; for these beds are from twenty to thirty feet in thickness, and they stretch for some miles along shore, attaining, however, only a very trifling breadth. At Valdivia there is some sandstone with imperfect casts of shells, which POSSIBLY may belong to the recent period: parts of the boulder formation and the shingle-beds on the lower plains of Patagonia probably belong to this same period, but neither are fossiliferous: it also so happens that the great Pampean formation does not include, with the exception of the Azara, any mollusca. There cannot be the smallest doubt that the upraised shells along the shores of the Atlantic and Pacific, whether lying on the bare surface, or embedded in mould or in sand-hillocks, will in the course of ages be destroyed by alluvial action: this probably will be the case even with the calcareous beds of Coquimbo, so liable to dissolution by rain-water. If we take into consideration the probability of oscillations of level and the consequent action of the tidal-waves at different heights, their destruction will appear almost certain. Looking to an epoch as far distant in futurity as we now are from the past Miocene period, there seems to me scarcely a chance, under existing conditions, of the numerous shells now living in those zones of depths most fertile in life, and found exclusively on the western and south-eastern coasts of South America, being preserved to this imaginary distant epoch. A whole conchological series will in time be swept away, with no memorials of their existence preserved in the earth's crust.

Can any light be thrown on this remarkable absence of recent conchiferous deposits on these coasts, on which, at an ancient tertiary epoch, strata abounding with organic remains were extensively accumulated? I think there can, namely, by considering the conditions necessary for the preservation of a formation to a distant age. Looking to the enormous amount of denudation which on all sides of us has been effected,—as evidenced by the lofty cliffs cutting off on so many coasts horizontal and once far-extended strata of no great antiquity (as in the case of Patagonia),—as evidenced by the level surface of the ground on both sides of great faults and dislocations,—by inland lines of escarpments, by outliers, and numberless other facts, and by that argument of high generality advanced by Mr. Lyell, namely, that every SEDIMENTARY formation, whatever its thickness may be, and over however many hundred square miles it may extend, is the result and the measure of an equal amount of wear and tear of pre-existing formations; considering these facts, we must conclude that, as an ordinary rule, a formation to resist such vast destroying powers, and to last to a distant epoch, must be of wide extent, and either in itself, or together with superincumbent strata, be of great thickness. In this discussion, we are considering only formations containing the remains of marine animals, which, as before mentioned, live, with some exceptions within (most of them much within) depths of 100 fathoms. How, then, can a thick and widely extended formation be accumulated, which shall include such organic remains? First, let us take the case of the bed of the sea long remaining at a stationary level: under these circumstances it is evident that CONCHIFEROUS strata can accumulate only to the same thickness with the depth at which the shells can live; on gently inclined coasts alone can they accumulate to any considerable width; and from the want of superincumbent pressure, it is probable that the sedimentary matter will seldom be much consolidated: such formations have no very good chance, when in the course of time they are upraised, of long resisting the powers of denudation. The chance will be less if the submarine surface, instead of having remained stationary, shall have gone on slowly rising during the deposition of the strata, for in this case their total thickness must be less, and each part, before being consolidated or thickly covered up by superincumbent matter, will have had successively to pass through the ordeal of the beach; and on most coasts, the waves on the beach tend to wear down and disperse every object exposed to their action. Now, both on the south-eastern and western shores of South America, we have had clear proofs that the land has been slowly rising, and in the long lines of lofty cliffs, we have seen that the tendency of the sea is almost everywhere to eat into the land. Considering these facts, it ceases, I think, to be surprising, that extensive recent conchiferous deposits are entirely absent on the southern and western shores of America.

Let us take the one remaining case, of the bed of the sea slowly subsiding during a length of time, whilst sediment has gone on being deposited. It is evident that strata might thus accumulate to any thickness, each stratum being deposited in shallow water, and consequently abounding with those shells which cannot live at great depths: the pressure, also, I may observe, of each fresh bed would aid in consolidating all the lower ones. Even on a rather steep coast, though such must ever be unfavourable to widely extended deposits, the formations would always tend to increase in breadth from the water encroaching on the land. Hence we may admit that periods of slow subsidence will commonly be most favourable to the accumulation of CONCHIFEROUS deposits, of sufficient thickness, extension, and hardness, to resist the average powers of denudation.

We have seen that at an ancient tertiary epoch, fossiliferous deposits were extensively deposited on the coasts of South America; and it is a very interesting fact, that there is evidence that these ancient tertiary beds were deposited during a period of subsidence. Thus, at Navidad, the strata are about eight hundred feet in thickness, and the fossil shells are abundant both at the level of the sea and some way up the cliffs; having sent a list of these fossils to Professor E. Forbes, he thinks they must have lived in water between one and ten fathoms in depth: hence the bottom of the sea on which these shells once lived must have subsided at least 700 feet to allow of the superincumbent matter being deposited. I must here remark, that, as all these and the following fossil shells are extinct species, Professor Forbes necessarily judges of the depths at which they lived only from their generic character, and from the analogical distribution of shells in the northern hemisphere; but there is no just cause from this to doubt the general results. At Huafo the strata are about the same thickness, namely, 800 feet, and Professor Forbes thinks the fossils found there cannot have lived at a greater depth than fifty fathoms, or 300 feet. These two points, namely, Navidad and Huafo, are 570 miles apart, but nearly halfway between them lies Mocha, an island 1,200 feet in height, apparently formed of tertiary strata up to its level summit, and with many shells, including the same Turritella with that found at Huafo, embedded close to the level of the sea. In Patagonia, shells are numerous at Santa Cruz, at the foot of the 350 feet plain, which has certainly been formed by the denudation of the 840 feet plain, and therefore was originally covered by strata that number of feet in thickness, and these shells, according to Professor Forbes, probably lived at a depth of between seven and fifteen fathoms: at Port S. Julian, sixty miles to the north, shells are numerous at the foot of the ninety feet plain (formed by the denudation of the 950 feet plain), and likewise occasionally at the height of several hundred feet in the upper strata; these shells must have lived in water somewhere between five and fifty fathoms in depth. Although in other parts of Patagonia I have no direct evidence of shoal-water shells having been buried under a great thickness of superincumbent submarine strata, yet it should be borne in mind that the lower fossiliferous strata with several of the same species of Mollusca, the upper tufaceous beds, and the high summit-plain, stretch for a considerable distance southward, and for hundreds of miles northward; seeing this uniformity of structure, I conceive it may be fairly concluded that the subsidence by which the shells at Santa Cruz and S. Julian were carried down and covered up, was not confined to these two points, but was co-extensive with a considerable portion of the Patagonian tertiary formation. In a succeeding chapter it will be seen, that we are led to a similar conclusion with respect to the secondary fossiliferous strata of the Cordillera, namely, that they also were deposited during a long- continued and great period of subsidence. From the foregoing reasoning, and from the facts just given, I think we must admit the probability of the following proposition: namely, that when the bed of the sea is either stationary or rising, circumstances are far less favourable, than when the level is sinking, to the accumulation of CONCHIFEROUS deposits of sufficient thickness and extension to resist, when upheaved, the average vast amount of denudation. This result appears to me, in several respects, very interesting: every one is at first inclined to believe that at innumerable points, wherever there is a supply of sediment, fossiliferous strata are now forming, which at some future distant epoch will be upheaved and preserved; but on the views above given, we must conclude that this is far from being the case; on the contrary, we require (1st), a long-continued supply of sediment; (2nd), an extensive shallow area; and (3rd), that this area shall slowly subside to a great depth, so as to admit the accumulation of a widely extended thick mass of superincumbent strata. In how few parts of the world, probably, do these conditions at the present day concur! We can thus, also, understand the general want of that close sequence in fossiliferous formations which we might theoretically have anticipated; for, without we suppose a subsiding movement to go on at the same spot during an enormous period, from one geological era to another, and during the whole of this period sediment to accumulate at the proper rate, so that the depth should not become too great for the continued existence of molluscous animals, it is scarcely possible that there should be a perfect sequence at the same spot in the fossil shells of the two geological formations. (Professor H.D. Rogers, in his excellent address to the Association of American Geologists ("Silliman's Journal" volume 47) makes the following remark: "I question if we are at all aware how COMPLETELY the whole history of all departed time lies indelibly recorded with the amplest minuteness of detail in the successive sediments of the globe, how effectually, in other words, every period of time HAS WRITTEN ITS OWN HISTORY, carefully preserving every created form and every trace of action." I think the correctness of such remarks is more than doubtful, even if we except (as I suppose he would) all those numerous organic forms which contain no hard parts.) So far from a very long-continued subsidence being probable, many facts lead to the belief that the earth's surface oscillates up and down; and we have seen that during the elevatory movements there is but a small chance of DURABLE fossiliferous deposits accumulating.

Lastly, these same considerations appear to throw some light on the fact that certain periods appear to have been favourable to the deposition, or at least to the preservation, of contemporaneous formations at very distant points. We have seen that in South America an enormous area has been rising within the recent period; and in other quarters of the globe immense spaces appear to have risen contemporaneously. From my examination of the coral- reefs of the great oceans, I have been led to conclude that the bed of the sea has gone on slowly sinking within the present era, over truly vast areas: this, indeed, is in itself probable, from the simple fact of the rising areas having been so large. In South America we have distinct evidence that at nearly the same tertiary period, the bed of the sea off parts of the coast of Chile and off Patagonia was sinking, though these regions are very remote from each other. If, then, it holds good, as a general rule, that in the same quarter of the globe the earth's crust tends to sink and rise contemporaneously over vast spaces, we can at once see, that we have at distant points, at the same period, those very conditions which appear to be requisite for the accumulation of fossiliferous masses of sufficient extension, thickness, and hardness, to resist denudation, and consequently to last unto an epoch distant in futurity. (Professor Forbes has some admirable remarks on this subject, in his "Report on the Shells of the Aegean Sea." In a letter to Mr. Maclaren ("Edinburgh New Philosophical Journal" January 1843), I partially entered into this discussion, and endeavoured to show that it was highly improbable, that upraised atolls or barrier-reefs, though of great thickness, should, owing to their small extension or breadth, be preserved to a distant future period.)

CHAPTER VI. PLUTONIC AND METAMORPHIC ROCKS:—CLEAVAGE AND FOLIATION.
Brazil, Bahia, gneiss with disjointed metamorphosed dikes.
Strike of foliation.
Rio de Janeiro, gneiss-granite, embedded fragment in, decomposition of.
La Plata, metamorphic and old volcanic rocks of.
S. Ventana.
Claystone porphyry formation of Patagonia; singular metamorphic rocks; pseudo-dikes.
Falkland Islands, Palaeozoic fossils of.
Tierra del Fuego, clay-slate formation, cretaceous fossils of; cleavage and foliation; form of land.
Chonos Archipelago, mica-schists, foliation disturbed by granitic axis; dikes.
Chiloe.
Concepcion, dikes, successive formation of.
Central and Northern Chile.
Concluding remarks on cleavage and foliation.
Their close analogy and similar origin.
Stratification of metamorphic schists.
Foliation of intrusive rocks.
Relation of cleavage and foliation to the lines of tension during metamorphosis.
The metamorphic and plutonic formations of the several districts visited by the "Beagle" will be here chiefly treated of, but only such cases as appear to me new, or of some special interest, will be described in detail; at the end of the chapter I will sum up all the facts on cleavage and foliation,— to which I particularly attended.
BAHIA, BRAZIL: latitude 13 degrees south.
The prevailing rock is gneiss, often passing, by the disappearance of the quartz and mica, and by the feldspar losing its red colour, into a brilliantly grey primitive greenstone. Not unfrequently quartz and hornblende are arranged in layers in almost amorphous feldspar. There is some fine-grained syenitic granite, orbicularly marked by ferruginous lines, and weathering into vertical, cylindrical holes, almost touching each other. In the gneiss, concretions of granular feldspar and others of garnets with mica occur. The gneiss is

traversed by numerous dikes composed of black, finely crystallised, hornblendic rock, containing a little glassy feldspar and sometimes mica, and varying in thickness from mere threads to ten feet: these threads, which are often curvilinear, could sometimes be traced running into the larger dikes. One of these dikes was remarkable from having been in two or three places laterally disjointed, with unbroken gneiss interposed between the broken ends, and in one part with a portion of the gneiss driven, apparently whilst in a softened state, into its side or wall. In several neighbouring places, the gneiss included angular, well- defined, sometimes bent, masses of hornblende rock, quite like, except in being more perfectly crystallised, that forming the dikes, and, at least in one instance, containing (as determined by Professor Miller) augite as well as hornblende. In one or two cases these angular masses, though now quite separate from each other by the solid gneiss, had, from their exact correspondence in size and shape, evidently once been united; hence I cannot doubt that most or all of the fragments have been derived from the breaking up of the dikes, of which we see the first stage in the above- mentioned laterally disjointed one. The gneiss close to the fragments generally contained many large crystals of hornblende, which are entirely absent or rare in other parts: its folia or laminae were gently bent round the fragments, in the same manner as they sometimes are round concretions. Hence the gneiss has certainly been softened, its composition modified, and its folia arranged, subsequently to the breaking up of the dikes, these latter also having been at the same time bent and softened. (Professor Hitchcock "Geology of Massachusetts" volume 2, gives a closely similar case of a greenstone dike in syenite.)

I must here take the opportunity of premising, that by the term CLEAVAGE I imply those planes of division which render a rock, appearing to the eye quite or nearly homogeneous, fissile. By the term FOLIATION, I refer to the layers or plates of different mineralogical nature of which most metamorphic schists are composed; there are, also, often included in such masses, alternating, homogeneous, fissile layers or folia, and in this case the rock is both foliated and has a cleavage. By STRATIFICATION, as applied to these formations, I mean those alternate, parallel, large masses of different composition, which are themselves frequently either foliated or fissile,—such as the alternating so-called strata of mica-slate, gneiss, glossy clay-slate, and marble.

The folia of the gneiss within a few miles round Bahia generally strike irregularly, and are often curvilinear, dipping in all directions at various angles: but where best defined, they extended most frequently in a N.E. by N. (or East 50 degrees N.) and S.W. by S. line, corresponding nearly with the coast-line northwards of the bay. I may add that Mr. Gardner found in several parts of the province of Ceara, which lies between four and five hundred miles north of Bahia, gneiss with the folia extending E. 45 degrees N.; and in Guyana according to Sir R. Schomburgk, the same rock strikes E. 57 degrees N. Again, Humboldt describes the gneiss-granite over an immense area in Venezuela and even in Colombia, as striking E. 50 degrees N., and dipping to the N.W. at an angle of fifty degrees. (Gardner "Geological Section of the British Association" 1840. For Sir R. Schomburgk's observations see "Geographical Journal" 1842. See also Humboldt's discussion on Loxodrism in the "Personal Narrative.") Hence all the observations hitherto made tend to show that the gneissic rocks over the whole of this part of the continent have their folia extending generally within almost a point of the compass of the same direction. (I landed at only one place north of Bahia, namely, at Pernambuco. I found there only soft, horizontally stratified matter, formed from disintegrated granitic rocks, and some yellowish impure limestone, probably of a tertiary epoch. I have described a most singular natural bar of hard sandstone, which protects the harbour, in the 19th volume 1841 of the "London and Edinburgh Philosophical Magazine.")

ABROLHOS ISLETS, Latitude 18 degrees S. off the coast of Brazil.

Although not strictly in place, I do not know where I can more conveniently describe this little group of small islands. The lowest bed is a sandstone with ferruginous veins; it weathers into an extraordinary honeycombed mass; above it there is a dark-coloured argillaceous shale; above this a coarser sandstone—making a total thickness of about sixty feet; and lastly, above these sedimentary beds, there is a fine conformable mass of greenstone, in some parts having a columnar structure. All the strata, as well as the surface of the land, dip at an angle of about 12 degrees to N. by W. Some of the islets are composed entirely of the sedimentary, others of the trappean rocks, generally, however, with the sandstone, cropping out on the southern shores.)

RIO DE JANEIRO.
This whole district is almost exclusively formed of gneiss, abounding with garnets, and porphyritic with large crystals, even three and four inches in length, of orthoclase feldspar: in these crystals mica and garnets are often enclosed. At the western base of the Corcovado, there is some ferruginous carious quartz-rock; and in the Tijeuka range, much fine- grained granite. I observed boulders of greenstone in several places; and on the islet of Villegagnon, and likewise on the coast some miles northward, two large trappean dikes. The porphyritic gneiss, or gneiss- granite as it has been called by Humboldt, is only so far foliated that the constituent minerals are arranged with a certain degree of regularity, and may be said to have a "GRAIN," but they are not separated into distinct folia or laminae. There are, however, several other varieties of gneiss regularly foliated, and alternating with each other in so-called strata. The stratification and foliation of the ordinary gneisses, and the foliation or "grain" of the gneiss-granite, are parallel to each other, and generally strike within a point of N.E. and S.W. dipping at a high angle (between 50 and 60 degrees) generally to S.E.: so that here again we meet with the strike so prevalent over the more northern parts of this continent. The mountains of gneiss-granite are to a remarkable degree abruptly conical, which seems caused by the rock tending to exfoliate in thick, conically concentric layers: these peaks resemble in shape those of phonolite and other injected rocks on volcanic islands; nor is the grain or foliation (as we shall afterwards see) any difficulty on the idea of the gneiss-granite having been an intrusive rather than a metamorphic formation. The lines of mountains, but not always each separate hill, range nearly in the same direction with the foliation and so-called stratification, but rather more easterly.

(FIGURE 22. FRAGMENT OF GNEISS EMBEDDED IN ANOTHER VARIETY OF THE SAME ROCK.)
On a bare gently inclined surface of the porphyritic gneiss in Botofogo Bay, I observed the appearance represented in Figure 22. A fragment seven yards long and two in width, with angular and distinctly defined edges, composed of a peculiar variety of gneiss with dark layers of mica and garnets, is surrounded on all sides by the ordinary gneiss- granite; both having been dislocated by a granitic vein. The folia in the fragment and in the surrounding rock strike in the same N.N.E. and S.S.W. line; but in the fragment they are vertical, whereas in the gneiss-granite they dip at a small angle, as shown by the arrows, to S.S.E. This fragment, considering its great size, its solitary position, and its foliated structure parallel to that of the surrounding rock, is, as far as I know, a unique case: and I will not attempt any explanation of its origin.

The numerous travellers in this country, have all been greatly surprised at the depth to which the gneiss and other granitic rocks, as well as the talcose slates of the interior, have been decomposed. (Spix and Martius have collected in an Appendix to their "Travels," the largest body of facts on this subject. See also some remarks by M. Lund in his communications to the Academy at Copenhagen; and others by M. Gaudichaud in Freycinet "Voyage.") Near Rio, every mineral except the quartz has been completely softened, in some places to a depth little less than one hundred feet. (Dr. Benza describes granitic rock, "Madras Journal of Literature" etc. October 183?), in the Neelgherries, decomposed to a depth of forty feet.) The minerals retain their positions in folia ranging in the usual direction; and fractured quartz veins may be traced from the solid rock, running for some distance into the softened, mottled, highly coloured, argillaceous mass. It is said that these decomposed rocks abound with gems of various kinds, often in a fractured state, owing, as some have supposed, to the collapse of geodes, and that they contain gold and diamonds. At Rio, it appeared to me that the gneiss had been softened before the excavation (no doubt by the sea) of the existing, broad, flat-bottomed valleys; for the depth of decomposition did not appear at all conformable with the present undulations of the surface. The porphyritic gneiss, where now exposed to the air, seems to withstand decomposition remarkably well; and I could see no signs of any tendency to the production of argillaceous masses like those here described. I was also struck with the fact, that where a bare surface of this rock sloped into one of the quiet bays, there were no marks of erosion at the level of the water, and the parts both beneath and above it preserved a uniform curve. At Bahia, the gneiss rocks are similarly decomposed, with the upper parts insensibly losing their foliation, and passing, without any distinct line of separation, into a bright red argillaceous earth, including partially rounded fragments of quartz and granite. From this circumstance, and from the rocks appearing to have suffered decomposition before the excavation of the valleys, I suspect that here, as at Rio, the decomposition took place under the sea. The subject appeared to me a curious one, and would probably well repay careful examination by an able mineralogist.

THE NORTHERN PROVINCES OF LA PLATA.
According to some observations communicated to me by Mr. Fox, the coast from Rio de Janeiro to the mouth of the Plata seems everywhere to be granitic, with a few trappean dikes. At Port Alegre, near the boundary of Brazil, there are porphyries and diorites. (M. Isabelle "Voyage a Buenos Ayres") At the mouth of the Plata, I examined the country for twenty-five miles west, and for about seventy miles north of Maldonado: near this town, there is some common gneiss, and much, in all parts of the country, of a coarse-grained mixture of quartz and reddish feldspar, often, however, assuming a little dark-green imperfect hornblende, and then immediately becoming foliated. The abrupt hillocks thus composed, as well as the highly inclined folia of the common varieties of gneiss, strike N.N.E. or a little more easterly, and S.S.W. Clay-slate is occasionally met with, and near the L. del Potrero, there is white marble, rendered fissile from the presence of hornblende, mica, and asbestus; the cleavage of these rocks and their stratification, that is the alternating masses thus composed, strike N.N.E. and S.S.W. like the foliated gneisses, and have an almost vertical dip. The Sierra Larga, a low range five miles west of Maldonado, consists of quartzite, often ferruginous, having an arenaceous feel, and divided into excessively thin, almost vertical laminae or folia by microscopically minute scales, apparently of mica, and striking in the usual N.N.E. and S.S.W. direction. The range itself is formed of one principal line with some subordinate ones; and it extends with remarkable uniformity far northward (it is said even to the confines of Brazil), in the same line with the vertically ribboned quartz rock of which it is composed. The S. de Las Animas is the highest range in the country; I estimated it at 1,000 feet; it runs north and south, and is formed of feldspathic porphyry; near its base there is a N.N.W. and S.S.E. ridge of a conglomerate in a highly porphyritic basis.

Northward of Maldonado, and south of Las Minas, there is an E. and W. hilly band of country, some miles in width, formed of siliceous clay-slate, with some quartz, rock, and limestone, having a tortuous irregular cleavage, generally ranging east and west. E. and S.E. of Las Minas there is a confused district of imperfect gneiss and laminated quartz, with the hills ranging in various directions, but with each separate hill generally running in the same line with the folia of the rocks of which it is composed: this confusion appears to have been caused by the intersection of the [E. and W.] and [N.N.E. and S.S.W.] strikes. Northward of Las Minas, the more regular northerly ranges predominate: from this place to near Polanco, we meet with the coarse-grained mixture of quartz and feldspar, often with the imperfect hornblende, and then becoming foliated in a N. and S. line—with imperfect clay-slate, including laminae of red crystallised feldspar—with white or black marble, sometimes containing asbestus and crystals of gypsum—with quartz-rock—with syenite—and lastly, with much granite. The marble and granite alternate repeatedly in apparently vertical masses: some miles northward of the Polanco, a wide district is said to be entirely composed of marble. It is remarkable, how rare mica is in the whole range of country north and westward of Maldonado. Throughout this district, the cleavage of the clay-slate and marble—the foliation of the gneiss and the quartz—the stratification or alternating masses of these several rocks—and the range of the hills, all coincide in direction; and although the country is only hilly, the planes of division are almost everywhere very highly inclined or vertical.

Some ancient submarine volcanic rocks are worth mentioning, from their rarity on this eastern side of the continent. In the valley of the Tapas (fifty or sixty miles N. of Maldonado) there is a tract three or four miles in length, composed of various trappean rocks with glassy feldspar—of apparently metamorphosed grit-stones—of purplish amygdaloids with large kernels of carbonate of lime (Near the Pan de Azucar there is some greenish porphyry, in one place amygdaloidal with agate.)—and much of a harshish rock with glassy feldspar intermediate in character between claystone porphyry and trachyte. This latter rock was in one spot remarkable from being full of drusy cavities, lined with quartz crystals, and arranged in planes, dipping at an angle of 50 degrees to the east, and striking parallel to the foliation of an adjoining hill composed of the common mixture of quartz, feldspar, and imperfect hornblende: this fact perhaps indicates that these volcanic rocks have been metamorphosed, and their constituent parts rearranged, at the same time and according to the same laws, with the granitic and metamorphic formations of this whole region. In the valley of the Marmaraya, a few miles south of the Tapas, a band of trappean and amygdaloidal rock is interposed between a hill of granite and an extensive surrounding formation of red conglomerate, which (like that at the foot of the S. Animas) has its basis porphyritic with crystals of feldspar, and which hence has certainly suffered metamorphosis.

MONTE VIDEO.

The rocks here consist of several varieties of gneiss, with the feldspar often yellowish, granular and imperfectly crystallised, alternating with, and passing insensibly into, beds, from a few yards to nearly a mile in thickness, of fine or coarse grained, dark-green hornblendic slate; this again often passing into chloritic schist. These passages seem chiefly due to changes in the mica, and its replacement by other minerals. At Rat Island I examined a mass of chloritic schist, only a few yards square, irregularly surrounded on all sides by the gneiss, and intricately penetrated by many curvilinear veins of quartz, which gradually BLEND into the gneiss: the cleavage of the chloritic schist and the foliation of the gneiss were exactly parallel. Eastward of the city there is much fine- grained, dark-coloured gneiss, almost assuming the character of hornblende- slate, which alternates in thin laminae with laminae of quartz, the whole mass being transversely intersected by numerous large veins of quartz: I particularly observed that these veins were absolutely continuous with the alternating laminae of quartz. In this case and at Rat Island, the passage of the gneiss into imperfect hornblendic or into chloritic slate, seemed to be connected with the segregation of the veins of quartz. (Mr. Greenough "Critical Examination" etc., observes that quartz in mica-slate sometimes appears in beds and sometimes in veins. Von Buch also in his "Travels in Norway", remarks on alternating laminae of quartz and hornblende-slate replacing mica-schist.)

The Mount, a hill believed to be 450 feet in height, from which the place takes its name, is much the highest land in this neighbourhood: it consists of hornblendic slate, which (except on the eastern and disturbed base) has an east and west nearly vertical cleavage; the longer axis of the hill also ranges in this same line. Near the summit the hornblende-slate gradually becomes more and more coarsely crystallised, and less plainly laminated, until it passes into a heavy, sonorous greenstone, with a slaty conchoidal fracture; the laminae on the north and south sides near the summit dip inwards, as if this upper part had expanded or bulged outwards. This greenstone must, I conceive, be considered as metamorphosed hornblende- slate. The Cerrito, the next highest, but much less elevated point, is almost similarly composed. In the more western parts of the province, besides gneiss, there is quartz-rock, syenite, and granite; and at Colla, I heard of marble.

Near M. Video, the space which I more accurately examined was about fifteen miles in an east and west line, and here I found the foliation of the gneiss and the cleavage of the slates generally well developed, and extending parallel to the alternating strata composed of the gneiss, hornblendic and chloritic schists. These planes of division all range within one point of east and west, frequently east by south and west by north; their dip is generally almost vertical, and scarcely anywhere under 45 degrees: this fact, considering how slightly undulatory the surface of the country is, deserves attention. Westward of M. Video, towards the Uruguay, wherever the gneiss is exposed, the highly inclined folia are seen striking in the same direction; I must except one spot where the strike was N.W. by W. The little Sierra de S. Juan, formed of gneiss and laminated quartz, must also be excepted, for it ranges between [N. to N.E.] and [S. to S.W.] and seems to belong to the same system with the hills in the Maldonado district. Finally, we have seen that, for many miles northward of Maldonado and for twenty-five miles westward of it, as far as the S. de las Animas, the foliation, cleavage, so-called stratification and lines of hills, all range N.N.E. and S.S.W., which is nearly coincident with the adjoining coast of the Atlantic. Westward of the S. de las Animas, as far as even the Uruguay, the foliation, cleavage, and stratification (but not lines of hills, for there are no defined ones) all range about E. by S. and W. by N., which is nearly coincident with the direction of the northern shore of the Plata; in the confused country near Las Minas, where these two great systems appear to intersect each other, the cleavage, foliation, and stratification run in various directions, but generally coincide with the line of each separate hill.

SOUTHERN LA PLATA.

The first ridge, south of the Plata, which projects through the Pampean formation, is the Sierra Tapalguen and Vulcan, situated 200 miles southward of the district just described. This ridge is only a few hundred feet in height, and runs from C. Corrientes in a W.N.W. line for at least 150 miles into the interior: at Tapalguen, it is composed of unstratified granular quartz, remarkable from forming tabular masses and small plains, surrounded by precipitous cliffs: other parts of the range are said to consist of granite: and marble is found at the S. Tinta. It appears from M. Parchappe's observations, that at Tandil there is a range of quartzose gneiss, very like the rocks of the S. Larga near Maldonado, running in the same N.N.E. and S.S.W. direction; so that the framework of the country here is very similar to that on the northern shore of the Plata. (M. d'Orbigny's "Voyage" Part. Geolog.. I have given a short account of the peculiar forms of the quartz hills of Tapalguen, so unusual in a metamorphic formation, in my "Journal of Researches" 2nd edition)

The Sierra Guitru-gueyu is situated sixty miles south of the S. Tapalguen: it consists of numerous parallel, sometimes blended together ridges, about twenty-three miles in width, and five hundred feet in height above the plain, and extending in a N.W. and S.E. direction. Skirting round the extreme S.E. termination, I ascended only a few points, which were composed of a fine-grained gneiss, almost composed of feldspar with a little mica, and passing in the upper parts of the hills into a rather compact purplish clay-slate. The cleavage was nearly vertical, striking in a N.W. by W. and S.E. by E. line, nearly, though not quite, coincident with the direction of the parallel ridges.

The Sierra Ventana lies close south of that of Guitru-gueyu; it is remarkable from attaining a height, very unusual on this side of the continent, of 3,340 feet. It consists up to its summit, of quartz, generally pure and white, but sometimes reddish, and divided into thick laminae or strata: in one part there is a little glossy clay-slate with a tortuous cleavage. The thick layers of quartz strike in a W. 30 degrees N. line, dipping southerly at an angle of 45 degrees and upwards. The principal line of mountains, with some quite subordinate parallel ridges, range about W. 45 degrees N.: but at their S.E. termination, only W. 25

degrees N. This Sierra is said to extend between twenty and thirty leagues into the interior.

PATAGONIA.

With the exception perhaps of the hill of S. Antonio (600 feet high) in the Gulf of S. Matias, which has never been visited by a geologist, crystalline rocks are not met with on the coast of Patagonia for a space of 380 miles south of the S. Ventana. At this point (latitude 43 degrees 50 minutes), at Points Union and Tombo, plutonic rocks are said to appear, and are found, at rather wide intervals, beneath the Patagonian tertiary formation for a space of about three hundred miles southward, to near Bird Island, in latitude 48 degrees 56 minutes. Judging from specimens kindly collected for me by Mr. Stokes, the prevailing rock at Ports St. Elena, Camerones, Malaspina, and as far south as the Paps of Pineda, is a purplish-pink or brownish claystone porphyry, sometimes laminated, sometimes slightly vesicular, with crystals of opaque feldspar and with a few grains of quartz; hence these porphyries resemble those immediately to be described at Port Desire, and likewise a series which I have seen from P. Alegre on the southern confines of Brazil. This porphyritic formation further resembles in a singularly close manner the lowest stratified formation of the Cordillera of Chile, which, as we shall hereafter see, has a vast range, and attains a great thickness. At the bottom of the Gulf of St. George, only tertiary deposits appear to be present. At Cape Blanco, there is quartz rock, very like that of the Falkland Islands, and some hard, blue siliceous clay-slate.

At Port Desire there is an extensive formation of the claystone porphyry, stretching at least twenty-five miles into the interior: it has been denuded and deeply worn into gullies before being covered up by the tertiary deposits, through which it here and there projects in hills; those north of the bay being 440 feet in height. The strata have in several places been tilted at small angles, generally either to N.N.W. or S.S.E. By gradual passages and alternations, the porphyries change incessantly in nature. I will describe only some of the principal mineralogical changes, which are highly instructive, and which I carefully examined. The prevailing rock has a compact purplish base, with crystals of earthy or opaque feldspar, and often with grains of quartz. There are other varieties, with an almost truly trachytic base, full of little angular vesicles and crystals of glassy feldspar; and there are beds of black perfect pitchstone, as well as of a concretionary imperfect variety. On a casual inspection, the whole series would be thought to be of the same plutonic or volcanic nature with the trachytic varieties and pitchstone; but this is far from being the case, as much of the porphyry is certainly of metamorphic origin. Besides the true porphyries, there are many beds of earthy, quite white or yellowish, friable, easily fusible matter, resembling chalk, which under the microscope is seen to consist of minute broken crystals, and which, as remarked in a former chapter, singularly resembles the upper tufaceous beds of the Patagonian tertiary formation. This earthy substance often becomes coarser, and contains minute rounded fragments of porphyries and rounded grains of quartz, and in one case so many of the latter as to resemble a common sandstone. These beds are sometimes marked with true lines of aqueous deposition, separating particles of different degrees of coarseness; in other cases there are parallel ferruginous lines not of true deposition, as shown by the arrangement of the particles, though singularly resembling them. The more indurated varieties often include many small and some larger angular cavities, which appear due to the removal of earthy matter: some varieties contain mica. All these earthy and generally white stones insensibly pass into more indurated sonorous varieties, breaking with a conchoidal fracture, yet of small specific gravity; many of these latter varieties assume a pale purple tint, being singularly banded and veined with different shades, and often become plainly porphyritic with crystals of feldspar. The formation of these crystals could be most clearly traced by minute angular and often partially hollow patches of earthy matter, first assuming a FIBROUS STRUCTURE, then passing into opaque imperfectly shaped crystals, and lastly, into perfect glassy crystals. When these crystals have appeared, and when the basis has become compact, the rock in many places could not be distinguished from a true claystone porphyry without a trace of mechanical structure.

In some parts, these earthy or tufaceous beds pass into jaspery and into beautifully mottled and banded porcelain rocks, which break into splinters, translucent at their edges, hard enough to scratch glass, and fusible into white transparent beads: grains of quartz included in the porcelainous varieties can be seen melting into the surrounding paste. In other parts, the earthy or tufaceous beds either insensibly pass into, or alternate with, breccias composed of large and small fragments of various purplish porphyries, with the matrix generally porphyritic: these breccias, though their subaqueous origin is in many places shown both by the arrangement of their smaller particles and by an oblique or current lamination, also pass into porphyries, in which every trace of mechanical origin and stratification has been obliterated.

Some highly porphyritic though coarse-grained masses, evidently of sedimentary origin, and divided into thin layers, differing from each other chiefly in the number of embedded grains of quartz, interested me much from the peculiar manner in which here and there some of the layers terminated in abrupt points, quite unlike those produced by a layer of sediment naturally thinning out, and apparently the result of a subsequent process of metamorphic aggregation. In another common variety of a finer texture, the aggregating process had gone further, for the whole mass consisted of quite short, parallel, often slightly curved layers or patches, of whitish or reddish finely granulo-crystalline feldspathic matter, generally terminating at both ends in blunt points; these layers or patches further tended to pass into wedge or almond-shaped little masses, and these finally into true crystals of feldspar, with their centres often slightly drusy. The series was so perfect that I could not doubt that these large crystals, which had their longer axes placed parallel to each other, had primarily originated in the metamorphosis and aggregation of alternating layers of tuff; and hence their parallel position must be attributed (unexpected though the conclusion may be), not to laws of chemical action, but to the original planes of deposition. I am tempted briefly to describe three other singular allied varieties of rock; the first without examination would have passed for a stratified porphyritic breccia, but all the included angular fragments consisted of a border of pinkish crystalline feldspathic matter, surrounding a dark translucent siliceous centre, in which grains of quartz not quite blended into the paste could be distinguished: this uniformity in the nature of the fragments shows that they are not of mechanical, but of concretionary origin, having resulted perhaps from the self-breaking up and aggregation of layers of indurated tuff containing numerous grains of quartz,—into which, indeed, the whole mass in one part passed. The second variety is a reddish non-porphyritic claystone, quite full of spherical cavities, about half an inch in diameter, each lined with a collapsed crust formed of crystals of quartz. The third variety also consists of a pale purple non-porphyritic claystone, almost wholly formed of concretionary balls, obscurely arranged in layers, of a less compact and paler coloured claystone; each ball being on one side partly hollow and lined with crystals of quartz.

PSEUDO-DIKES.

Some miles up the harbour, in a line of cliffs formed of slightly metamorphosed tufaceous and porphyritic claystone beds, I observed three vertical dikes, so closely resembling in general appearance ordinary volcanic dikes, that I did not doubt, until closely examining their composition, that they had been injected from below. The first is straight, with parallel sides, and about four feet wide; it consists of whitish, indurated tufaceous matter, precisely like some of the beds intersected by it. The second dike is more remarkable; it is slightly tortuous, about eighteen inches thick, and can be traced for a considerable distance along the beach; it is of a purplish-red or brown colour, and is formed chiefly of ROUNDED grains of quartz, with broken crystals of earthy feldspar, scales of black mica, and minute fragments of claystone porphyry, all firmly united together in a hard sparing base. The structure of this dike shows obviously that it is of mechanical and sedimentary origin; yet it thinned out upwards, and did not cut through the uppermost strata in the cliffs. This fact at first appears to indicate that the matter could not have been washed in from above (Upfilled fissures are known to occur both in volcanic and in ordinary sedimentary formations. At the Galapagos Archipelago "Volcanic Islands" etc., there are some striking examples of pseudo-dikes composed of hard tuff.); but if we reflect on the suction which would result from a deep-seated fissure being formed, we may admit that if the fissure were in any part open to the surface, mud and water might well be drawn into it along its whole course. The third dike consisted of a hard, rough, white rock, almost composed of broken crystals of glassy feldspar, with numerous scales of black mica, cemented in a scanty base; there was little in the appearance of this rock, to preclude the idea of its having been a true injected feldspathic dike. The matter composing these three pseudo-dikes, especially the second one, appears to have suffered, like the surrounding strata, a certain degree of metamorphic action; and this has much aided the deceptive appearance. At Bahia, in Brazil, we have seen that a true injected hornblendic dike, not only has suffered metamorphosis, but has been dislocated and even diffused in the surrounding gneiss, under the form of separate crystals and of fragments.

FALKLAND ISLANDS.

I have described these islands in a paper published in the third volume of the "Geological Journal." The mountain-ridges consist of quartz, and the lower country of clay-slate and sandstone, the latter containing Palaeozoic fossils. These fossils have been separately described by Messrs. Morris and Sharpe: some of them resemble Silurian, and others Devonian forms. In the eastern part of the group the several parallel ridges of quartz extend in a west and east line; but further westward the line becomes W.N.W. and E.S.E., and even still more northerly. The cleavage-planes of the clay-slate are highly inclined, generally at an angle of above 50 degrees, and often vertical; they strike almost invariably in the same direction with the quartz ranges. The outline of the indented shores of the two main islands, and the relative positions of the smaller islets, accord with the strike both of the main axes of elevation and of the cleavage of the clay-slate.

TIERRA DEL FUEGO.

My notes on the geology of this country are copious, but as they are unimportant, and as fossils were found only in one district, a brief sketch will be here sufficient. The east coast from the S. of Magellan (where the boulder formation is largely developed) to St. Polycarp's Bay is formed of horizontal tertiary strata, bounded some way towards the interior by a broad mountainous band of clay-slate. This great clay-slate formation extends from St. Le Maire westward for 140 miles, along both sides of the Beagle Channel to near its bifurcation. South of this channel, it forms all Navarin Island, and the eastern half of Hoste Island and of Hardy Peninsula; north of the Beagle Channel it extends in a north-west line on both sides of Admiralty Sound to Brunswick Peninsula in the St. of Magellan, and I have reason to believe, stretches far up the eastern side of the Cordillera. The western and broken side of Tierra del Fuego towards the Pacific is formed of metamorphic schists, granite and various trappean rocks: the line of separation between the crystalline and clay-slate formations can generally be distinguished, as remarked by Captain King, by the parallelism in the clay-slate districts of the shores and channels, ranging in a line between [W. 20 degrees to 40 degrees N.] and [E. 20 degrees to 40 degrees S.]. ("Geographical Journal" volume 1)

The clay-slate is generally fissile, sometimes siliceous or ferruginous, with veins of quartz and calcareous spar; it often assumes, especially on the loftier mountains, an altered feldspathic character, passing into feldspathic porphyry: occasionally it is associated with breccia and grauwacke. At Good Success Bay, there is a little intercalated black crystalline limestone. At Port Famine much of the clay-slate is calcareous, and passes either into a mudstone or into grauwacke, including odd-shaped concretions of dark argillaceous limestone. Here alone, on the shore a few miles north of Port Famine, and on the summit of Mount Tarn (2,600 feet high), I found organic remains; they consist of:—

1. Ancyloceras simplex, d'Orbigny "Pal Franc" Mount Tarn. 2. Fusus (in imperfect state), d'Orbigny "Pal Franc" Mount Tarn. 3. Natica, d'Orbigny "Pal Franc" Mount Tarn. 4. Pentacrimus, d'Orbigny "Pal Franc" Mount Tarn. 5. Lucina excentrica, G.B. Sowerby, Port

Famine. 6. Venus (in imperfect state), G.B. Sowerby, Port Famine. 7. Turbinolia (?), G.B. Sowerby, Port Famine. 8. Hamites elatior, G.B. Sowerby, Port Famine.

M. d'Orbigny states that MM. Hombron and Grange found in this neighbourhood an Ancyloceras, perhaps A. simplex, an Ammonite, a Plicatula and Modiola. ("Voyage" Part Geolog..) M. d'Orbigny believes from the general character of these fossils, and from the Ancyloceras being identical (as far as its imperfect condition allows of comparison) with the A. simplex of Europe, that the formation belongs to an early stage of the Cretaceous system. Professor E. Forbes, judging only from my specimens, concurs in the probability of this conclusion. The Hamites elatior of the above list, of which a description has been given by Mr. Sowerby, and which is remarkable from its large size, has not been seen either by M. d'Orbigny or Professor E. Forbes, as, since my return to England, the specimens have been lost. The great clay-slate formation of Tierra del Fuego being cretaceous, is certainly a very interesting fact,—whether we consider the appearance of the country, which, without the evidence afforded by the fossils, would form the analogy of most known districts, probably have been considered as belonging to the Palaeozoic series,—or whether we view it as showing that the age of this terminal portion of the great axis of South America, is the same (as will hereafter be seen) with the Cordillera of Chile and Peru.

The clay-slate in many parts of Tierra del Fuego, is broken by dikes and by great masses of greenstone, often highly hornblendic (In a greenstone-dike in the Magdalen Channel, the feldspar cleaved with the angle of albite. This dike was crossed, as well as the surrounding slate, by a large vein of quartz, a circumstance of unusual occurrence.): almost all the small islets within the clay-slate districts are thus composed. The slate near the dikes generally becomes paler-coloured, harder, less fissile, of a feldspathic nature, and passes into a porphyry or greenstone: in one case, however, it became more fissile, of a red colour, and contained minute scales of mica, which were absent in the unaltered rock. On the east side of Ponsonby Sound some dikes composed of a pale sonorous feldspathic rock, porphyritic with a little feldspar, were remarkable from their number,—there being within the space of a mile at least one hundred,—from their nearly equalling in bulk the intermediate slate,—and more especially from the excessive fineness (like the finest inlaid carpentry) and perfect parallelism of their junctions with the almost vertical laminae of clay-slate. I was unable to persuade myself that these great parallel masses had been injected, until I found one dike which abruptly thinned out to half its thickness, and had one of its walls jagged, with fragments of the slate embedded in it.

In Southern Tierra del Fuego, the clay-slate towards its S.W. boundary, becomes much altered and feldspathic. Thus on Wollaston Island slate and grauwacke can be distinctly traced passing into feldspathic rocks and greenstones, including iron pyrites and epidote, but still retaining traces of cleavage with the usual strike and dip. One such metamorphosed mass was traversed by large vein-like masses of a beautiful mixture (as ascertained by Professor Miller) of green epidote, garnets, and white calcareous spar. On the northern point of this same island, there were various ancient submarine volcanic rocks, consisting of amygdaloids with dark bole and agate,—of basalt with decomposed olivine—of compact lava with glassy feldspar,—and of a coarse conglomerate of red scoriae, parts being amygdaloidal with carbonate of lime. The southern part of Wollaston Island and the whole of Hermite and Horn Islands, seem formed of cones of greenstone; the outlying islets of Il Defenso and D. Raminez are said to consist of porphyritic lava. (Determined by Professor Jameson. Weddell's "Voyage") In crossing Hardy Peninsula, the slate still retaining traces of its usual cleavage, passes into columnar feldspathic rocks, which are succeeded by an irregular tract of trappean and basaltic rocks, containing glassy feldspar and much iron pyrites: there is, also, some harsh red claystone porphyry, and an almost true trachyte, with needles of hornblende, and in one spot a curious slaty rock divided into quadrangular columns, having a base almost like trachyte, with drusy cavities lined by crystals, too imperfect, according to Professor Miller, to be measured, but resembling Zeagonite. (See Mr. Brooke's Paper in the "London Philosophical Magazine" volume 10. This mineral occurs in an ancient volcanic rock near Rome.) In the midst of these singular rocks, no doubt of ancient submarine volcanic origin, a high hill of feldspathic clay-slate projected, retaining its usual cleavage. Near this point, there was a small hillock, having the aspect of granite, but formed of white albite, brilliant crystals of hornblende (both ascertained by the reflecting goniometer) and mica; but with no quartz. No recent volcanic district has been observed in any part of Tierra del Fuego.

Five miles west of the bifurcation of the Beagle Channel, the slate- formation, instead of becoming, as in the more southern parts of Tierra del Fuego, feldspathic, and associated with trappean or old volcanic rocks, passes by alternations into a great underlying mass of fine gneiss and glossy clay-slate, which at no great distance is succeeded by a grand formation of mica-slate containing garnets. The folia of these metamorphic schists strike parallel to the cleavage-planes of the clay-slate, which have a very uniform direction over the whole of this part of the country: the folia, however, are undulatory and tortuous, whilst the cleavage- laminae of the slate are straight. These schists compose the chief mountain-chain of Southern Tierra del Fuego, ranging along the north side of the northern arm of the Beagle Channel, in a short W.N.W. and E.S.E. line, with two points (Mounts Sarmiento and Darwin) rising to heights of 6,800 and 6,900 feet. On the south-western side of this northern arm of the Beagle Channel, the clay-slate is seen with its STRATA dipping from the great chain, so that the metamorphic schists here form a ridge bordered on each side by clay-slate. Further north, however, to the west of this great range, there is no clay-slate, but only gneiss, mica, and hornblendic slates, resting on great barren hills of true granite, and forming a tract about sixty miles in width. Again, westward of these rocks, the outermost islands are of trappean formation, which, from information obtained during the voyages of the "Adventure" and "Beagle," seem, together with granite, chiefly to prevail along the western coast as far north as the entrance of the St. of Magellan (See the Paper by Captain King in the "Geographical Journal"; also a Letter to Dr. Fitton in "Geological Proceedings" volume 1; also some observations by Captain Fitzroy "Voyages" volume 1. I am indebted also to Mr. Lyell for a series of specimens collected by Lieutenant Graves.): a little more inland, on the eastern side of Clarence Island and S. Desolation, granite, greenstone, mica-slate, and gneiss appear to predominate. I am tempted to believe, that where the clay-slate has been metamorphosed at great depths beneath the surface, gneiss, mica- slate, and other allied rocks have been formed, but where the action has taken place nearer the surface, feldspathic porphyries, greenstones, etc., have resulted, often accompanied by submarine volcanic eruptions.

Only one other rock, met with in both arms of the Beagle Channel, deserves any notice, namely a granulo-crystalline mixture of white albite, black hornblende (ascertained by measurement of the crystals, and confirmed by Professor Miller), and more or less of brown mica, but without any quartz. This rock occurs in large masses, closely resembling in external form granite or syenite: in the southern arm of the Channel, one such mass underlies the mica-slate, on which clay-slate was superimposed: this peculiar plutonic rock which, as we have seen, occurs also in Hardy Peninsula, is interesting, from its perfect similarity with that (hereafter often to be referred to under the name of andesite) forming the great injected axes of the Cordillera of Chile.

The stratification of the clay-slate is generally very obscure, whereas the cleavage is remarkably well defined: to begin with the extreme eastern parts of Tierra del Fuego; the cleavage-planes near the St. of Le Maire strike either W. and E. or W.S.W. and E.N.E., and are highly inclined; the form of the land, including Staten Island, indicates that the axes of elevation have run in this same line, though I was unable to distinguish the planes of stratification. Proceeding westward, I accurately examined the cleavage of the clay-slate on the northern, eastern, and western sides (thirty-five miles apart) of Navarin Island, and everywhere found the laminae ranging with extreme regularity, W.N.W. and E.S.E., seldom varying more than one point of the compass from this direction. (The clay-slate in this island was in many places crossed by parallel smooth joints. Out of five cases, the angle of intersection between the strike of these joints and that of the cleavage-laminae was in two cases 45 degrees and in two others 79 degrees.) Both on the east and west coasts, I crossed at right angles the cleavage-planes for a space of about eight miles, and found them dipping at an angle of between 45 degrees and 90 degrees, generally to S.S.W., sometimes to N.N.E., and often quite vertically. The S.S.W. dip was occasionally succeeded abruptly by a N.N.E. dip, and this by a vertical cleavage, or again by the S.S.W. dip; as in a lofty cliff on the eastern end of the island the laminae of slate were seen to be folded into very large steep curves, ranging in the usual W.N.W. line, I suspect that the varying and opposite dips may possibly be accounted for by the cleavage- laminae, though to the eye appearing straight, being parts of large abrupt curves, with their summits cut off and worn down.

In several places I was particularly struck with the fact, that the fine laminae of the clay-slate, where cutting straight through the bands of stratification, and therefore indisputably true cleavage-planes, differed slightly in their greyish and greenish tints of colour, in compactness, and in some of the laminae having a rather more jaspery appearance than others. I have not seen this fact recorded, and it appears to me important, for it shows that the same cause which has produced the highly fissile structure, has altered in a slight degree the mineralogical character of the rock in the same planes. The bands of stratification, just alluded to, can be distinguished in many places, especially in Navarin Island, but only on the weathered surfaces of the slate; they consist of slightly undulatory zones of different shades of colour and of thicknesses, and resemble the marks (more closely than anything else to which I can compare them) left on the inside of a vessel by the draining away of some dirty slightly agitated liquid: no difference in composition, corresponding with these zones, could be seen in freshly fractured surfaces. In the more level parts of Navarin Island, these bands of stratification were nearly horizontal; but on the flanks of the mountains they were inclined from them, but in no instance that I saw at a very high angle. There can, I think, be no doubt that these zones, which appear only on the weathered surfaces, are the last vestiges of the original planes of stratification, now almost obliterated by the highly fissile and altered structure which the mass has assumed.

The clay-slate cleaves in the same W.N.W. and E.S.E. direction, as on Navarin Island, on both sides of the Beagle Channel, on the eastern side of Hoste Island, on the N.E. side of Hardy Peninsula, and on the northern point of Wollaston Island; although in these two latter localities the cleavage has been much obscured by the metamorphosed and feldspathic condition of the slate. Within the area of these several islands, including Navarin Island, the direction of the stratification and of the mountain- chains is very obscure; though the mountains in several places appeared to range in the same W.N.W. line with the cleavage: the outline of the coast, however, does not correspond with this line. Near the bifurcation of the Beagle Channel, where the underlying metamorphic schists are first seen, they are foliated (with some irregularities), in this same W.N.W. line, and parallel, as before stated, to the main mountain-axis of this part of the country. Westward of this main range, the metamorphic schists are foliated, though less plainly, in the same direction, which is likewise common to the zone of old erupted trappean rocks, forming the outermost islets. Hence the area, over which the cleavage of the slate and the foliation of the metamorphic schists extends with an average W.N.W. and E.S.E. strike, is about forty miles in a north and south line, and ninety miles in an east and west line.

Further northward, near Port Famine, the stratification of the clay-slate and of the associated rocks, is well defined, and there alone the cleavage and strata-planes are parallel. A little north of this port there is an anticlinal axis ranging N.W. (or a little more westerly) and S.E.: south of the port, as far as Admiralty Sound and Gabriel Channel, the outline of the land clearly indicates the existence of several lines of elevation in this same N.W. direction, which, I may add, is so uniform in the western half of the St. of Magellan, that, as Captain King has remarked, "a parallel ruler placed on the map upon the projecting points of the south shore, and extended across the strait, will also touch the headlands on the opposite coast." ("Geographical Journal" volume 1.) It would appear, from Captain King's observations, that over all this area the cleavage extends in the same line. Deep-water channels, however, in all parts of Tierra del Fuego have burst through the trammels both of stratification and cleavage; most of them may have been formed during the elevation of the land by long- continued erosion, but others, for instance the Beagle Channel, which stretches like a narrow canal for 120 miles obliquely through the mountains, can hardly have thus originated.

Finally, we have seen that in the extreme eastern point of Tierra del Fuego, the cleavage and coast-lines extend W. and E. and even W.S.W. and E.N.E.: over a large area westward, the cleavage, the main range of mountains, and some subordinate ranges, but not the outlines of the coast, strike W.N.W., and E.S.E.: in the central and western parts of the St. of Magellan, the stratification, the mountain-ranges, the outlines of the coast, and the cleavage all strike nearly N.W. and S.E. North of the strait, the outline of the coast, and the mountains on the mainland, run nearly north and south. Hence we see, at this southern point of the continent, how gradually the Cordillera bend, from their north and south course of so many thousand miles in length, into an E. and even E.N.E. direction.

WEST COAST, FROM THE SOUTHERN CHONOS ISLANDS TO NORTHERN CHILE.

The first place at which we landed north of the St. of Magellan was near Cape Tres Montes, in latitude 47 degrees S. Between this point and the Northern Chonos Islands, a distance of 200 miles, the "Beagle" visited several points, and specimens were collected for me from the intermediate spaces by Lieutenant Stokes. The predominant rock is mica-slate, with thick folia of quartz, very frequently alternating with and passing into a chloritic, or into a black, glossy, often striated, slightly anthracitic schist, which soils paper, and becomes white under a great heat, and then fuses. Thin layers of feldspar, swelling at intervals into well crystallised kernels, are sometimes included in these black schists; and I observed one mass of the ordinary black variety insensibly lose its fissile structure, and pass into a singular mixture of chlorite, epidote, feldspar, and mica. Great veins of quartz are numerous in the mica-schists; wherever these occur the folia are much convoluted. In the southern part of the Peninsula of Tres Montes, a compact altered feldspathic rock with crystals of feldspar and grains of quartz is the commonest variety; this rock exhibits occasionally traces of an original brecciated structure, and often presents (like the altered state of Tierra del Fuego) traces of cleavage- planes, which strike in the same direction with the folia of mica-schist further northward. (The peculiar, abruptly conical form of the hills in this neighbourhood, would have led any one at first to have supposed that they had been formed of injected or intrusive rocks. At Inchemo Island, a similar rock gradually becomes granulo-crystalline and acquires scales of mica; and this variety at S. Estevan becomes highly laminated, and though still exhibiting some rounded grains of quartz, passes into the black, glossy, slightly anthracitic schist, which, as we have seen, repeatedly alternates with and passes into the micaceous and chloritic schists. Hence all the rocks on this line of coast belong to one series, and insensibly vary from an altered feldspathic clay-slate into largely foliated, true mica-schist.

The cleavage of the homogeneous schists, the foliation of those composed of more or less distinct minerals in layers, and the planes of alternation of the different varieties or so-called stratification, are all parallel, and preserve over this 200 miles of coast a remarkable degree of uniformity in direction. At the northern end of the group, at Low's Harbour, the well- defined folia of mica-schist everywhere ranged within eight degrees (or less than one point of the compass) of N. 19 degrees W. and S. 19 degrees E.; and even the point of dip varied very little, being always directed to the west and generally at an angle of forty degrees; I should mention that I had here good opportunities of observation, for I followed the naked rock on the beach, transversely to the strike, for a distance of four miles and a half, and all the way attended to the dip. Along the outer islands for 100 miles south of Low's Harbour, Lieutenant Stokes, during his boat- survey, kindly observed for me the strike of the foliation, and he assures me that it was invariably northerly, and the dip with one single exception to the west. Further south at Vallenar Bay, the strike was almost universally N. 25 degrees W. and the dip, generally at an angle of about 40 degrees to W. 25 degrees S., but in some places almost vertical. Still farther south, in the neighbourhood of the harbours of Anna Pink, S. Estevan and S. Andres, and (judging from a distance) along the southern part of Tres Montes, the foliation and cleavage extended in a line between [N. 11 degrees to 22 degrees W.] and [S. 11 degrees to 22 degrees E.]; and the planes dipped generally westerly, but often easterly, at angles varying from a gentle inclination to vertical. At A. Pink's Harbour, where the schists generally dipped easterly, wherever the angle became very high, the strike changed from N. 11 degrees W. to even as much as N. 45 degrees W.: in an analogous manner at Vallenar Bay, where the dip was westerly (viz. on an average directed to W. 25 degrees S.), as soon as the angle became very high, the planes struck in a line more than 25 degrees west of north. The average result from all the observations on this 200 miles of coast, is a strike of N. 19 degrees W. and S. 19 degrees E.: considering that in each specified place my examination extended over an area of several miles, and that Lieutenant Stokes' observations apply to a length of 100 miles, I think this remarkable uniformity is pretty well established. The prevalence, throughout the northern half of this line of coast, of a dip in one direction, that is to the west, instead of being sometimes west and sometimes east, is, judging from what I have elsewhere seen, an unusual circumstance. In Brazil, La Plata, the Falkland Islands, and Tierra del Fuego, there is generally an obvious relation between the axis of elevation, the outline of the coast, and the strike of the cleavage or foliation: in the Chonos Archipelago, however, neither the minor details of the coast-line, nor the chain of the Cordillera, nor the subordinate transverse mountain-axes, accord with the strike of the foliation and cleavage: the seaward face of the numerous islands composing this Archipelago, and apparently the line of the Cordillera, range N. 11 degrees E., whereas, as we have just seen, the average strike of the foliation is N. 19 degrees W.

There is one interesting exception to the uniformity in the strike of the foliation. At the northern point of Tres Montes (latitude 45 degrees 52 minutes) a bold chain of granite, between two and three thousand feet in height, runs from the coast far into the interior, in an E.S.E. line, or more strictly E. 28 degrees S. and W. 28 degrees N. (In the distance, other mountains could be seen apparently ranging N.N.E. and S.S.W. at right angles to this one. I may add, that not far from Vallenar Bay there is a fine range, apparently of granite, which has burst through the mica-slate in a N.E. by E. and S.W. by S. line.) In a bay, at the northern foot of this range, there are a few islets of mica-slate, with the folia in some parts horizontal, but mostly inclined at an average angle of 20 degrees to the north. On the northern steep flank of the range, there are a few patches (some quite isolated, and not larger than half a-crown!) of the mica-schist, foliated with the same northerly dip. On the broad summit, as far as the southern crest, there is much mica-slate, in some places even 400 feet in thickness, with the folia all dipping north, at angles varying from 5 degrees to 20 degrees, but sometimes mounting up to 30 degrees. The southern flank consists of bare granite. The mica-slate is penetrated by small veins of granite, branching from the main body. (The granite within these veins, as well as generally at the junction with the mica-slate, is more quartzose than elsewhere. The granite, I may add, is traversed by dikes running for a very great length in the line of the mountains; they are composed of a somewhat laminated eurite, containing crystals of feldspar, hornblende, and octagons of quartz.) Leaving out of view the prevalent strike of the folia in other parts of this Archipelago, it might have been expected that they would have dipped N. 28 degrees E., that is directly from the ridge, and, considering its abruptness, at a high inclination; but the real dip, as we have just seen, both at the foot and on the northern flank, and over the entire summit, is at a small angle, and directed nearly due north. From these considerations it occurred to me, that perhaps we here had the novel and curious case of already inclined laminae obliquely tilted at a subsequent period by the granitic axis. Mr. Hopkins, so well known from his mathematical investigations, has most kindly calculated the problem: the proposition sent was,—Take a district composed of laminae, dipping at an angle of 40 degrees to W. 19 degrees S., and let an axis of elevation traverse it in an E. 28 degrees S. line, what will the position of the laminae be on the northern flank after a tilt, we will first suppose, of 45 degrees? Mr. Hopkins informs me, that the angle of the dip will be 28 degrees 31 minutes, and its direction to north 30 degrees 33 minutes west. (On the south side of the axis (where, however, I did not see any mica-slate) the dip of the folia would be at an angle of 77 degrees 55 minutes, directed to west 35 degrees 33 minutes south. Hence the two points of dip on the opposite sides of the range, instead of being as in ordinary cases directly opposed to each other at an angle of 180 degrees, would here be only 86 degrees 50 minutes apart.) By varying the supposed angle of the tilt, our previously inclined folia can be thrown into any angle between 26 degrees, which is the least possible angle, and 90 degrees; but if a small inclination be thus given to them, their point of dip will depart far from the north, and therefore not accord with the actual position of the folia of mica-schist on our granitic range. Hence it appears very difficult, without varying considerably the elements of the problem, thus to explain the anomalous strike and dip of the foliated mica- schist, especially in those parts, namely, at the base of the range, where the folia are almost horizontal. Mr. Hopkins, however, adds, that great irregularities and lateral thrusts might be expected in every great line of elevation, and that these would account for considerable deviations from the calculated results: considering that the granitic axis, as shown by the veins, has indisputably been injected after the perfect formation of the mica-slate, and considering the uniformity of the strike of the folia throughout the rest of the Archipelago, I cannot but still think that their anomalous position at this one point is someway directly and mechanically related to the intrusion of this W.N.W. and E.S.E. mountain-chain of granite.

Dikes are frequent in the metamorphic schists of the Chonos Islands, and seem feebly to represent that great band of trappean and ancient volcanic rocks on the south-western coast of Tierra del Fuego. At S. Andres I observed in the space of half-a-mile, seven broad, parallel dikes, composed of three varieties of trap, running in a N.W. and S.E. line, parallel to the neighbouring mountain-ranges of altered clay-slate; but they must be of long subsequent origin to these mountains; for they intersected the volcanic formation described in the last chapter. North of Tres Montes, I noticed three dikes differing from each other in composition, one of them having a euritic base including large octagons of quartz; these dikes, as well as several of porphyritic greenstone at Vallenar Bay, extended N.E. and S.W., nearly at right angles to the foliation of the schists, but in the line of their joints. At Low's Harbour, however, a set of great parallel dikes, one ninety yards and another sixty yards in width, have been guided by the foliation of the mica-schist, and hence are inclined westward at an angle of 45 degrees: these dikes are formed of various porphyritic traps, some of which are remarkable from containing numerous rounded grains of quartz. A porphyritic trap of this latter kind, passed in one of the dikes into a most curious hornstone, perfectly white, with a waxy fracture and pellucid edges, fusible, and containing many grains of quartz and specks of iron pyrites. In the ninety-yard dike several large, apparently now quite isolated, fragments of mica-slate were embedded: but as their foliation was exactly parallel to that of the surrounding solid rock, no doubt these new separate fragments originally formed wedge-shaped depending portions of a continuous vault or crust, once extending over the dike, but since worn down and denuded.

CHILOE, VALDIVIA, CONCEPCION.

In Chiloe, a great formation of mica-schist strikingly resembles that of the Chonos Islands. For a space of eleven miles on the S.E. coast, the folia were very distinct, though slightly convoluted, and ranged within a point of N.N.W. and S.S.E., dipping either E.N.E. or more commonly W.S.W., at an average angle of 22 degrees (in one spot, however, at 60 degrees), and therefore decidedly at a lesser inclination than amongst the Chonos Islands. On the west and north-western shores, the foliation was often obscure, though, where best defined, it ranged within a point of N. by W. and S. by E., dipping either easterly or westerly, at varying and generally very small angles. Hence, from the southern part of Tres Montes to the northern end of Chiloe, a distance of 300 miles, we have closely allied rocks with their folia striking on an average in the same direction, namely between N. 11 degrees and 22 degrees W. Again, at Valdivia, we meet with the same mica-schist, exhibiting nearly the same mineralogical passages as in the Chonos Archipelago, often, however, becoming more ferruginous, and containing so much feldspar as to pass into gneiss. The folia were generally well defined; but nowhere else in South America did I see them varying so much in direction: this seemed chiefly caused by their forming parts, as I could sometimes distinctly trace, of large flat curves: nevertheless, both near the settlement and towards the interior, a N.W. and S.E. strike seemed more frequent than any other direction; the angle of the dip was generally small. At Concepcion, a highly glossy clay-slate had its cleavage often slightly curvilinear, and inclined, seldom at a high angle, towards various points of the compass: but here, as at Valdivia, a N.W. and S.E. strike seemed to be the most frequent one. ((FIGURE 23.) I observed in some parts that the tops of the laminae of the clay-slate (b in Figure 23) under the superficial detritus and soil (a) were bent, sometimes without being broken, as represented in Figure 23, which is copied from one given by Sir H. De la Beche "Geological Manual") of an exactly similar phenomenon in Devonshire. Mr. R.A.C. Austen, also, in his excellent paper on S.E. Devon ("Geological Transactions" volume 6), has described this phenomenon; he attributes it to the action of frosts, but at the same time doubts whether the frosts of the present day penetrate to a sufficient depth. As it is known that earthquakes particularly affect the surface of the ground, it occurred to me that this appearance might perhaps be due, at least at Concepcion, to their frequent occurrence; the superficial layers of detritus being either jerked in one direction, or, where the surface was inclined, pushed a little downwards during each strong vibration. In North Wales I have seen a somewhat analogous but less regular appearance, though on a greater scale ("London Philosophical Magazine" volume 21), and produced by a quite different cause, namely, by the stranding of great icebergs; this latter appearance has also been observed in N. America.)

In certain spots large quartz veins were numerous, and near them, the cleavage, as was the case with the foliation of the schists in the Chonos Archipelago, became extremely tortuous.

At the northern end of Quiriquina Island, in the Bay of Concepcion, at least eight rudely parallel dikes, which have been guided to a certain extent by the cleavage of the slate, occur within the space of a quarter of a mile. They vary much in composition, resembling in many respects the dikes at Low's Harbour: the greater number consist of feldspathic porphyries, sometimes containing grains of quartz: one, however, was black and brilliant, like an augitic rock, but really formed of feldspar; others of a feldspathic nature were perfectly white, with either an earthy or crystalline fracture, and including grains and regular octagons of quartz; these white varieties passed into ordinary greenstones. Although, both here and at Low's Harbour, the nature of the rock varied considerably in the same dike, yet I cannot but think that at these two places and in other parts of the Chonos group, where the dikes, though close to each other and running parallel, are of different composition, that they must have been formed at different periods. In the case of Quiriquina this is a rather interesting conclusion, for these eight parallel dikes cut through the metamorphic schists in a N.W. and S.E. line, and since their injection the overlying cretaceous or tertiary strata have been tilted (whilst still under the sea) from a N.W. by N. and S.E. by S. line; and again, during the great earthquake of February 1835, the ground in this neighbourhood was fissured in N.W. and S.E. lines; and from the manner in which buildings were thrown down, it was evident that the surface undulated in this same direction. ("Geological Transactions" volume 6 . "Journal of Researches" 2nd edition.)

CENTRAL AND NORTHERN CHILE.

Northward of Concepcion, as far as Copiapo, the shores of the Pacific consist, with the exception of some small tertiary basins, of gneiss, mica- schist, altered clay-slate, granite, greenstone and syenite: hence the coast from Tres Montes to Copiapo, a distance of 1,200 miles, and I have reason to believe for a much greater space, is almost similarly constituted.

Near Valparaiso the prevailing rock is gneiss, generally including much hornblende: concretionary balls formed of feldspar, hornblende and mica, from two or three feet in diameter, are in very many places conformably enfolded by the foliated gneiss: veins of quartz and feldspar, including black schorl and well-crystallised epidote, are numerous. Epidote likewise occurs in the gneiss in thin layers, parallel to the foliation of the mass. One large vein of a coarse granitic character was remarkable from in one part quite changing its character, and insensibly passing into a blackish porphyry, including acicular crystals of glassy feldspar and of hornblende: I have never seen any other such case. (Humboldt "Personal Narrative" volume 4, has described with much surprise, concretionary balls, with concentric divisions, composed of partially vitreous feldspar, hornblende, and garnets, included within great veins of gneiss, which cut across the mica-slate near Venezuela.)

I shall in the few following remarks on the rocks of Chile allude exclusively to their foliation and cleavage. In the gneiss round Valparaiso the strike of the foliation is very variable, but I think about N. by W. and S. by E. is the commonest direction; this likewise holds good with the cleavage of the altered feldspathic clay-slates, occasionally met with on the coast for ninety miles north of Valparaiso. Some feldspathic slate, alternating with strata of claystone porphyry in the Bell of Quillota and at Jajuel, and therefore, perhaps, belonging to a later period than the metamorphic schists on the coast, cleaved in this same direction. In the Eastern Cordillera, in the Portillo Pass, there is a grand mass of mica- slate, foliated in a north and south line, and with a high westerly dip: in the Uspallata range, clay-slate and grauwacke have a highly inclined, nearly north and south cleavage, though in some parts the strike is irregular: in the main or Cumbre range, the direction of the cleavage in the feldspathic clay-slate is N.W. and S.E.

Between Coquimbo and Guasco there are two considerable formations of mica- slate, in one of which the rock passed sometimes into common clay-slate and sometimes into a glossy black variety, very like that in the Chonos Archipelago. The folia and cleavage of these rocks ranged between [N. and N.W. by N.] and [S. and S.W. by S.]. Near the Port of Guasco several varieties of altered clay-slate have a quite irregular cleavage. Between Guasco and Copiapo, there are some siliceous and talcaceous slates cleaving in a north and south line, with an easterly dip of between 60 and 70 degrees: high up, also, the main valley of Copiapo, there is mica-slate with a high easterly dip. In the whole space between Valparaiso and Copiapo an easterly dip is much more common than an opposite or westerly one.

CONCLUDING REMARKS ON CLEAVAGE AND FOLIATION.

In this southern part of the southern hemisphere, we have seen that the cleavage-laminae range over wide areas with remarkable uniformity, cutting straight through the planes of stratification, but yet being parallel in strike to the main axes of elevation, and generally to the outlines of the coast. (In my paper on the Falkland Islands "Geological Journal" volume 3, I have given a curious case on the authority of Captain Sulivan, R.N., of much folded beds of clay-slate, in some of which the cleavage is perpendicular to the horizon, and in others it is perpendicular to each curvature or fold of the bed: this appears a new case.) The dip, however, is as variable, both in angle and in direction (that is, sometimes being inclined to the one side and sometimes to the directly opposite side), as the strike is uniform. In all these respects there is a close agreement with the facts given by Professor Sedgwick in his celebrated memoir in the "Geological Transactions," and by Sir R.I. Murchison in his various excellent discussions on this subject. The Falkland Islands, and more especially Tierra del Fuego, offer striking instances of the lines of cleavage, the principle axes of elevation, and the outlines of the coast, gradually changing together their courses. The direction which prevails throughout Tierra del Fuego and the Falkland Islands, namely, from west with some northing to east with some southing, is also common to the several ridges in Northern Patagonia and in the western parts of Banda Oriental: in this latter province, in the Sierra Tapalguen, and in the Western Falkland Island, the W. by N., or

W.N.W. and E.S.E., ridges, are crossed at right angles by others ranging N.N.E. and S.S.W.

The fact of the cleavage-laminae in the clay-slate of Tierra del Fuego, where seen cutting straight through the planes of stratification, and where consequently there could be no doubt about their nature, differing slightly in colour, texture, and hardness, appears to me very interesting. In a thick mass of laminated, feldspathic and altered clay-slate, interposed between two great strata of porphyritic conglomerate in Central Chile, and where there could be but little doubt about the bedding, I observed similar slight differences in composition, and likewise some distinct thin layers of epidote, parallel to the highly inclined cleavage of the mass. Again, I incidentally noticed in North Wales, where glaciers had passed over the truncated edges of the highly inclined laminae of clay-slate, that the surface, though smooth, was worn into small parallel undulations, caused by the competent laminae being of slightly different degrees of hardness. ("London Philosophical Magazine" volume 21.) With reference to the slates of North Wales, Professor Sedgwick describes the planes of cleavage, as "coated over with chlorite and semi-crystalline matter, which not only merely define the planes in question, but strike in parallel flakes through the whole mass of the rock." ("Geological Transactions" volume 3.) In some of those glossy and hard varieties of clay-slate, which may often be seen passing into mica-schist, it has appeared to me that the cleavage- planes were formed of excessively thin, generally slighted convoluted, folia, composed of microscopically minute scales of mica. From these several facts, and more especially from the case of the clay-slate in Tierra del Fuego, it must, I think, be concluded, that the same power which has impressed on the slate its fissile structure or cleavage has tended to modify its mineralogical character in parallel planes.

Let us now turn to the foliation of the metamorphic schists, a subject which has been much less attended to. As in the case of cleavage-laminae, the folia preserve over very large areas a uniform strike: thus Humboldt found for a distance of 300 miles in Venezuela, and indeed over a much larger space, gneiss, granite, mica, and clay-slate, striking very uniformly N.E. and S.W., and dipping at an angle of between 60 and 70 degrees to N.W. ("Personal Narrative" volume 6 et seq.); it would even appear from the facts given in this chapter, that the metamorphic rocks throughout the north-eastern part of South America are generally foliated within two points of N.E. and S.W. Over the eastern parts of Banda Oriental, the foliation strikes with a high inclination, very uniformly N.N.E. to S.S.W., and over the western parts, in a W. by N. and E. by S. line. For a space of 300 miles on the shores of the Chonos and Chiloe Islands, we have seen that the foliation seldom deviates more than a point of the compass from a N. 19 degrees W. and S. 19 degrees E. strike. As in the case of cleavage, the angle of the dip in foliated rocks is generally high but variable, and alternates from one side of the line of strike to the other side, sometimes being vertical: in the Northern Chonos Islands, however, the folia are inclined almost always to the west; in nearly the same manner, the cleavage-laminae in Southern Tierra del Fuego certainly dip much more frequently to S.S.W. than to the opposite point. In Eastern Banda Oriental, in parts of Brazil, and in some other districts, the foliation runs in the same direction with the mountain-ranges and adjoining coast-lines: amongst the Chonos Islands, however, this coincidence fails, and I have given my reasons for suspecting that one granitic axis has burst through and tilted the already inclined folia of mica-schist: in the case of cleavage, the coincidence between its strike and that of the main stratification seems sometimes to fail. (Cases are given by Mr. Jukes in his "Geology of Newfoundland") Foliation and cleavage resemble each other in the planes winding round concretions, and in becoming tortuous where veins of quartz abound. (I have seen in Brazil and Chile concretions thus enfolded by foliated gneiss; and Macculloch "Highlands" volume 1, has described a similar case. For analogous cases in clay-slate, see Professor Henslow's Memoir in "Cambridge Philosophical Transactions" volume 1, and Macculloch's "Classification of Rocks". With respect to both foliation and cleavage becoming tortuous where quartz-veins abound, I have seen instances near Monte Video, at Concepcion, and in the Chonos Islands. See also Mr. Greenough's "Critical Examination") On the flanks of the mountains both in Tierra del Fuego and in other countries, I have observed that the cleavage-planes frequently dip at a high angle inwards; and this was long ago observed by Von Buch to be the case in Norway: this fact is perhaps analogous to the folded, fan-like or radiating structure in the metamorphic schists of the Alps, in which the folia in the central crests are vertical and on the two flanks inclined inwards. (Studer in "Edinburgh New Philosophical Journal" volume 23) Where masses of fissile and foliated rocks alternate together, the cleavage and foliation, in all cases which I have seen, are parallel. Where in one district the rocks are fissile, and in another adjoining district they are foliated, the planes of cleavage and foliation are likewise generally parallel: this is the case with the feldspathic homogeneous slates in the southern part of the Chonos group, compared with the fine foliated mica-schists of the northern part; so again the clay-slate of the whole eastern side of Tierra del Fuego cleaves in exactly the same line with the foliated gneiss and mica-slate of the western coast; other analogous instances might have been adduced. (I have given a case in Australia. See my "Volcanic Islands.")

With respect to the origin of the folia of quartz, mica, feldspar, and other minerals composing the metamorphic schists, Professor Sedgwick, Mr. Lyell, and most authors believe, that the constituent parts of each layer were separately deposited as sediment, and then metamorphosed. This view, in the majority of cases, I believe to be quite untenable. In those not uncommon instances, where a mass of clay-slate, in approaching granite, gradually passes into gneiss, we clearly see that folia of distinct minerals can originate through the metamorphosis of a homogeneous fissile rock. (I have described in "Volcanic Islands" a good instance of such a passage at the Cape of Good Hope.) The deposition, it may be remarked, of numberless alternations of pure quartz, and of the elements of mica or feldspar does not appear a probable event. (See some excellent remarks on this subject, in D'Aubuisson's "Traite de Geog." tome 1. Also some remarks by Mr. Dana in "Silliman's American Journal" volume 45) In those districts in which the metamorphic schists are foliated in planes parallel to the cleavage of the rocks in an adjoining district, are we to believe that the folia are due to sedimentary layers, whilst the cleavage- laminae, though parallel, have no relation whatever to such planes of deposition? On this view, how can we reconcile the vastness of the areas over which the strike of the foliation is uniform, with what we see in disturbed districts composed of true strata: and especially, how can we understand the high and even vertical dip throughout many wide districts, which are not mountainous, and throughout some, as in Western Banda Oriental, which are not even hilly? Are we to admit that in the northern part of the Chonos Archipelago, mica-slate was first accumulated in parallel horizontal folia to a thickness of about four geographical miles, and then upturned at an angle of forty degrees; whilst, in the southern part of this same Archipelago, the cleavage-laminae of closely allied rocks, which none would imagine had ever been horizontal, dip at nearly the same angle, to nearly the same point?

Seeing, then, that foliated schists indisputably are sometimes produced by the metamorphosis of homogeneous fissile rocks; seeing that foliation and cleavage are so closely analogous in the several above-enumerated respects; seeing that some fissile and almost homogeneous rocks show incipient mineralogical changes along the planes of their cleavage, and that other rocks with a fissile structure alternate with, and pass into varieties with a foliated structure, I cannot doubt that in most cases foliation and cleavage are parts of the same process: in cleavage there being only an incipient separation of the constituent minerals; in foliation a much more complete separation and crystallisation.

The fact often referred to in this chapter, of the foliation and the so- called strata in the metamorphic series,—that is, the alternating masses of different varieties of gneiss, mica-schist, and hornblende-slate, etc.,- -being parallel to each other, at first appears quite opposed to the view, that the folia have no relation to the planes of original deposition. Where the so-called beds are not very thick and of widely different mineralogical composition from each other, I do not think that there is any difficulty in supposing that they have originated in an analogous manner with the separate folia. We should bear in mind what thick strata, in ordinary sedimentary masses, have obviously been formed by a concretionary process. In a pile of volcanic rocks on the Island of Ascension, there are strata, differing quite as much in appearance as the ordinary varieties of the metamorphic schists, which undoubtedly have been produced, not by successive flowings of lava, but by internal molecular changes. Near Monte Video, where the stratification, as it would be called, of the metamorphic series is, in most parts, particularly well developed, being as usual, parallel to the foliation, we have seen that a mass of chloritic schist, netted with quartz-veins, is entangled in gneiss, in such a manner as to show that it had certainly originated in some process of segregation: again, in another spot, the gneiss tended to pass into hornblendic schist by alternating with layers of quartz; but these layers of quartz almost certainly had never been separately deposited, for they were absolutely continuous with the numerous intersecting veins of quartz. I have never had an opportunity of tracing for any distance, along the line both of strike and of dip, the so-called beds in the metamorphic schists, but I strongly suspect that they would not be found to extend with the same character, very far in the line either of their dip or strike. Hence I am led to believe, that most of the so-called beds are of the nature of complex folia, and have not been separately deposited. Of course, this view cannot be extended to THICK masses included in the metamorphic series, which are of totally different composition from the adjoining schists, and which are far extended, as is sometimes the case with quartz and marble; these must generally be of the nature of true strata. (Macculloch "Classification of Rocks", states that primary limestones are often found in irregular masses or great nodules, "which can scarcely be said to possess a stratified shape!") Such strata, however, will almost always strike in the same direction with the folia, owing to the axes of elevation being in most countries parallel to the strike of the foliation; but they will generally dip at a different angle from that of the foliation; and the angle of the foliation in itself almost always varies much: hence, in crossing a metamorphosed schistose district, it would require especial attention to discriminate between true strata of deposition and complex foliated masses. The mere presence of true strata in the midst of a set of metamorphic schists, is no argument that the foliation is of sedimentary origin, without it be further shown in each case, that the folia not only strike, but dip throughout in parallel planes with those of the true stratification.

As in some cases it appears that where a fissile rock has been exposed to partial metamorphic action, for instance from the irruption of granite, the foliation has supervened on the already existing cleavage-planes; so perhaps in some instances, the foliation of a rock may have been determined by the original planes of deposition or of oblique current-laminae: I have, however, myself, never seen such a case, and I must maintain that in most extensive metamorphic areas, the foliation is the extreme result of that process, of which cleavage is the first effect. That foliation may arise without any previous structural arrangement in the mass, we may infer from injected, and therefore once liquified, rocks, both of volcanic and plutonic origin, sometimes having a "grain" (as expressed by Professor Sedgwick), and sometimes being composed of distinct folia or laminae of different compositions. In my work on "Volcanic Islands," I have given several instances of this structure in volcanic rocks, and it is not uncommonly seen in plutonic masses—thus, in the Cordillera of Chile, there are gigantic mountain-like masses of red granite, which have been injected whilst liquified, and which, nevertheless, display in parts a decidedly laminar structure. (As remarked in a former part of this chapter, I suspect that the boldly conical mountains of gneiss-granite, near Rio de Janeiro, in which the constituent minerals are arranged in parallel planes, are of intrusive origin. We must not, however, forget the lesson of caution taught by the curious claystone porphyries of Port Desire, in which we have seen that the breaking up and aggregation of a thinly stratified tufaceous mass, has yielded a rock semi-porphyritic with crystals of feldspar, arranged in the planes of original deposition.)

Finally, we have seen that the planes of cleavage and of foliation, that is, of the incipient process and of the final result, generally strike parallel to the principal axes of elevation, and to the outline of the land: the strike of the axes of elevation (that is, of the lines of fissures with the strata on their edges upturned), according to the reasoning of Mr. Hopkins, is determined by the form of the area undergoing changes of level, and the consequent direction of the lines of tension and fissure. Now, in that remarkable pile of volcanic rocks at Ascension, which has several times been alluded to (and in some other cases), I have endeavoured to show, that the lamination of the several varieties, and their alternations, have been caused by the moving mass, just before its final consolidation, having been subjected (as in a glacier) to planes of different tension; this difference in the tension affecting the crystalline and concretionary processes. (In "Volcanic Islands.") One of the varieties of rock thus produced at Ascension, at first sight, singularly resembles a fine-grained gneiss; it consists of quite straight and parallel zones of excessive tenuity, of more or less coloured crystallised feldspar, of distinct crystals of quartz, diopside, and oxide of iron. These considerations, notwithstanding the experiments made by Mr. Fox, showing the influence of electrical currents in producing a structure like that of cleavage, and notwithstanding the apparently inexplicable variation, both in the inclination of the cleavage-laminae and in their dipping first to one side and then to the other side of the line of strike, lead me to suspect that the planes of cleavage and foliation are intimately connected with the planes of different tension, to which the area was long subjected, AFTER the main fissures or axes of upheavement had been formed, but BEFORE the final consolidation of the mass and the total cessation of all molecular movement.

CHAPTER VII. CENTRAL CHILE:—STRUCTURE OF THE CORDILLERA.
Central Chile.
Basal formations of the Cordillera.
Origin of the porphyritic clay-stone conglomerate.
Andesite.
Volcanic rocks.
Section of the Cordillera by the Peuquenes are Portillo Pass.
Great gypseous formation.
Peuquenes line; thickness of strata, fossils of.
Portillo line.
Conglomerate, orthitic granite, mica-schist, volcanic rocks of.
Concluding remarks on the denudation and elevation of the Portillo line.
Section by the Cumbre, or Uspallata Pass.
Porphyries.
Gypseous strata.
Section near the Puente del Inca; fossils of.
Great subsidence.
Intrusive porphyries.
Plain of Uspallata.
Section of the Uspallata chain.
Structure and nature of the strata.
Silicified vertical trees.
Great subsidence.
Granitic rocks of axis.
Concluding remarks on the Uspallata range; origin subsequent to that of the
main Cordillera; two periods of subsidence; comparison with the Portillo
chain.

The district between the Cordillera and the Pacific, on a rude average, is from about eighty to one hundred miles in width. It is crossed by many chains of mountains, of which the principal ones, in the latitude of Valparaiso and southward of it, range nearly north and south; but in the more northern parts of the province, they run in almost every possible direction. Near the Pacific, the mountain-ranges are generally formed of syenite or granite, and or of an allied euritic porphyry; in the low country, besides these granitic rocks and greenstone, and much gneiss, there are, especially northward of Valparaiso, some considerable districts of true clay-slate with quartz veins, passing into a feldspathic and porphyritic slate; there is also some grauwacke and quartzose and jaspery rocks, the latter occasionally assuming the character of the basis of claystone porphyry: trap-dikes are numerous. Nearer the Cordillera the ranges (such as those of S. Fernando, the Prado (Meyen "Reise um Erde" th. 1 s. 235.), and Aconcagua) are formed partly of granitic rocks, and partly of purple porphyritic conglomerates, claystone porphyry, greenstone porphyry, and other rocks, such as we shall immediately see, form the basal strata of the main Cordillera. In the more northern parts of Chile, this porphyritic series extends over large tracts of country far from the Cordillera; and even in Central Chile such occasionally occur in outlying positions.

I will describe the Campana of Quillota, which stands only fifteen miles from the Pacific, as an instance of one of these outlying masses. This hill is conspicuous from rising to the height of 6,400 feet: its summit shows a nucleus, uncovered for a height of 800 feet, of fine greenstone, including epidote and octahedral magnetic iron ore; its flanks are formed of great strata of porphyritic claystone conglomerate associated with various true porphyries and amygdaloids, alternating with thick masses of a highly feldspathic, sometimes porphyritic, pale-coloured slaty rock, with its cleavage-laminae dipping inwards at a high angle. At the base of the hill there are syenites, a granular mixture of quartz and feldspar, and harsh quartzose rocks, all belonging to the basal metamorphic series. I may observe that at the foot of several hills of this class, where the porphyries are first seen (as near S. Fernando, the Prado, Las Vacas, etc.), similar harsh quartzose rocks and granular mixtures of quartz and feldspar occur, as if the more fusible constituent parts of the granitic series had been drawn off to form the overlying porphyries.

In Central Chile, the flanks of the main Cordillera, into which I penetrated by four different valleys, generally consist of distinctly stratified rocks. The strata are inclined at angles varying from sometimes even under ten, to twenty degrees, very rarely exceeding forty degrees: in some, however, of the quite small, exterior, spur-like ridges, the inclination was not unfrequently greater. The dip of the strata in the main outer lines was usually outwards or from the Cordillera, but in Northern Chile frequently inwards,—that is, their basset-edges fronted the Pacific. Dikes occur in extraordinary numbers. In the great, central, loftiest ridges, the strata, as we shall presently see, are almost always highly inclined and often vertical. Before giving a detailed account of my two sections across the Cordillera, it will, I think, be convenient to describe the basal strata as seen, often to a thickness of four or five thousand feet, on the flanks of the outer lines.

BASAL STRATA OF THE CORDILLERA.

The prevailing rock is a purplish or greenish, porphyritic claystone conglomerate. The embedded fragments vary in size from mere particles to blocks as much as six or eight inches (rarely more) in diameter; in many places, where the fragments were minute, the signs of aqueous deposition were unequivocally distinct; where they were large, such evidence could rarely be detected. The basis is generally porphyritic with perfect crystals of feldspar, and resembles that of a true injected claystone porphyry: often, however, it has a mechanical or sedimentary aspect, and sometimes (as at Jajuel) is jaspery. The included fragments are either angular, or partially or quite rounded (Some of the rounded fragments in the porphyritic conglomerate near the Baths of Cauquenes, were marked with radii and concentric zones of different shades of colour: any one who did not know that pebbles, for instance flint pebbles from the chalk, are sometimes zoned concentrically with their worn and rounded surfaces, might have been led to infer, that these balls of porphyry were not true pebbles, but had originated in concretionary action.); in some parts the rounded, in others the angular fragments prevail, and usually both kinds are mixed together: hence the word BRECCIA ought strictly to be appended to the term PORPHYRITIC CONGLOMERATE. The fragments consist of many varieties of claystone porphyry, usually of nearly the same colour with the surrounding basis, namely, purplish-reddish, brownish, mottled or bright green; occasionally fragments of a laminated, pale-coloured, feldspathic rock, like altered clay-slate are included; as are sometimes grains of quartz, but only in one instance in Central Chile (namely, at the mines of Jajuel) a few pebbles of quartz. I nowhere observed mica in this formation, and rarely hornblende; where the latter mineral did occur, I was generally in doubt whether the mass really belonged to this formation, or was of intrusive origin. Calcareous spar occasionally occurs in small cavities; and nests and layers of epidote are common. In some few places in the finer-grained varieties (for instance, at Quillota), there were short, interrupted layers of earthy feldspar, which could be traced, exactly as at Port Desire, passing into large crystals of feldspar: I doubt, however, whether in this instance the layers had ever been separately deposited as tufaceous sediment.

All the varieties of porphyritic conglomerates and breccias pass into each other, and by innumerable gradations into porphyries no longer retaining the least trace of mechanical origin: the transition appears to have been effected much more easily in the finer-grained, than in the coarser-grained varieties. In one instance, near Cauquenes, I noticed that a porphyritic conglomerate assumed a spheroidal structure, and tended to become columnar. Besides the porphyritic conglomerates and the perfectly characterised porphyries, of metamorphic origin, there are other porphyries, which, though differing not at all or only slightly in composition, certainly have had a different origin: these consist of pink or purple claystone porphyries, sometimes including grains of quartz,—of greenstone porphyry, and of other dusky rocks, all generally porphyritic with fine, large, tabular, opaque crystals, often placed crosswise, of feldspar cleaving like albite (judging from several measurements), and often amygdaloidal with silex, agate, carbonate of lime, green and brown bole. (This bole is a very common mineral in the amygdaloidal rocks; it is generally of a greenish- brown colour, with a radiating structure; externally it is black with an almost metallic lustre, but often coated by a bright green film. It is soft and can be scratched by a quill; under the blowpipe swells greatly and becomes scaly, then fuses easily into a black magnetic bead. This substance is evidently similar to that which often occurs in submarine volcanic rocks. An examination of some very curious specimens of a fine porphyry (from Jajuel) leads me to suspect that some of these amygdaloidal balls, instead of having been deposited in pre-existing air-vesicles, are of concretionary origin; for in these specimens, some of the pea-shaped little masses (often externally marked with minute pits) are formed of a mixture of green earth with stony matter, like the basis of the porphyry, including minute imperfect crystals of feldspar; and these pea-shaped little masses are themselves amygdaloidal with minute spheres of the green earth, each enveloped by a film of white, apparently feldspathic, earthy matter: so that the porphyry is doubly amygdaloidal. It should not, however, be overlooked, that all the strata here have undergone metamorphic action, which may have caused crystals of feldspar to appear, and other changes to be effected, in the originally simple amygdaloidal balls. Mr. J.D. Dana, in an excellent paper on Trap-rocks "Edinburgh New Philosophical Journal" volume 41, has argued with great force, that all amygdaloidal minerals have been deposited by aqueous infiltration. I may take this opportunity of alluding to a curious case, described in my work on "Volcanic Islands," of an amygdaloid with many of its cells only half filled up with a mesotypic mineral. M. Rose has described an amygdaloid, brought by Dr. Meyen "Reise um Erde" Th. 1. s. 316, from Chile, as consisting of crystallised quartz, with crystals of stilbite within, and lined externally by green earth.) These several porphyritic and amygdaloidal varieties never show any signs of passing into masses of sedimentary origin: they occur both in great and small intrusive masses, and likewise in strata alternating with those of the porphyritic conglomerate, and with the planes of junction often quite distinct, yet not seldom blended together. In some of these intrusive masses, the porphyries exhibit, more or less plainly, a brecciated structure, like that often seen in volcanic masses. These brecciated porphyries could generally be distinguished at once from the metamorphosed, porphyritic breccia- conglomerates, by all the fragments being angular and being formed of the same variety, and by the absence of every trace of aqueous deposition. One of the porphyries above specified, namely, the greenstone porphyry with large tabular crystals of albite, is particularly abundant, and in some parts of the Cordillera (as near St. Jago) seemed more common even than the purplish porphyritic conglomerate. Numerous dikes likewise consist of this greenstone porphyry; others are formed of various fine-grained trappean rocks; but very few of claystone porphyry: I saw no true basaltic dikes.

In several places in the lower part of the series, but not everywhere, thick masses of a highly feldspathic, often porphyritic, slaty rock occur interstratified with the porphyritic conglomerate; I believe in one or two cases blackish limestone has been found in a similar position. The feldspathic rock is of a pale grey or greenish colour; it is easily fusible; where porphyritic, the crystals of feldspar are generally small and vitreous: it is distinctly laminated, and sometimes includes parallel layers of epidote (This mineral is extremely common in all the formations of Chile; in the gneiss near Valparaiso and in the granitic veins crossing it, in the injected greenstone crowning the C. of Quillota, in some granitic porphyries, in the porphyritic conglomerate, and in the feldspathic clay-slates.); the lamination appears to be distinct from stratification. Occasionally this rock is somewhat curious; and at one spot, namely, at the C. of Quillota, it had a brecciated structure. Near the mines of Jajuel, in a thick stratum of this feldspathic, porphyritic slate, there was a layer of hard, blackish, siliceous, infusible, compact clay-slate, such as I saw nowhere else; at the same place I was able to follow for a considerable distance the junction between the slate and the conformably underlying porphyritic conglomerate, and they certainly passed gradually into each other. Wherever these slaty feldspathic rocks abound, greenstone seems common; at the C. of Quillota a bed of well-crystallised greenstone lay conformably in the midst of the feldspathic slate, with the upper and lower junctions passing insensibly into it. From this point, and from the frequently porphyritic condition of the slate, I should perhaps have considered this rock as an erupted one (like certain laminated feldspathic lavas in the trachytic series), had I not seen in Tierra del Fuego how readily true clay-slate becomes feldspathic and porphyritic, and had I not seen at Jajuel the included layer of black, siliceous clay-slate, which no one could have thought of igneous origin. The gentle passage of the feldspathic slate, at Jajuel, into the porphyritic conglomerate, which is certainly of aqueous origin, should also be taken in account.

The alternating strata of porphyries and porphyritic conglomerate, and with the occasionally included beds of feldspathic slate, together make a grand formation; in several places within the Cordillera, I estimated its thickness at from six to seven thousand feet. It extends for many hundred miles, forming the western flank of the Chilean Cordillera; and even at Iquique in Peru, 850 miles north of the southernmost point examined by me in Chile, the coast-escarpment which rises to a height of between two and three thousand feet is thus composed. In several parts of Northern Chile this formation extends much further towards the Pacific, over the granitic and metamorphic lower rocks, than it does in Central Chile; but the main Cordillera may be considered as its central line, and its breadth in an east and west direction is never great. At first the origin of this thick, massive, long but narrow formation, appeared to me very anomalous: whence were derived, and how were dispersed the innumerable fragments, often of large size, sometimes angular and sometimes rounded, and almost invariably composed of porphyritic rocks? Seeing that the interstratified porphyries are never vesicular and often not even amygdaloidal, we must conclude that the pile was formed in deep water; how then came so many fragments to be well rounded and so many to remain angular, sometimes the two kinds being equally mingled, sometimes one and sometimes the other preponderating? That the claystone, greenstone, and other porphyries and amygdaloids, which lie CONFORMABLY between the beds of conglomerate, are ancient submarine lavas, I think there can be no doubt; and I believe we must look to the craters whence these streams were erupted, as the source of the breccia- conglomerate; after the great explosion, we may fairly imagine that the water in the heated and scarcely quiescent crater would remain for a considerable time sufficiently agitated to triturate and round the loose fragments, few or many in number, would be shot forth at the next eruption, associated with few or many angular fragments, according to the strength of the explosion. (This certainly seems to have taken place in some recent volcanic archipelagos, as at the Galapagos, where numerous craters are exclusively formed of tuff and fragments of lava.)

The porphyritic conglomerate being purple or reddish, even when alternating with dusty-coloured or bright green porphyries and amygdaloids, is probably an analogous circumstance to the scoriae of the blackish basalts being often bright red. The ancient submarine orifices whence the porphyries and their fragments were ejected having been arranged in a band, like most still active volcanoes, accounts for the thickness, the narrowness, and linear extension of this formation.

This whole great pile of rock has suffered much metamorphic action, as is very obvious in the gradual formation and appearance of the crystals of albitic feldspar and of epidote—in the bending together of the fragments— in the appearance of a laminated structure in the feldspathic slate—and, lastly, in the disappearance of the planes of stratification, which could sometimes be seen on the same mountain quite distinct in the upper part, less and less plain on the flanks, and quite obliterated at the base. Partly owing to this metamorphic action, and partly to the close relationship in origin, I have seen fragments of porphyries—taken from a metamorphosed conglomerate—from a neighbouring stream of lava—from the nucleus or centre (as it appeared to me) of the whole submarine volcano— and lastly from an intrusive mass of quite subsequent origin, all of which were absolutely undistinguishable in external characters.

One other rock, of plutonic origin, and highly important in the history of the Cordillera, from having been injected in most of the great axes of elevation, and from having apparently been instrumental in metamorphosing the superincumbent strata, may be conveniently described in this preliminary discussion. It has been called by some authors ANDESITE: it mainly consists of well-crystallised white albite (as determined with the goniometer in numerous specimens both by Professor Miller and myself), of less perfectly crystallised green hornblende, often associated with much mica, with chlorite and epidote, and occasionally with a few grains of quartz: in one instance in Northern Chile, I found crystals of orthitic or potash feldspar, mingled with those of albite. (I here, and elsewhere, call by this name, those feldspathic minerals which cleave like albite: but it now appears ("Edinburgh New Philosophical Journal" volume 24) that Abich has analysed a mineral from the Cordillera, associated with hornblende and quartz (probably the same rock with that here under discussion), which cleaves like albite, but which is a new and distinct kind, called by him ANDESINE. It is allied to leucite, with the greater proportion of its potash replaced by lime and soda. This mineral seems scarcely distinguishable from albite, except by analysis.) Where the mica and quartz are abundant, the rock cannot be distinguished from granite; and it may be called andesitic granite. Where these two minerals are quite absent, and when, as often then happens, the crystals of albite are imperfect and blend together, the rock may be called andesitic porphyry, which bears nearly the same relation to andesitic granite that euritic porphyry does to common granite. These andesitic rocks form mountain masses of a white colour, which, in their general outline and appearance—in their joints—in their occasionally including dark-coloured, angular fragments, apparently of some pre-existing rock—and in the great dikes branching from them into the superincumbent strata, manifest a close and striking resemblance to masses of common granite and syenite: I never, however, saw in these andesitic rocks, those granitic veins of segregation which are so common in true granites. We have seen that andesite occurs in three places in Tierra del Fuego; in Chile, from S. Fernando to Copiapo, a distance of 450 miles, I found it under most of the axes of elevation; in a collection of specimens from the Cordillera of Lima in Peru, I immediately recognised it; and Erman states that it occurs in Eastern Kamtschatka. ("Geographical Journal" volume 9) From its wide range, and from the important part it has played in the history of the Cordillera, I think this rock has well deserved its distinct name of Andesite.

The few still active volcanoes in Chile are confined to the central and loftiest ranges of the Cordillera; and volcanic matter, such as appears to have been of subaerial eruption, is everywhere rare. According to Meyen, there is a hill of pumice high up the valley of the Maypu, and likewise a trachytic formation at Colina, a village situated north of St. Jago. ("Reise um Erde" Th. 1 ss. 338 and 362.) Close to this latter city, there are two hills formed of a pale feldspathic porphyry, remarkable from being doubly columnar, great cylindrical columns being subdivided into smaller four- or five-sided ones; and a third hillock (Cerro Blanco) is formed of a fragmentary mass of rock, which I believed to be of volcanic origin, intermediate in character between the above feldspathic porphyry and common trachyte, and containing needles of hornblende and granular oxide of iron. Near the Baths of Cauquenes, between two short parallel lines of elevation, where they are intersected by the valley, there is a small, though distinct volcanic district; the rock is a dark grey (andesitic) trachyte, which fuses into a greenish-grey bead, and is formed of long crystals of fractured glassy albite (judging from one measurement) mingled with well- formed crystals, often twin, of augite. The whole mass is vesicular, but the surface is darker coloured and much more vesicular than any other part. This trachyte forms a cliff-bounded, horizontal, narrow strip on the steep southern side of the valley, at the height of four or five hundred feet above the river-bed; judging from an apparently corresponding line of cliff on the northern side, the valley must once have been filled up to this height by a field of lava. On the summit of a lofty mountain some leagues higher up this same valley of the Cachapual, I found columnar pitchstone porphyritic with feldspar; I do not suppose this rock to be of volcanic origin, and only mention it here, from its being intersected by masses and dikes of a VESICULAR rock, approaching in character to trachyte; in no other part of Chile did I observe vesicular or amygdaloidal dikes, though these are so common in ordinary volcanic districts.

PASSAGE OF THE ANDES BY THE PORTILLO OR PEQUENES PASS.

Although I crossed the Cordillera only once by this pass, and only once by that of the Cumbre or Uspallata (presently to be described), riding slowly and halting occasionally to ascend the mountains, there are many circumstances favourable to obtaining a more faithful sketch of their structure than would at first be thought possible from so short an examination. The mountains are steep and absolutely bare of vegetation; the atmosphere is resplendently clear; the stratification distinct; and the rocks brightly and variously coloured: some of the natural sections might be truly compared for distinctness to those coloured ones in geological works. Considering how little is known of the structure of this gigantic range, to which I particularly attended, most travellers having collected only specimens of the rocks, I think my sketch-sections, though necessarily imperfect, possess some interest. Section 1/1 in Plate 1 which I will now describe in detail, is on a horizontal scale of a third of an inch to a nautical mile, and on a vertical scale of one inch to a mile (or 6,000 feet). The width of the range (excluding a few outlying hillocks), from the plain on which St. Jago the capital of Chile stands, to the Pampas, is sixty miles, as far as I can judge from the maps, which differ from each other and are all EXCEEDINGLY imperfect. The St. Jago plain at the mouth of the Maypu, I estimate from adjoining known points at 2,300 feet, and the Pampas at 3,500 feet, both above the level of the sea. The height of the Pequenes line, according to Dr. Gillies, is 13,210 feet ("Journal of Natural and Geographical Science" August 1830.); and that of the Portillo line (both in the gaps where the road crosses them) is 14,345 feet; the lowest part of the intermediate valley of Tenuyan is 7,530 feet—all above the level of the sea.

The Cordillera here, and indeed I believe throughout Chile, consist of several parallel, anticlinal and uniclinal mountain-lines, ranging north, or north with a little westing, and south. Some exterior and much lower ridges often vary considerably from this course, projecting like oblique spurs from the main ranges: in the district towards the Pacific, the mountains, as before remarked, extend in various directions, even east and west. In the main exterior lines, the strata, as also before remarked, are seldom inclined at a high angle; but in the central lofty ridges they are almost always highly inclined, broken by many great faults, and often vertical. As far as I could judge, few of the ranges are of great length: and in the central parts of the Cordillera, I was frequently able to follow with my eye a ridge gradually becoming higher and higher, as the stratification increased in inclination, from one end where its height was trifling and its strata gently inclined to the other end where vertical strata formed snow-clad pinnacles. Even outside the main Cordillera, near the baths of Cauquenes, I observed one such case, where a north and south ridge had its strata in the valley inclined at 37 degrees, and less than a mile south of it at 67 degrees: another parallel and similarly inclined ridge rose at the distance of about five miles, into a lofty mountain with absolutely vertical strata. Within the Cordillera, the height of the ridges and the inclination of the strata often became doubled and trebled in much shorter distances than five miles; this peculiar form of upheaval probably indicates that the stratified crust was thin, and hence yielded to the underlying intrusive masses unequally, at certain points on the lines of fissure.

The valleys, by which the Cordillera are drained, follow the anticlinal or rarely synclinal troughs, which deviate most from the usual north and south course; or still more commonly those lines of faults or of unequal curvature (that is, lines with the strata on both hands dipping in the same direction, but at a somewhat different angle) which deviate most from a northerly course. Occasionally the torrents run for some distance in the north and south valleys, and then recover their eastern or western course by bursting through the ranges at those points where the strata have been least inclined and the height consequently is less. Hence the valleys, along which the roads run, are generally zigzag; and, in drawing an east and west section, it is necessary to contract greatly that which is actually seen on the road.

Commencing at the western end of Section 1/1 where the R. Maypu debouches on the plain of St. Jago, we immediately enter on the porphyritic conglomerate formation, and in the midst of it find some hummocks [A] of granite and syenite, which probably (for I neglected to collect specimens) belong to the andesitic class. These are succeeded by some rugged hills [B] of dark-green, crystalline, feldspathic and in some parts slaty rocks, which I believe belong to the altered clay-slate formation. From this point, great mountains of purplish and greenish, generally thinly stratified, highly porphyritic conglomerates, including many strata of amygdaloidal and greenstone porphyries, extend up the valley to the junction of the rivers Yeso and Volcan. As the valley here runs in a very southerly course, the width of the porphyritic conglomerate formation is quite conjectural; and from the same cause, I was unable to make out much about the stratification. In most of the exterior mountains the dip was gentle and directed inwards; and at only one spot I observed an inclination as high as 50 degrees. Near the junction of the R. Colorado with the main stream, there is a hill of whitish, brecciated, partially decomposed feldspathic porphyry, having a volcanic aspect but not being really of that nature: at Tolla, however, in this valley, Dr. Meyen met with a hill of pumice containing mica. ("Reise um Erde" Th.1 ss. 338, 341.) At the junction of the Yeso and Volcan [D] there is an extensive mass, in white conical hillocks, of andesite, containing some mica, and passing either into andesitic granite, or into a spotted, semi-granular mixture of albitic (?) feldspar and hornblende: in the midst of this formation Dr. Meyen found true trachyte. The andesite is covered by strata of dark-coloured, crystalline, obscurely porphyritic rocks, and above them by the ordinary porphyritic conglomerates,—the strata all dipping away at a small angle from the underlying mass. The surrounding lofty mountains appear to be entirely composed of the porphyritic conglomerate, and I estimated its thickness here at between six and seven thousand feet. Beyond the junction of the Yeso and Volcan, the porphyritic strata appear to dip towards the hillocks of andesite at an angle of 40 degrees; but at some distant points on the same ridge they are bent up and vertical. Following the valley of the Yeso, trending N.E. (and therefore still unfavourable for our transverse section), the same porphyritic conglomerate formation is prolonged to near the Cuestadel Indio, situated at the western end of the basin (like a drained lake) of Yeso. Some way before arriving at this point, distant lofty pinnacles capped by coloured strata belonging to the great gypseous formation could first be seen. From the summit of the Cuesta, looking southward, there is a magnificent sectional view of a mountain-mass, at least 2,000 feet in thickness [E], of fine andesite granite (containing much black mica, a little chlorite and quartz), which sends great white dikes far into the superincumbent, dark-coloured, porphyritic conglomerates. At the line of junction the two formations are wonderfully interlaced together: in the lower part of the porphyritic conglomerate, the stratification has been quite obliterated, whilst in the upper part it is very distinct, the beds composing the crests of the surrounding mountains being inclined at angles of between 70 and 80 degrees, and some being even vertical. On the northern side of the valley, there is a great corresponding mass of andesitic granite, which is encased by porphyritic conglomerate, dipping both on the western and eastern sides, at about 80 degrees to west, but on the eastern side with the tips of the strata bent in such a manner, as to render it probable that the whole mass has been on that side thrown over and inverted.

In the valley basin of the Yeso, which I estimated at 7,000 feet above the level of the sea, we first reach at [F] the gypseous formation. Its thickness is very great. It consists in most parts of snow-white, hard, compact gypsum, which breaks with a saccharine fracture, having translucent edges; under the blowpipe gives out much vapour; it frequently includes nests and exceedingly thin layers of crystallised, blackish carbonate of lime. Large, irregularly shaped concretions (externally still exhibiting lines of aqueous deposition) of blackish-grey, but sometimes white, coarsely and brilliantly crystallised, hard anhydrite, abound within the common gypsum. Hillocks, formed of the hardest and purest varieties of the white gypsum, stand up above the surrounding parts, and have their surfaces cracked and marked, just like newly baked bread. There is much pale brown, soft argillaceous gypsum; and there were some intercalated green beds which I had not time to reach. I saw only one fragment of selenite or transparent gypsum, and that perhaps may have come from some subsequently formed vein. From the mineralogical characters here given, it is probable that these gypseous beds have undergone some metamorphic action. The strata are much hidden by detritus, but they appeared in most parts to be highly inclined; and in an adjoining lofty pinnacle they could be distinctly seen bending up, and becoming vertical, conformably with the underlying porphyritic conglomerate. In very many parts of the great mountain-face [F], composed of thin gypseous beds, there were innumerable masses, irregularly shaped and not like dikes, yet with well-defined edges, of an imperfectly granular, pale greenish, or yellowish-white rock, essentially composed of feldspar, with a little chlorite or hornblende, epidote, iron-pyrites, and ferruginous powder: I believe that these curious trappean masses have been injected from the not far distant mountain-mass [E] of andesite whilst still fluid, and that owing to the softness of the gypseous strata they have not acquired the ordinary forms of dikes. Subsequently to the injection of these feldspathic rocks, a great dislocation has taken place; and the much shattered gypseous strata here overlie a hillock [G], composed of vertical strata of impure limestone and of black highly calcareous shale including threads of gypsum: these rocks, as we shall presently see, belong to the upper parts of the gypseous series, and hence must here have been thrown down by a vast fault.

Proceeding up the valley-basin of the Yeso, and taking our section sometimes on one hand and sometimes on the other, we come to a great hill of stratified porphyritic conglomerate [H] dipping at 45 degrees to the west; and a few hundred yards farther on, we have a bed between three or four hundred feet thick of gypsum [I] dipping eastward at a very high angle: here then we have a fault and anticlinal axis. On the opposite side of the valley, a vertical mass of red conglomerate, conformably underlying the gypsum, appears gradually to lose its stratification and passes into a mountain of porphyry. The gypsum [I] is covered by a bed [K], at least 1,000 feet in thickness, of a purplish-red, compact, heavy, fine-grained sandstone or mudstone, which fuses easily into a white enamel, and is seen under a lens to contain triturated crystals. This is succeeded by a bed [L], 1,000 feet thick (I believe I understate the thickness) of gypsum, exactly like the beds before described; and this again is capped by another great bed [M] of purplish-red sandstone. All these strata dip eastward; but the inclination becomes less and less, as we leave the first and almost vertical bed [I] of gypsum.

Leaving the basin-plain of Yeso, the road rapidly ascends, passing by mountains composed of the gypseous and associated beds, with their stratification greatly disturbed and therefore not easily intelligible: hence this part of the section has been left uncoloured.

Shortly before reaching the great Pequenes ridge, the lowest stratum visible [N] is a red sandstone or mudstone, capped by a vast thickness of black, compact, calcareous, shaly rock [O], which has been thrown into four lofty, though small ridges: looking northward, the strata in these ridges are seen gradually to rise in inclination, becoming in some distant pinnacles absolutely vertical.

The ridge of Pequenes, which divides the waters flowing into the Pacific and Atlantic Oceans, extends in a nearly N.N.W. and S.S.E. line; its strata dip eastward at an angle of between 30 and 45 degrees, but in the higher peaks bending up and becoming almost vertical. Where the road crosses this range, the height is 13,210 feet above the sea-level, and I estimated the neighbouring pinnacles at from fourteen to fifteen thousand feet. The lowest stratum visible in this ridge is a red stratified sandstone [P]; on it are superimposed two great masses [Q and S] of black, hard, compact, even having a conchoidal fracture, calcareous, more or less laminated shale, passing into limestone: this rock contains organic remains, presently to be enumerated. The compacter varieties fuse easily in a white glass; and this I may add is a very general character with all the sedimentary beds in the Cordillera: although this rock when broken is generally quite black, it everywhere weathers into an ash-grey tint. Between these two great masses [Q and S], a bed [R] of gypsum is interposed, about three hundred feet in thickness, and having the same characters as heretofore described. I estimated the total thickness of these three beds [Q, R, S] at nearly three thousand feet; and to this must be added, as will be immediately seen, a great overlying mass of red sandstone.

In descending the eastern slope of this great central range, the strata, which in the upper part dip eastward at about an angle of 40 degrees, become more and more curved, till they are nearly vertical; and a little further onwards there is seen on the further side of a ravine, a thick mass of strata of bright red sandstone [T], with their upper extremities slightly curved, showing that they were once conformably prolonged over the beds [S]: on the southern and opposite side of the road, this red sandstone and the underlying black shaly rocks stand vertical, and in actual juxtaposition. Continuing to descend, we come to a synclinal valley filled with rubbish, beyond which we have the red sandstone [T2] corresponding with [T], and now dipping, as is seen both north and south of the road, at 45 degrees to the west; and under it, the beds [S2, R2, Q2, and I believe P2] in corresponding order and of similar composition, with those on the western flank of the Pequenes range, but dipping westward. Close to the synclinal valley the dip of these strata is 45 degrees, but at the eastern or farther end of the series it increases to 60 degrees. Here the great gypseous formation abruptly terminates, and is succeeded eastward by a pile of more modern strata. Considering how violently these central ranges have been dislocated, and how very numerous dikes are in the exterior and lower parts of the Cordillera, it is remarkable that I did not here notice a single dike. The prevailing rock in this neighbourhood is the black, calcareous, compact shale, whilst in the valley-basin of the Yeso the purplish red sandstone or mudstone predominates,—both being associated with gypseous strata of exactly the same nature. It would be very difficult to ascertain the relative superposition of these several masses, for we shall afterwards see in the Cumbre Pass that the gypseous and intercalated beds are lens-shaped, and that they thin out, even where very thick, and disappear in short horizontal distances: it is quite possible that the black shales and red sandstones may be contemporaneous, but it is more probable that the former compose the uppermost parts of the series.

The fossils above alluded to in the black calcareous shales are few in number, and are in an imperfect condition; they consist, as named for me by M. d'Orbigny, of:—

1. Ammonite, indeterminable, near to A. recticostatus, d'Orbigny, "Pal. Franc." (Neocomian formation). 2. Gryphaea, near to G. Couloni (Neocomian formations of France and Neufchatel). 3. Natica, indeterminable. 4. Cyprina rostrata, d'Orbigny, "Pal. Franc." (Neocomian formation). 5. Rostellaria angulosa (?), d'Orbigny, "Pal. de l'Amer. Mer." 6. Terebratula (?).

Some of the fragments of Ammonites were as thick as a man's arm: the Gryphaea is much the most abundant shell. These fossils M. d'Orbigny considers as belonging to the Neocomian stage of the Cretaceous system. Dr. Meyen, who ascended the valley of the Rio Volcan, a branch of the Yeso, found a nearly similar, but apparently more calcareous formation, with much gypsum, and no doubt the equivalent of that here described ("Reise um Erde" etc. Th. 1 s. 355.): the beds were vertical, and were prolonged up to the limits of perpetual snow; at the height of 9,000 feet above the sea, they abounded with fossils, consisting, according to Von Buch ("Descript. Phys. des Iles Canaries"), of:—

1. Exogyra (Gryphaea) Couloni, absolutely identical with specimens from the Jura and South of France. 2. Trigonia costata, identical with those found in the upper Jurassic beds at Hildesheim. 3. Pecten striatus, identical with those found in the upper Jurassic beds at Hildesheim. 4. Cucullaea, corresponding in form to C. longirostris, so frequent in the upper Jurassic beds of Westphalia. 5. Ammonites resembling A. biplex.

Von Buch concludes that this formation is intermediate between the limestone of the Jura and the chalk, and that it is analogous with the uppermost Jurassic beds forming the plains of Switzerland. Hence M. D'Orbigny and Von Buch, under different terms, compare these fossils to those from the same late stage in the secondary formations of Europe.

Some of the fossils which I collected were found a good way down the western slope of the main ridge, and hence must originally have been covered up by a great thickness of the black shaly rock, independently of the now denuded, thick, overlying masses of red sandstone. I neglected at the time to estimate how many hundred or rather thousand feet thick the superincumbent strata must have been: and I will not now attempt to do so. This, however, would have been a highly interesting point, as indicative of a great amount of subsidence, of which we shall hereafter find in other parts of the Cordillera analogous evidence during this same period. The altitude of the Peuquenes Range, considering its not great antiquity, is very remarkable; many of the fossils were embedded at the height of 13,210 feet, and the same beds are prolonged up to at least from fourteen to fifteen thousand feet above the level of the sea.

THE PORTILLO OR EASTERN CHAIN.

The valley of Tenuyan, separating the Peuquenes and Portillo lines, is, as estimated by Dr. Gillies and myself, about twenty miles in width; the lowest part, where the road crosses the river, being 7,500 feet above the sea-level. The pass on the Portillo line is 14,365 feet high (1,100 feet higher than that on the Peuquenes), and the neighbouring pinnacles must, I conceive, rise to nearly 16,000 feet above the sea. The river draining the intermediate valley of Tenuyan, passes through the Portillo line. To return to our section:—shortly after leaving the lower beds [P2] of the gypseous formation, we come to grand masses of a coarse, red conglomerate [V], totally unlike any strata hitherto seen in the Cordillera. This conglomerate is distinctly stratified, some of the beds being well defined by the greater size of the pebbles: the cement is calcareous and sometimes crystalline, though the mass shows no signs of having been metamorphosed. The included pebbles are either perfectly or only partially rounded: they consist of purplish sandstones, of various porphyries, of brownish limestone, of black calcareous, compact shale precisely like that in situ in the Peuquenes range, and CONTAINING SOME OF THE SAME FOSSIL SHELLS; also very many pebbles of quartz, some of micaceous schist, and numerous, broken, rounded crystals of a reddish orthitic or potash feldspar (as determined by Professor Miller), and these from their size must have been derived from a coarse-grained rock, probably granite. From this feldspar being orthitic, and even from its external appearance, I venture positively to affirm that it has not been derived from the rocks of the western ranges; but, on the other hand, it may well have come, together with the quartz and metamorphic schists, from the eastern or Portillo line, for this line mainly consists of coarse orthitic granite. The pebbles of the fossiliferous slate and of the purple sandstone, certainly have been derived from the Peuquenes or western ranges.

The road crosses the valley of Tenuyan in a nearly east and west line, and for several miles we have on both hands the conglomerate, everywhere dipping west and forming separate great mountains. The strata, where first met with, after leaving the gypseous formation, are inclined westward at an angle of only 20 degrees, which further on increases to about 45 degrees. The gypseous strata, as we have seen, are also inclined westward: hence, when looking from the eastern side of the valley towards the Peuquenes range, a most deceptive appearance is presented, as if the newer beds of conglomerate dipped directly under the much older beds of the gypseous formation. In the middle of the valley, a bold mountain of unstratified lilac-coloured porphyry (with crystals of hornblende) projects; and further on, a little south of the road, there is another mountain, with its strata inclined at a small angle eastwards, which in its general aspect and colour, resembles the porphyritic conglomerate formation, so rare on this side of the Peuquenes line and so grandly developed throughout the western ranges.

The conglomerate is of great thickness: I do not suppose that the strata forming the separate mountain-masses [V,V,V] have ever been prolonged over each other, but that one mass has been broken up by several, distinct, parallel, uniclinal lines of elevation. Judging therefore of the thickness of the conglomerate, as seen in the separate mountain-masses, I estimated it at least from one thousand five hundred to two thousand feet. The lower beds rest conformably on some singularly coloured, soft strata [W], which I could not reach to examine; and these again rest conformably on a thick mass of micaceous, thinly laminated, siliceous sandstone [X], associated with a little black clay-slate. These lower beds are traversed by several dikes of decomposing porphyry. The laminated sandstone is directly superimposed on the vast masses of granite [Y,Y] which mainly compose the Portillo range. The line of junction between this latter rock, which is of a bright red colour, and the whitish sandstone was beautifully distinct; the sandstone being penetrated by numerous, great, tortuous dikes branching from the granite, and having been converted into a granular quartz rock (singularly like that of the Falkland Islands), containing specks of an ochrey powder, and black crystalline atoms, apparently of imperfect mica. The quartzose strata in one spot were folded into a regular dome.

The granite which composes the magnificent bare pinnacles and the steep western flank of the Portillo chain, is of a brick-red colour, coarsely crystallised, and composed of orthitic or potash feldspar, quartz, and imperfect mica in small quantity, sometimes passing into chlorite. These minerals occasionally assume a laminar or foliated arrangement. The fact of the feldspar being orthitic in this range, is very remarkable, considering how rare, or rather, as I believe, entirely absent, this mineral is throughout the western ranges, in which soda-feldspar, or at least a variety cleaving like albite, is so extremely abundant. In one spot on the western flank, and on the eastern flank near Los Manantiales and near the crest, I noticed some great masses of a whitish granite, parts of it fine- grained, and parts containing large crystals of feldspar; I neglected to collect specimens, so I do not know whether this feldspar is also orthitic, though I am inclined to think so from its general appearance. I saw also some syenite and one mass which resembled andesite, but of which I likewise neglected to collect specimens. From the manner in which the whitish granites formed separate mountain-masses in the midst of the brick-red variety, and from one such mass near the crest being traversed by numerous veins of flesh-coloured and greenish eurite (into which I occasionally observed the brick-red granite insensibly passing), I conclude that the white granites probably belong to an older formation, almost overwhelmed and penetrated by the red granite.

On the crest I saw also, at a short distance, some coloured stratified beds, apparently like those [W] at the western base, but was prevented examining them by a snowstorm: Mr. Caldcleugh, however, collected here specimens of ribboned jasper, magnesian limestone, and other minerals. ("Travels" etc. volume 1) A little way down the eastern slope a few fragments of quartz and mica-slate are met with; but the great formation of this latter rock [Z], which covers up much of the eastern flank and base of the Portillo range, cannot be conveniently examined until much lower down at a place called Mal Paso. The mica-schist here consists of thick layers of quartz, with intervening folia of finely-scaly mica, often passing into a substance like black glossy clay-slate: in one spot, the layers of the quartz having disappeared, the whole mass became converted into glossy clay-slate. Where the folia were best defined, they were inclined at a high angle westward, that is, towards the range. The line of junction between the dark mica-slate and the coarse red granite was most clearly distinguishable from a vast distance: the granite sent many small veins into the mica-slate, and included some angular fragments of it. As the sandstone on the western base has been converted by the red granite into a granular quartz-rock, so this great formation of mica-schist may possibly have been metamorphosed at the same time and by the same means; but I think it more probable, considering its more perfect metamorphic character and its well-pronounced foliation, that it belongs to an anterior epoch, connected with the white granites: I am the more inclined to this view, from having found at the foot of the range the mica-schist surrounding a hummock [Y2], exclusively composed of white granite. Near Los Arenales, the mountains on all sides are composed of the mica-slate; and looking backwards from this point up to the bare gigantic peaks above, the view was eminently interesting. The colours of the red granite and the black mica-slate are so distinct, that with a bright light these rocks could be readily distinguished even from the Pampas, at a level of at least 9,000 feet below. The red granite, from being divided by parallel joints, has weathered into sharp pinnacles, on some of which, even on some of the loftiest, little caps of mica-schist could be clearly seen: here and there isolated patches of this rock adhered to the mountain-flanks, and these often corresponded in height and position on the opposite sides of the immense valleys. Lower down the schist prevailed more and more, with only a few quite small points of granite projecting through. Looking at the entire eastern face of the Portillo range, the red colour far exceeds in area the black; yet it was scarcely possible to doubt that the granite had once been almost wholly encased by the mica-schist.

At Los Arenales, low down on the eastern flank, the mica-slate is traversed by several closely adjoining, broad dikes, parallel to each other and to the foliation of the schist. The dikes are formed of three different varieties of rock, of which a pale brown feldspathic porphyry with grains of quartz was much the most abundant. These dikes with their granules of quartz, as well as the mica-schist itself, strikingly resemble the rocks of the Chonos Archipelago. At a height of about twelve hundred feet above the dikes, and perhaps connected with them, there is a range of cliffs formed of successive lava-streams [AA], between three and four hundred feet in thickness, and in places finely columnar. The lava consists of dark- greyish, harsh rocks, intermediate in character between trachyte and basalt, containing glassy feldspar, olivine, and a little mica, and sometimes amygdaloidal with zeolite: the basis is either quite compact, or crenulated with air-vesicles arranged in laminae. The streams are separated from each other by beds of fragmentary brown scoriae, firmly cemented together, and including a few well-rounded pebbles of lava. From their general appearance, I suspect that these lava-streams flowed at an ancient period under the pressure of the sea, when the Atlantic covered the Pampas and washed the eastern foot of the Cordillera. (This conclusion might, perhaps, even have been anticipated, from the general rarity of volcanic action, except near the sea or large bodies of water. Conformably with this rule, at the present day, there are no active volcanoes on this eastern side of the Cordillera; nor are severe earthquakes experienced here.) On the opposite and northern side of the valley there is another line of lava- cliffs at a corresponding height; the valley between being of considerable breadth, and as nearly as I could estimate 1,500 feet in depth. This field of lava is confined on both sides by the mountains of mica-schist, and slopes down rapidly but irregularly to the edge of the Pampas, where, having a thickness of about two hundred feet, it terminates against a little range of claystone porphyry. The valley in this lower part expands into a bay-like, gentle slope, bordered by the cliffs of lava, which must certainly once have extended across this wide expanse. The inclination of the streams from Los Arenales to the mouth of the valley is so great, that at the time (though ignorant of M. Elie de Beaumont's researches on the extremely small slope over which lava can flow, and yet retain a compact structure and considerable thickness) I concluded that they must subsequently to their flowing have been upheaved and tilted from the mountains; of this conclusion I can now entertain not the smallest doubt.

At the mouth of the valley, within the cliffs of the above lava-field, there are remnants, in the form of separate small hillocks and of lines of low cliffs, of a considerable deposit of compact white tuff (quarried for filtering-stones), composed of broken pumice, volcanic crystals, scales of mica, and fragments of lava. This mass has suffered much denudation; and the hard mica-schist has been deeply worn, since the period of its deposition; and this period must have been subsequent to the denudation of the basaltic lava-streams, as attested by their encircling cliffs standing at a higher level. At the present day, under the existing arid climate, ages might roll past without a square yard of rock of any kind being denuded, except perhaps in the rarely moistened drainage-channel of the valley. Must we then look back to that ancient period, when the waves of the sea beat against the eastern foot of the Cordillera, for a power sufficient to denude extensively, though superficially, this tufaceous deposit, soft although it be?

There remains only to mention some little water-worn hillocks [BB], a few hundred feet in height, and mere mole-hills compared with the gigantic mountains behind them, which rise out of the sloping, shingle-covered margin of the Pampas. The first little range is composed of a brecciated purple porphyritic claystone, with obscurely marked strata dipping at 70 degrees to the S.W.; the other ranges consist of—a pale-coloured feldspathic porphyry,—a purple claystone porphyry with grains of quartz,— and a rock almost exclusively composed of brick-red crystals of feldspar. These outermost small lines of elevation extend in a N.W. by W. and S.E. by S. direction.

CONCLUDING REMARKS ON THE PORTILLO RANGE.

When on the Pampas and looking southward, and whilst travelling northward, I could see for very many leagues the red granite and dark mica-schist forming the crest and eastern flank of the Portillo line. This great range, according to Dr. Gillies, can be traced with little interruption for 140 miles southward to the R. Diamante, where it unites with the western ranges: northward, according to this same author, it terminates where the R. Mendoza debouches from the mountains; but a little further north in the eastern part of the Cumbre section, there are, as we shall hereafter see, some mountain-masses of a brick-red porphyry, the last injected amidst many other porphyries, and having so close an analogy with the coarse red granite of the Portillo line, that I am tempted to believe that they belong to the same axis of injection; if so, the Portillo line is at least 200 miles in length. Its height, even in the lowest gap in the road, is 14,365 feet, and some of the pinnacles apparently attain an elevation of about 16,000 feet above the sea. The geological history of this grand chain appears to me eminently interesting. We may safely conclude, that at a former period the valley of Tenuyan existed as an arm of the sea, about twenty-miles in width, bordered on one hand by a ridge or chain of islets of the black calcareous shales and purple sandstones of the gypseous formation; and on the other hand, by a ridge or chain of islets composed of mica-slate, white granite, and perhaps to a partial extent of red granite. These two chains, whilst thus bordering the old sea-channel, must have been exposed for a vast lapse of time to alluvial and littoral action, during which the rocks were shattered, the fragments rounded, and the strata of conglomerate accumulated to a thickness of at least fifteen hundred or two thousand feet. The red orthitic granite now forms, as we have seen, the main part of the Portillo chain: it is injected in dikes not only into the mica-schist and white granites, but into the laminated sandstone, which it has metamorphosed, and which it has thrown off, together with the conformably overlying coloured beds and stratified conglomerate, at an angle of forty-five degrees. To have thrown off so vast a pile of strata at this angle, is a proof that the main part of the red granite (whether or not portions, as perhaps is probable, previously existed) was injected in a liquified state after the accumulation both of the laminated sandstone and of the conglomerate; this conglomerate, we know, was accumulated, not only after the deposition of the fossiliferous strata of the Peuquenes line, but after their elevation and long-continued denudation: and these fossiliferous strata belong to the early part of the Cretaceous system. Late, therefore, in a geological sense, as must be the age of the main part of the red granite, I can conceive nothing more impressive than the eastern view of this great range, as forcing the mind to grapple with the idea of the thousands of thousands of years requisite for the denudation of the strata which originally encased it,—for that the fluidified granite was once encased, its mineralogical composition and structure, and the bold conical shape of the mountain-masses, yield sufficient evidence. Of the encasing strata we see the last vestiges in the coloured beds on the crest, in the little caps of mica-schist on some of the loftiest pinnacles, and in the isolated patches of this same rock at corresponding heights on the now bare and steep flanks.

The lava-streams at the eastern foot of the Portillo are interesting, not so much from the great denudation which they have suffered at a comparatively late period as from the evidence they afford by their inclination taken conjointly with their thickness and compactness, that after the great range had assumed its present general outline, it continued to rise as an axis of elevation. The plains extending from the base of the Cordillera to the Atlantic show that the continent has been upraised in mass to a height of 3,500 feet, and probably to a much greater height, for the smooth shingle-covered margin of the Pampas is prolonged in a gentle unbroken slope far up many of the great valleys. Nor let it be assumed that the Peuquenes and Portillo ranges have undergone only movements of elevation; for we shall hereafter see, that the bottom of the sea subsided several thousand feet during the deposition of strata, occupying the same relative place in the Cordillera, with those of the Peuquenes ridge; moreover, we shall see from the unequivocal evidence of buried upright trees, that at a somewhat later period, during the formation of the Uspallata chain, which corresponds geographically with that of the Portillo, there was another subsidence of many thousand feet: here, indeed, in the valley of Tenuyan, the accumulation of the coarse stratified conglomerate to a thickness of fifteen hundred or two thousand feet, offers strong presumptive evidence of subsidence; for all existing analogies lead to the belief that large pebbles can be transported only in shallow water, liable to be affected by currents and movements of undulation—and if so, the shallow bed of the sea on which the pebbles were first deposited must necessarily have sunk to allow of the accumulation of the superincumbent strata. What a history of changes of level, and of wear and tear, all since the age of the latter secondary formations of Europe, does the structure of this one great mountain-chain reveal!

PASSAGE OF THE ANDES BY THE CUMBRE OR USPALLATA PASS.

This Pass crosses the Andes about sixty miles north of that just described: the section given in Plate 1, Section 1/2, is on the same scale as before, namely, at one-third of an inch to a mile in distance, and one inch to a mile (or 6,000 feet) in height. Like the last section, it is a mere sketch, and cannot pretend to accuracy, though made under favourable circumstances. We will commence as before, with the western half, of which the main range bears the name of the Cumbre (that is the Ridge), and corresponds to the Peuquenes line in the former section; as does the Uspallata range, though on a much smaller scale, to that of the Portillo. Near the point where the river Aconcagua debouches on the basin plain of the same name, at a height of about two thousand three hundred feet above the sea, we meet with the usual purple and greenish porphyritic claystone conglomerate. Beds of this nature, alternating with numerous compact and amygdaloidal porphyries, which have flowed as submarine lavas, and associated with great mountain- masses of various, injected, non-stratified porphyries, are prolonged the whole distance up to the Cumbre or central ridge. One of the commonest stratified porphyries is of a green colour, highly amygdaloidal with the various minerals described in the preliminary discussion, and including fine tabular crystals of albite. The mountain-range north (often with a little westing) and south. The stratification, wherever I could clearly distinguish it, was inclined westward or towards the Pacific, and, except near the Cumbre, seldom at angles above 25 degrees. Only at one spot on this western side, on a lofty pinnacle not far from the Cumbre, I saw strata apparently belonging to the gypseous formation, and conformably capping a pile of stratified porphyries. Hence, both in composition and in stratification, the structure of the mountains on this western side of the divortium aquarum, is far more simple than in the corresponding part of the Peuquenes section. In the porphyritic claystone conglomerate, the mechanical structure and the planes of stratification have generally been much obscured and even quite obliterated towards the base of the series, whilst in the upper parts, near the summits of the mountains, both are distinctly displayed. In these upper portions the porphyries are generally lighter coloured. In three places [X, Y, Z] masses of andesite are exposed: at [Y], this rock contained some quartz, but the greater part consisted of andesitic porphyry, with only a few well-developed crystals of albite, and forming a great white mass, having the external aspect of granite, capped by much dark unstratified porphyry. In many parts of the mountains, there are dikes of a green colour, and other white ones, which latter probably spring from underlying masses of andesite.

The Cumbre, where the road crosses it, is, according to Mr. Pentland, 12,454 feet above the sea; and the neighbouring peaks, composed of dark purple and whitish porphyries, some obscurely stratified with a westerly dip, and others without a trace of stratification, must exceed 13,000 feet in height. Descending the eastern slope of the Cumbre, the structure becomes very complicated, and generally differs on the two sides of the east and west line of road and section. First we come to a great mass [A] of nearly vertical, singularly contorted strata, composed of highly compact red sandstones, and of often calcareous conglomerates, and penetrated by green, yellow, and reddish dikes; but I shall presently have an opportunity of describing in some detail an analogous pile of strata. These vertical beds are abruptly succeeded by others [B], of apparently nearly the same nature but more metamorphosed, alternating with porphyries and limestones; these dip for a short space westward, but there has been here an extraordinary dislocation, which, on the north side of the road, appears to have determined the excavation of the north and south valley of the R. de las Cuevas. On this northern side of the road, the strata [B] are prolonged till they come in close contact with a jagged lofty mountain [D] of dark-coloured, unstratified, intrusive porphyry, where the beds have been more highly inclined and still more metamorphosed. This mountain of porphyry seems to form a short axis of elevation, for south of the road in its line there is a hill [C] of porphyritic conglomerate with absolutely vertical strata.

We now come to the gypseous formation: I will first describe the structure of the several mountains, and then give in one section a detailed account of the nature of the rocks. On the north side of the road, which here runs in an east and west valley, the mountain of porphyry [D] is succeeded by a hill [E] formed of the upper gypseous strata tilted, at an angle of between 70 and 80 degrees to the west, by a uniclinal axis of elevation which does not run parallel to the other neighbouring ranges, and which is of short length; for on the south side of the valley its prolongation is marked only by a small flexure in a pile of strata inclined by a quite separate axis. A little further on the north and south valley of Horcones enters at right angles our line of section; its western side is bounded by a hill of gypseous strata [F] dipping westward at about 45 degrees, and its eastern side by a mountain of similar strata [G] inclined westward at 70 degrees, and superimposed by an oblique fault on another mass of the same strata [H], also inclined westward, but at an angle of about 30 degrees: the complicated relation of these three masses [F, G, H] is explained by the structure of a great mountain-range lying some way to the north, in which a regular anticlinal axis (represented in the section by dotted lines) is seen, with the strata on its eastern side again bending up and forming a distinct uniclinal axis, of which the beds marked [H] form the lower part. This great uniclinal line is intersected, near the Puente del Inca, by the valley along which the road runs, and the strata composing it will be immediately described. On the south side of the road, in the space corresponding with the mountains [E, F, and G], the strata everywhere dip westward generally at an angle of 30 degrees, occasionally mounting up to 45 degrees, but not in an unbroken line, for there are several vertical faults, forming separate uniclinal masses, all dipping in the same direction,—a form of elevation common in the Cordillera. We thus see that within a narrow space, the gypseous strata have been upheaved and crushed together by a great uniclinal, anticlinal, and one lesser uniclinal line [E] of elevation; and that between these three lines and the Cumbre, in the sandstones, conglomerates and porphyritic formation, there have been at least two or three other great elevatory axes.

The uniclinal axis [I] intersected near the Puente del Inca (of which the strata at [H] form a part) ranges N. by W. and S. by E., forming a chain of mountains, apparently little inferior in height to the Cumbre: the strata, as we have seen, dip at an average angle of 30 degrees to the west. (At this place, there are some hot and cold springs, the warmest having a temperature, according to Lieutenant Brand "Travels,", of 91 degrees; they emit much gas. According to Mr. Brande, of the Royal Institution, ten cubical inches contain forty-five grains of solid matter, consisting chiefly of salt, gypsum, carbonate of lime, and oxide of iron. The water is charged with carbonic acid and sulphuretted hydrogen. These springs deposit much tufa in the form of spherical balls. They burst forth, as do those of Cauquenes, and probably those of Villa Vicencio, on a line of elevation.) The flanks of the mountains are here quite bare and steep, affording an excellent section; so that I was able to inspect the strata to a thickness of about 4,000 feet, and could clearly distinguish their general nature for 1,000 feet higher, making a total thickness of 5,000 feet, to which must be added about 1,000 feet of the inferior strata seen a little lower down the valley, I will describe this one section in detail, beginning at the bottom.

1st. The lowest mass is the altered clay-slate described in the preliminary discussion, and which in this line of section was here first met with. Lower down the valley, at the R. de las Vacas, I had a better opportunity of examining it; it is there in some parts well characterised, having a distinct, nearly vertical, tortuous cleavage, ranging N.W. and S.E., and intersected by quartz veins: in most parts, however, it is crystalline and feldspathic, and passes into a true greenstone often including grains of quartz. The clay-slate, in its upper half, is frequently brecciated, the embedded angular fragments being of nearly the same nature with the paste.

2nd. Several strata of purplish porphyritic conglomerate, of no very great thickness, rest conformably upon the feldspathic slate. A thick bed of fine, purple, claystone porphyry, obscurely brecciated (but not of metamorphosed sedimentary origin), and capped by porphyritic conglomerate, was the lowest bed actually examined in this section at the Puente del Inca.

3rd. A stratum, eighty feet thick, of hard and very compact impure whitish limestone, weathering bright red, with included layers brecciated and re- cemented. Obscure marks of shell are distinguishable in it.

4th. A red, quartzose, fine-grained conglomerate, with grains of quartz, and with patches of white earthy feldspar, apparently due to some process of concretionary crystalline action; this bed is more compact and metamorphosed than any of the overlying conglomerates.

5th. A whitish cherty limestone, with nodules of bluish argillaceous limestone.

6th. A white conglomerate, with many particles of quartz, almost blending into the paste.

7th. Highly siliceous, fine-grained white sandstone.

8th and 9th. Red and white beds not examined.

10th. Yellow, fine-grained, thinly stratified, magnesian (judging from its slow dissolution in acids) limestone: it includes some white quartz pebbles, and little cavities, lined with calcareous spar, some retaining the form of shells.

11th. A bed between twenty and thirty feet thick, quite conformable with the underlying ones, composed of a hard basis, tinged lilac-grey porphyritic with NUMEROUS crystals of whitish feldspar, with black mica and little spots of soft ferruginous matter: evidently a submarine lava.

12th. Yellow magnesian limestone, as before, part-stained purple.

13th. A most singular rock; basis purplish grey, obscurely crystalline, easily fusible into a dark green glass, not hard, thickly speckled with crystals more or less perfect of white carbonate of lime, of red hydrous oxide of iron, of a white and transparent mineral like analcime, and of a green opaque mineral like soap-stone; the basis is moreover amygdaloidal with many spherical balls of white crystallised carbonate of lime, of which some are coated with the red oxide of iron. I have no doubt, from the examination of a superincumbent stratum (19), that this is a submarine lava; though in Northern Chile, some of the metamorphosed sedimentary beds are almost as crystalline, and of as varied composition.

14th. Red sandstone, passing in the upper part into a coarse, hard, red conglomerate, 300 feet thick, having a calcareous cement, and including grains of quartz and broken crystals of feldspar; basis infusible; the pebbles consist of dull purplish porphyries, with some of quartz, from the size of a nut to a man's head. This is the coarsest conglomerate in this part of the Cordillera: in the middle there was a white layer not examined.

15th. Grand thick bed, of a very hard, yellowish-white rock, with a crystalline feldspathic base, including large crystals of white feldspar, many little cavities mostly full of soft ferruginous matter, and numerous hexagonal plates of black mica. The upper part of this great bed is slightly cellular; the lower part compact: the thickness varied a little in different parts. Manifestly a submarine lava; and is allied to bed 11.

16th and 17th. Dull purplish, calcareous, fine-grained, compact sandstones, which pass into coarse white conglomerates with numerous particles of quartz.

18th. Several alternations of red conglomerate, purplish sandstone, and submarine lava, like that singular rock forming bed 13.

19th. A very heavy, compact, greenish-black stone, with a fine-grained obviously crystalline basis, containing a few specks of white calcareous spar, many specks of the crystallised hydrous red oxide of iron, and some specks of a green mineral; there are veins and nests filled with epidote: certainly a submarine lava.

20th. Many thin strata of compact, fine-grained, pale purple sandstone.

21st. Gypsum in a nearly pure state, about three hundred feet in thickness: this bed, in its concretions of anhydrite and layers of small blackish crystals of carbonate of lime, exactly resembles the great gypseous beds in the Peuquenes range.

22nd. Pale purple and reddish sandstone, as in bed 20: about three hundred feet in thickness.

23rd. A thick mass composed of layers, often as thin as paper and convoluted, of pure gypsum with others very impure, of a purplish colour.

24th. Pure gypsum, thick mass.

25th. Red sandstones, of great thickness.

26th. Pure gypsum, of great thickness.

27th. Alternating layers of pure and impure gypsum, of great thickness.

I was not able to ascend to these few last great strata, which compose the neighbouring loftiest pinnacles. The thickness, from the lowest to the uppermost bed of gypsum, cannot be less than 2,000 feet: the beds beneath I estimated at 3,000 feet, and this does not include either the lower parts of the porphyritic conglomerate, or the altered clay-slate; I conceive the total thickness must be about six thousand feet. I distinctly observed that not only the gypsum, but the alternating sandstones and conglomerates were lens-shaped, and repeatedly thinned out and replaced each other: thus in the distance of about a mile, a bed 300 feet thick of sandstone between two beds of gypsum, thinned out to nothing and disappeared. The lower part of this section differs remarkably,—in the much greater diversity of its mineralogical composition,—in the abundance of calcareous matter,—in the greater coarseness of some of the conglomerates,—and in the numerous particles and well-rounded pebbles, sometimes of large size, of quartz,— from any other section hitherto described in Chile. From these peculiarities and from the lens-form of the strata, it is probable that this great pile of strata was accumulated on a shallow and very uneven bottom, near some pre-existing land formed of various porphyries and quartz-rock. The formation of porphyritic claystone conglomerate does not in this section attain nearly its ordinary thickness; this may be PARTLY attributed to the metamorphic action having been here much less energetic than usual, though the lower beds have been affected to a certain degree. If it had been as energetic as in most other parts of Chile, many of the beds of sandstone and conglomerate, containing rounded masses of porphyry, would doubtless have been converted into porphyritic conglomerate; and these would have alternated with, and even blended into, crystalline and porphyritic strata without a trace of mechanical structure,—namely, into those which, in the present state of the section, we see are unquestionably submarine lavas.

The beds of gypsum, together with the red alternating sandstones and conglomerates, present so perfect and curious a resemblance with those seen in our former section in the basin-valley of Yeso, that I cannot doubt the identity of the two formations: I may add, that a little westward of the P. del Inca, a mass of gypsum passed into a fine-grained, hard, brown sandstone, which contained some layers of black, calcareous, compact, shaly rock, precisely like that seen in such vast masses on the Peuquenes range.

Near the Puente del Inca, numerous fragments of limestone, containing some fossil remains, were scattered on the ground: these fragments so perfectly resemble the limestone of bed No. 3, in which I saw impressions of shells, that I have no doubt they have fallen from it. The yellow magnesian limestone of bed No. 10, which also includes traces of shells, has a different appearance. These fossils (as named by M. d'Orbigny) consist of:- -

Gryphaea, near to G. Couloni (Neocomian formation).

Arca, perhaps A. Gabrielis, d'Orbigny, "Pal. Franc." (Neocomian formation).

Mr. Pentland made a collection of shells from this same spot, and Von Buch considers them as consisting of:—

Trigonia, resembling in form T. costata.

Pholadomya, like one found by M. Dufresnoy near Alencon.

Isocardi excentrica, Voltz., identical with that from the Jura.

("Description Phys. des Iles Can.".)

Two of these shells, namely, the Gryphaea and Trigonia, appear to be identical with species collected by Meyen and myself on the Peuquenes range; and in the opinion of Von Buch and M. d'Orbigny, the two formations belong to the same age. I must here add, that Professor E. Forbes, who has examined my specimens from this place and from the Peuquenes range, has likewise a strong impression that they indicate the Cretaceous period, and probably an early epoch in it: so that all the palaeontologists who have seen these fossils nearly coincide in opinion regarding their age. The limestone, however, with these fossils here lies at the very base of the formation, just above the porphyritic conglomerate, and certainly several thousand feet lower in the series, than the equivalent, fossiliferous, black, shaly rocks high up on the Peuquenes range.

It is well worthy of remark that these shells, or at least those of which I saw impressions in the limestone (bed No. 3), must have been covered up, on the LEAST computation, by 4,000 feet of strata: now we know from Professor E. Forbes's researches, that the sea at greater depths than 600 feet becomes exceedingly barren of organic beings,—a result quite in accordance with what little I have seen of deep-sea soundings. Hence, after this limestone with its shells was deposited, the bottom of the sea where the main line of the Cordillera now stands, must have subsided some thousand feet to allow of the deposition of the superincumbent submarine strata. Without supposing a movement of this kind, it would, moreover, be impossible to understand the accumulation of the several lower strata of COARSE, well-rounded conglomerates, which it is scarcely possible to believe were spread out in profoundly deep water, and which, especially those containing pebbles of quartz, could hardly have been rounded in submarine craters and afterwards ejected from them, as I believe to have been the case with much of the porphyritic conglomerate formation. I may add that, in Professor Forbes's opinion, the above-enumerated species of mollusca probably did not live at a much greater depth than twenty fathoms, that is only 120 feet.

To return to our section down the valley; standing on the great N. by W. and S. by E. uniclinal axis of the Puente del Inca, of which a section has just been given, and looking north-east, greater tabular masses of gypseous formation (KK) could be seen in the distance, very slightly inclined towards the east. Lower down the valley, the mountains are almost exclusively composed of porphyries, many of them of intrusive origin and non-stratified, others stratified, but with the stratification seldom distinguishable except in the upper parts. Disregarding local disturbances, the beds are either horizontal or inclined at a small angle eastwards: hence, when standing on the plain of Uspallata and looking to the west or backwards, the Cordillera appear composed of huge, square, nearly horizontal, tabular masses: so wide a space, with such lofty mountains so equably elevated, is rarely met with within the Cordillera. In this line of section, the interval between the Puente del Inca and the neighbourhood of the Cumbre, includes all the chief axes of dislocation.

The altered clay-slate formation, already described, is seen in several parts of the valley as far down as Las Vacas, underlying the porphyritic conglomerate. At the Casa de Pujios [L], there is a hummock of (andesitic?) granite; and the stratification of the surrounding mountains here changes from W. by S. to S.W. Again, near the R. Vacas there is a larger formation of (andesitic?) granite [M], which sends a meshwork of veins into the superincumbent clay-slate, and which locally throws off the strata, on one side to N.W. and on the other to S.E. but not at a high angle: at the junction, the clay-slate is altered into fine-grained greenstone. This granitic axis is intersected by a green dike, which I mention, because I do not remember having elsewhere seen dikes in this lowest and latest intrusive rock. From the R. Vacas to the plain of Uspallata, the valley runs N.E., so that I have had to contract my section; it runs exclusively through porphyritic rocks. As far as the Pass of Jaula, the claystone conglomerate formation, in most parts highly porphyritic, and crossed by numerous dikes of greenstone porphyry, attains a great thickness: there is also much intrusive porphyry. From the Jaula to the plain, the stratification has been in most places obliterated, except near the tops of some of the mountains; and the metamorphic action has been extremely great. In this space, the number and bulk of the intrusive masses of differently coloured porphyries, injected one into another and intersected by dikes, is truly extraordinary. I saw one mountain of whitish porphyry, from which two huge dikes, thinning out, branched DOWNWARDS into an adjoining blackish porphyry. Another hill of white porphyry, which had burst through dark-coloured strata, was itself injected by a purple, brecciated, and recemented porphyry, both being crossed by a green dike, and both having been upheaved and injected by a granitic dome. One brick-red porphyry, which above the Jaula forms an isolated mass in the midst of the porphyritic conglomerate formation, and lower down the valley a magnificent group of peaked mountains, differs remarkably from all the other porphyries. It consists of a red feldspathic base, including some rather large crystals of red feldspar, numerous large angular grains of quartz, and little bits of a soft green mineral answering in most of its characters to soapstone. The crystals of red feldspar resemble in external appearance those of orthite, though, from being partially decomposed, I was unable to measure them; and they certainly are quite unlike the variety, so abundantly met with in almost all the other rocks of this line of section, and which, wherever I tried it, cleaved like albite. This brick-red porphyry appears to have burst through all the other porphyries, and numerous red dikes traversing the neighbouring mountains have proceeded from it: in some few places, however, it was intersected by white dikes. From this posteriority of intrusive origin,—from the close general resemblance between this red porphyry and the red granite of the Portillo line, the only difference being that the feldspar here is less perfectly granular, and that soapstone replaces the mica, which is there imperfect and passes into chlorite,—and from the Portillo line a little southward of this point appearing to blend (according to Dr. Gillies) into the western ranges,—I am strongly urged to believe (as formerly remarked) that the grand mountain-masses composed of this brick-red porphyry belong to the same axis of injection with the granite of the Portillo line; if so, the injection of this porphyry probably took place, as long subsequently to the several axes of elevation in the gypseous formation near the Cumbre, as the injection of the Portillo granite has been shown to have been subsequent to the elevation of the gypseous strata composing the Peuquenes range; and this interval, we have seen, must have been a very long one.

The Plain of Uspallata has been briefly described in Chapter 3; it resembles the basin-plains of Chile; it is ten or fifteen miles wide, and is said to extend for 180 miles northward; its surface is nearly six thousand feet above the sea; it is composed, to a thickness of some hundred feet of loosely aggregated, stratified shingle, which is prolonged with a gently sloping surface up the valleys in the mountains on both sides. One section in this plain [Z] is interesting, from the unusual circumstance of alternating layers of almost loose red and white sand with lines of pebbles (from the size of a nut to that of an apple), and beds of gravel, being inclined at an angle of 45 degrees, and in some spots even at a higher angle. (I find that Mr. Smith of Jordan Hill has described ("Edinburgh New Philosophical Journal" volume 25) beds of sand and gravel, near Edinburgh, tilted at an angle of 60 degrees, and dislocated by miniature faults.) These beds are dislocated by small faults: and are capped by a thick mass of horizontally stratified gravel, evidently of subaqueous origin. Having been accustomed to observe the irregularities of beds accumulated under currents, I feel sure that the inclination here has not been thus produced. The pebbles consist chiefly of the brick-red porphyry just described and of white granite, both probably derived from the ranges to the west, and of altered clay-slate and of certain porphyries, apparently belonging to the rocks of the Uspallata chain. This plain corresponds geographically with the valley of Tenuyan between the Portillo and Peuquenes ranges; but in that valley the shingle, which likewise has been derived both from the eastern and western ranges, has been cemented into a hard conglomerate, and has been throughout tilted at a considerable inclination; the gravel there apparently attains a much greater thickness, and is probably of higher antiquity.

THE USPALLATA RANGE.

The road by the Villa Vicencio Pass does not strike directly across the range, but runs for some leagues northward along its western base: and I must briefly describe the rocks here seen, before continuing with the coloured east and west section. At the mouth of the valley of Canota, and at several points northwards, there is an extensive formation of a glossy and harsh, and of a feldspathic clay-slate, including strata of grauwacke, and having a tortuous, nearly vertical cleavage, traversed by numerous metalliferous veins and others of quartz. The clay-slate is in many parts capped by a thick mass of fragments of the same rock, firmly recemented; and both together have been injected and broken up by very numerous hillocks, ranging north and south, of lilac, white, dark and salmon- coloured porphyries: one steep, now denuded, hillock of porphyry had its face as distinctly impressed with the angles of a fragmentary mass of the slate, with some of the points still remaining embedded, as sealing-wax could be by a seal. At the mouth of this same valley of Canota, in a fine escarpment having the strata dipping from 50 to 60 degrees to the N.E. (Nearly opposite to this escarpment, there is another corresponding one, with the strata dipping not to the exactly opposite point, or S.W., but to S.S.W.: consequently the two escarpments trend towards each other, and some miles southward they become actually united: this is a form of elevation which I have not elsewhere seen.), the clay-slate formation is seen to be covered by—(1st) a purple, claystone porphyry resting unconformably in some parts on the solid slate, and in others on a thick fragmentary mass; (2nd), a conformable stratum of compact blackish rock, having a spheroidal structure, full of minute acicular crystals of glassy feldspar, with red spots of oxide of iron; (3rd), a great stratum of purplish-red claystone porphyry, abounding with crystals of opaque feldspar, and laminated with thin, parallel, often short, layers, and likewise with great irregular patches of white, earthy, semi-crystalline feldspar; this rock (which I noticed in other neighbouring places) perfectly resembles a curious variety described at Port Desire, and occasionally occurs in the great porphyritic conglomerate formation of Chile; (4th), a thin stratum of greenish white, indurated tuff, fusible and containing broken crystals and particles of porphyries; (5th), a grand mass, imperfectly columnar and divided into three parallel and closely joined strata, of cream-coloured claystone porphyry; (6th), a thick stratum of lilac-coloured porphyry, which I could see was capped by another bed of the cream-coloured variety; I was unable to examine the still higher parts of the escarpment. These conformably stratified porphyries, though none are either vesicular are amygdaloidal, have evidently flowed as submarine lavas: some of them are separated from each other by seams of indurated tuff, which, however, are quite insignificant in thickness compared with the porphyries. This whole pile resembles, but not very closely, some of the less brecciated parts of the great porphyritic conglomerate formation of Chile; but it does not probably belong to the same age, as the porphyries here rest unconformably on the altered feldspathic clay-slate, whereas the porphyritic conglomerate formation alternates with and rests conformably on it. These porphyries, moreover, with the exception of the one blackish stratum, and of the one indurated, white tufaceous bed, differ from the beds composing the Uspallata range in the line of the Villa Vicencio Pass.

I will now give, first, a sketch of the structure of the range, as represented in the section, and will then describe its composition and interesting history. At its western foot, a hillock [N] is seen to rise out of the plain, with its strata dipping at 70 degrees to the west, fronted by strata [O] inclined at 45 degrees to the east, thus forming a little north and south anticlinal axis. Some other little hillocks of similar composition, with their strata highly inclined, range N.E. and S.W., obliquely to the main Uspallata line. The cause of these dislocations, which, though on a small scale, have been violent and complicated, is seen to lie in hummocks of lilac, purple and red porphyries, which have been injected in a liquified state through and into the underlying clay-slate formation. Several dykes were exposed here, but in no other part, that I saw of this range. As the strata consist of black, white, greenish and brown-coloured rocks, and as the intrusive porphyries are so brightly tinted, a most extraordinary view was presented, like a coloured geological drawing. On the gently inclined main western slope [PP], above the little anticlinal ridges just mentioned, the strata dip at an average angle of 25 degrees to the west; the inclination in some places being only 19 degrees, in some few others as much as 45 degrees. The masses having these different inclinations, are separated from each other by parallel vertical faults [as represented at Pa], often giving rise to separate, parallel, uniclinal ridges. The summit of the main range is broad and undulatory, with the stratification undulatory and irregular: in a few places granitic and porphyritic masses [Q] protrude, which, from the small effect they have locally produced in deranging the strata, probably form the upper points of a regular, great underlying dome. These denuded granitic points, I estimated at about nine thousand feet in height above the sea. On the eastern slope, the strata in the upper part are regularly inclined at about 25 degrees to the east, so that the summit of this chain, neglecting small irregularities, forms a broad anticlinal axis. Lower down, however, near Los Hornillos [R], there is a well-marked synclinal axis, beyond which the strata are inclined at nearly the same angle, namely from 20 to 30 degrees, inwards or westward. Owing to the amount of denudation which this chain has suffered, the outline of the gently inclined eastern flank scarcely offers the slightest indication of this synclinal axis. The stratified beds, which we have hitherto followed across the range, a little further down are seen to lie, I believe unconformably, on a broad mountainous band of clay-slate and grauwacke. The strata and laminae of this latter formation, on the extreme eastern flank, are generally nearly vertical; further inwards they become inclined from 45 to 80 degrees to the west: near Villa Vicencio [S] there is apparently an anticlinal axis, but the structure of this outer part of the clay-slate formation is so obscure, that I have not marked the planes of stratification in the section. On the margin of the Pampas, some low, much dislocated spurs of this same formation, project in a north- easterly line, in the same oblique manner as do the ridges on the western foot, and as is so frequently the case with those at the base of the main Cordillera.

I will now describe the nature of the beds, beginning at the base on the eastern side. First, for the clay-slate formation: the slate is generally hard and bluish, with the laminae coated by minute micaceous scales; it alternates many times with a coarse-grained, greenish grauwacke, containing rounded fragments of quartz and bits of slate in a slightly calcareous basis. The slate in the upper part generally becomes purplish, and the cleavage so irregular that the whole consists of mere splinters. Transverse veins of quartz are numerous. At the Calera, some leagues distant, there is a dark crystalline limestone, apparently included in this formation. With the exception of the grauwacke being here more abundant, and the clay-slate less altered, this formation closely resembles that unconformably underlying the porphyries at the western foot of this same range; and likewise that alternating with the porphyritic conglomerate in the main Cordillera. This formation is a considerable one, and extends several leagues southward to near Mendoza: the mountains composed of it rise to a height of about two thousand feet above the edge of the Pampas, or about seven thousand feet above the sea. (I infer this from the height of V. Vicencio, which was ascertained by Mr. Miers to be 5,328 feet above the sea.)

Secondly: the most usual bed on the clay-slate is a coarse, white, slightly calcareous conglomerate, of no great thickness, including broken crystals of feldspar, grains of quartz, and numerous pebbles of brecciated claystone porphyry, but without any pebbles of the underlying clay-slate. I nowhere saw the actual junction between this bed and the clay-slate, though I spent a whole day in endeavouring to discover their relations. In some places I distinctly saw the white conglomerate and overlying beds inclined at from 25 to 30 degrees to the west, and at the bottom of the same mountain, the clay-slate and grauwacke inclined to the same point, but at an angle from 70 to 80 degrees: in one instance, the clay-slate dipped not only at a different angle, but to a different point from the overlying formation. In these cases the two formations certainly appeared quite unconformable: moreover, I found in the clay-slate one great, vertical, dike-like fissure, filled up with an indurated whitish tuff, quite similar to some of the upper beds presently to be described; and this shows that the clay-slate must have been consolidated and dislocated before their deposition. On the other hand, the stratification of the slate and grauwacke, in some cases gradually and entirely disappeared in approaching the overlying white conglomerate; in other cases the stratification of the two formations became strictly conformable; and again in other cases, there was some tolerably well characterised clay-slate lying above the conglomerate. (The coarse, mechanical structure of many grauwackes has always appeared to me a difficulty; for the texture of the associated clay-slate and the nature of the embedded organic remains where present, indicate that the whole has been a deep-water deposit. Whence have the sometimes included angular fragments of clay-slate, and the rounded masses of quartz and other rocks, been derived? Many deep-water limestones, it is well known, have been brecciated, and then firmly recemented.) The most probable conclusion appears to be, that after the clay-slate formation had been dislocated and tilted, but whilst under the sea, a fresh and more recent deposition of clay-slate took place, on which the white conglomerate was conformably deposited, with here and there a thin intercalated bed of clay-slate. On this view the white conglomerates and the presently to be described tuffs and lavas are really unconformable to the main part of the clay-slate; and this, as we have seen, certainly is the case with the clay-stone lavas in the valley of Canota, at the western and opposite base of the range.

Thirdly: on the white conglomerate, strata several hundred feet in thickness are superimposed, varying much in nature in short distances: the commonest variety is a white, much indurated tuff, sometimes slightly calcareous, with ferruginous spots and water-lines, often passing into whitish or purplish compact, fine-grained grit or sandstones; other varieties become semi-porcellanic, and tinted faint green or blue; others pass into an indurated shale: most of these varieties are easily fusible.

Fourthly: a bed, about one hundred feet thick of a compact, partially columnar, pale-grey, feldspathic lava, stained with iron, including very numerous crystals of opaque feldspar, and with some crystallised and disseminated calcareous matter. The tufaceous stratum on which this feldspathic lava rests is much hardened, stained purple, and has a spherico-concretionary structure; it here contains a good many pebbles of claystone porphyry.

Fifthly: thin beds, 400 feet in thickness, varying much in nature, consisting of white and ferruginous tuffs, in some parts having a concretionary structure, in others containing rounded grains and a few pebbles of quartz; also passing into hard gritstones and into greenish mudstones: there is, also, much of a bluish-grey and green semi-porcellanic stone.

Sixthly: a volcanic stratum, 250 feet in thickness, of so varying a nature that I do not believe a score of specimens would show all the varieties; much is highly amygdaloidal, much compact; there are greenish, blackish, purplish, and grey varieties, rarely including crystals of green augite and minute acicular ones of feldspar, but often crystals and amygdaloidal masses of white, red, and black carbonate of lime. Some of the blackish varieties of this rock have a conchoidal fracture and resemble basalt; others have an irregular fracture. Some of the grey and purplish varieties are thickly speckled with green earth and with white crystalline carbonate of lime; others are largely amygdaloidal with green earth and calcareous spar. Again, other earthy varieties, of greenish, purplish and grey tints, contain much iron, and are almost half composed of amygdaloidal balls of dark brown bole, of a whitish indurated feldspathic matter, of bright green earth, of agate, and of black and white crystallised carbonate of lime. All these varieties are easily fusible. Viewed from a distance, the line of junction with the underlying semi-porcellanic strata was distinct; but when examined closely, it was impossible to point out within a foot where the lava ended and where the sedimentary mass began: the rock at the time of junction was in most places hard, of a bright green colour, and abounded with irregular amygdaloidal masses of ferruginous and pure calcareous spar, and of agate.

Seventhly: strata, eighty feet in thickness, of various indurated tuffs, as before; many of the varieties have a fine basis including rather coarse extraneous particles; some of them are compact and semi-porcellanic, and include vegetable impressions.

Eighthly: a bed, about fifty feet thick, of greenish-grey, compact, feldspathic lava, with numerous small crystals of opaque feldspar, black augite, and oxide of iron. The junction with the bed on which it rested, was ill defined; balls and masses of the feldspathic rock being enclosed in much altered tuff.

Ninthly: indurated tuffs, as before.

Tenthly: a conformable layer, less than two feet in thickness, of pitchstone, generally brecciated, and traversed by veins of agate and of carbonate of lime: parts are composed of apparently concretionary fragments of a more perfect variety, arranged in horizontal lines in a less perfectly characterised variety. I have much difficulty in believing that this thin layer of pitchstone flowed as lava.

Eleventhly: sedimentary and tufaceous beds as before, passing into sandstone, including some conglomerate: the pebbles in the latter are of claystone porphyry, well rounded, and some as large as cricket-balls.

Twelfthly: a bed of compact, sonorous, feldspathic lava, like that of bed No. 8, divided by numerous joints into large angular blocks.

Thirteenthly: sedimentary beds as before.

Fourteenthly: a thick bed of greenish or greyish black, compact basalt (fusing into a black enamel), with small crystals, occasionally distinguishable, of feldspar and augite: the junction with the underlying sedimentary bed, differently from that in most of the foregoing streams, here was quite distinct:—the lava and tufaceous matter preserving their perfect characters within two inches of each other. This rock closely resembles certain parts of that varied and singular lava-stream No. 6; it likewise resembles, as we shall immediately see, many of the great upper beds on the western flank and on the summit of this range.

The pile of strata here described attains a great thickness; and above the last-mentioned volcanic stratum, there were several other great tufaceous beds alternating with submarine lavas, which I had not time to examine; but a corresponding series, several thousand feet in thickness, is well exhibited on the crest and western flank of the range. Most of the lava-streams on the western side are of a jet-black colour and basaltic nature; they are either compact and fine-grained, including minute crystals of augite and feldspar, or they are coarse-grained and abound with rather large coppery-brown crystals of an augitic mineral. (Very easily fusible into a jet-black bead, attracted by the magnet: the crystals are too much tarnished to be measured by the goniometer.) Another variety was of a dull-red colour, having a claystone brecciated basis, including specks of oxide of iron and of calcareous spar, and amygdaloidal with green earth: there were apparently several other varieties. These submarine lavas often exhibit a spheroidal, and sometimes an imperfect columnar structure: their upper junctions are much more clearly defined than their lower junctions; but the latter are not so much blended into the underlying sedimentary beds as is the case in the eastern flank. On the crest and western flank of the range, the streams, viewed as a whole, are mostly basaltic; whilst those on the eastern side, which stand lower in the series, are, as we have seen, mostly feldspathic.

The sedimentary strata alternating with the lavas on the crest and western side, are of an almost infinitely varying nature; but a large proportion of them closely resemble those already described on the eastern flank: there are white and brown, indurated, easily fusible tuffs,—some passing into pale blue and green semi-porcellanic rocks,—others into brownish and purplish sandstones and gritstones, often including grains of quartz,— others into mudstone containing broken crystals and particles of rock, and occasionally single large pebbles. There was one stratum of a bright red, coarse, volcanic gritstone; another of conglomerate; another of a black, indurated, carbonaceous shale marked with imperfect vegetable impressions; this latter bed, which was thin, rested on a submarine lava, and followed all the considerable inequalities of its upper surface. Mr. Miers states that coal has been found in this range. Lastly, there was a bed (like No. 10 on the eastern flank) evidently of sedimentary origin, and remarkable from closely approaching in character to an imperfect pitchstone, and from including extremely thin layers of perfect pitchstone, as well as nodules and irregular fragments (but not resembling extraneous fragments) of this same rock arranged in horizontal lines: I conceive that this bed, which is only a few feet in thickness, must have assumed its present state through metamorphic and concretionary action. Most of these sedimentary strata are much indurated, and no doubt have been partially metamorphosed: many of them are extraordinarily heavy and compact; others have agate and crystalline carbonate of lime disseminated throughout them. Some of the beds exhibit a singular concretionary arrangement, with the curves determined by the lines of fissure. There are many veins of agate and calcareous spar, and innumerable ones of iron and other metals, which have blackened and curiously affected the strata to considerable distances on both sides.

Many of these tufaceous beds resemble, with the exception of being more indurated, the upper beds of the Great Patagonian tertiary formation, especially those variously coloured layers high up the River Santa Cruz, and in a remarkable degree the tufaceous formation at the northern end of Chiloe. I was so much struck with this resemblance, that I particularly looked out for silicified wood, and found it under the following extraordinary circumstances. High up on this western flank, at a height estimated at 7,000 feet above the sea, in a broken escarpment of thin strata, composed of compact green gritstone passing into a fine mudstone, and alternating with layers of coarser, brownish, very heavy mudstone, including broken crystals and particles of rock almost blended together, I counted the stumps of fifty-two trees. (For the information of any future traveller, I will describe the spot in detail. Proceeding eastward from the Agua del Zorro, and afterwards leaving on the north side of the road a rancho attached to some old goldmines, you pass through a gully with low but steep rocks on each hand: the road then bends, and the ascent becomes steeper. A few hundred yards farther on, a stone's throw on the south side of the road, the white calcareous stumps may be seen. The spot is about half a mile east of the Agua del Zorro.) They projected between two and five feet above the ground, and stood at exactly right angles to the strata, which were here inclined at an angle of about 25 degrees to the west. Eleven of these trees were silicified and well preserved; Mr. R. Brown has been so kind as to examine the wood when sliced and polished; he says it is coniferous, partaking of the characters of the Araucarian tribe, with some curious points of affinity with the Yew. The bark round the trunks must have been circularly furrowed with irregular lines, for the mudstone round them is thus plainly marked. One cast consisted of dark argillaceous limestone; and forty of them of coarsely crystallised carbonate of lime, with cavities lined by quartz crystals: these latter white calcareous columns do not retain any internal structure, but their external form plainly shows their origin. All the stumps have nearly the same diameter, varying from one foot to eighteen inches; some of them stand within a yard of each other; they are grouped in a clump within a space of about sixty yards across, with a few scattered round at the distance of 150 yards. They all stand at about the same level. The longest stump stood seven feet out of the ground: the roots, if they are still preserved, are buried and concealed. No one layer of the mudstone appeared much darker than the others, as if it had formerly existed as soil, nor could this be expected, for the same agents which replaced with silex and lime the wood of the trees, would naturally have removed all vegetable matter from the soil. Besides the fifty-two upright trees, there were a few fragments, like broken branches, horizontally embedded. The surrounding strata are crossed by veins of carbonate of lime, agate, and oxide of iron; and a poor gold vein has been worked not far from the trees.

The green and brown mudstone beds including the trees, are conformably covered by much indurated, compact, white or ferruginous tuffs, which pass upwards into a fine-grained, purplish sedimentary rock: these strata, which, together, are from four to five hundred feet in thickness, rest on a thick bed of submarine lava, and are conformably covered by another great mass of fine-grained basalt, which I estimated at 1,000 feet in thickness, and which probably has been formed by more than one stream. (This rock is quite black, and fuses into a black bead, attracted strongly by the magnet; it breaks with a conchoidal fracture; the included crystals of augite are distinguishable by the naked eye, but are not perfect enough to be measured: there are many minute acicular crystals of glassy feldspar.) Above this mass I could clearly distinguish five conformable alternations, each several hundred feet in thickness, of stratified sedimentary rocks and lavas, such as have been previously described. Certainly the upright trees have been buried under several thousand feet in thickness of matter, accumulated under the sea. As the trees obviously must once have grown on dry land, what an enormous amount of subsidence is thus indicated! Nevertheless, had it not been for the trees there was no appearance which would have led any one even to have conjectured that these strata had subsided. As the land, moreover, on which the trees grew, is formed of subaqueous deposits, of nearly if not quite equal thickness with the superincumbent strata, and as these deposits are regularly stratified and fine-grained, not like the matter thrown up on a sea-beach, a previous upward movement, aided no doubt by the great accumulation of lavas and sediment, is also indicated. (At first I imagined, that the strata with the trees might have been accumulated in a lake: but this seems highly improbable; for, first, a very deep lake was necessary to receive the matter below the trees, then it must have been drained for their growth, and afterwards re-formed and made profoundly deep, so as to receive a subsequent accumulation of matter SEVERAL THOUSAND feet in thickness. And all this must have taken place necessarily before the formation of the Uspallata range, and therefore on the margin of the wide level expanse of the Pampas! Hence I conclude, that it is infinitely more probable that the strata were accumulated under the sea: the vast amount of denudation, moreover, which this range has suffered, as shown by the wide valleys, by the exposure of the very trees and by other appearances, could have been effected, I conceive, only by the long-continued action of the sea; and this shows that the range was either upheaved from under the sea, or subsequently let down into it. From the natural manner in which the stumps (fifty-two in number) are GROUPED IN A CLUMP, and from their all standing vertically to the strata, it is superfluous to speculate on the chance of the trees having been drifted from adjoining land, and deposited upright: I may, however, mention that the late Dr. Malcolmson assured me, that he once met in the Indian Ocean, fifty miles from land, several cocoa-nut trees floating upright, owing to their roots being loaded with earth.)

In nearly the middle of the range, there are some hills [Q], before alluded to, formed of a kind of granite externally resembling andesite, and consisting of a white, imperfectly granular, feldspathic basis, including some perfect crystals apparently of albite (but I was unable to measure them), much black mica, epidote in veins, and very little or no quartz. Numerous small veins branch from this rock into the surrounding strata; and it is a singular fact that these veins, though composed of the same kind of feldspar and small scales of mica as in the solid rock, abound with innumerable minute ROUNDED grains of quartz: in the veins or dikes also, branching from the great granitic axis in the peninsula of Tres Montes, I observed that quartz was more abundant in them than in the main rock: I have heard of other analogous cases: can we account for this fact, by the long-continued vicinity of quartz when cooling, and by its having been thus more easily sucked into fissures than the other constituent minerals of granite? (See a paper by M. Elie de Beaumont, "Soc. Philomath." May 1839 "L'Institut." 1839) The strata encasing the flanks of these granitic or andesite masses, and forming a thick cap on one of their summits, appear originally to have been of the same tufaceous nature with the beds already described, but they are now changed into porcellanic, jaspery, and crystalline rocks, and into others of a white colour with a harsh texture, and having a siliceous aspect, though really of a feldspathic nature and fusible. Both the granitic intrusive masses and the encasing strata are penetrated by innumerable metallic veins, mostly ferruginous and auriferous, but some containing copper-pyrites and a few silver: near the veins, the rocks are blackened as if blasted by gunpowder. The strata are only slightly dislocated close round these hills, and hence, perhaps, it may be inferred that the granitic masses form only the projecting points of a broad continuous axis-dome, which has given to the upper parts of this range its anticlinal structure.

CONCLUDING REMARKS ON THE USPALLATA RANGE.

I will not attempt to estimate the total thickness of the pile of strata forming this range, but it must amount to many thousand feet. The sedimentary and tufaceous beds have throughout a general similarity, though with infinite variations. The submarine lavas in the lower part of the series are mostly feldspathic, whilst in the upper part, on the summit and western flank, they are mostly basaltic. We are thus reminded of the relative position in most recent volcanic districts of the trachytic and basaltic lavas,—the latter from their greater weight having sunk to a lower level in the earth's crust, and having consequently been erupted at a later period over the lighter and upper lavas of the trachytic series. (See on this subject, "Volcanic Islands" etc. by the Author.) Both the basaltic and feldspathic submarine streams are very compact; none being vesicular, and only a few amygdaloidal: the effects which some of them, especially those low in the series, have produced on the tufaceous beds over which they have flowed is highly curious. Independently of this local metamorphic action, all the strata undoubtedly display an indurated and altered character; and all the rocks of this range—the lavas, the alternating sediments, the intrusive granite and porphyries, and the underlying clay- slate—are intersected by metalliferous veins. The lava-strata can often be seen extending for great distances, conformably with the under and overlying beds; and it was obvious that they thickened towards the west. Hence the points of eruption must have been situated westward of the present range, in the direction of the main Cordillera: as, however, the flanks of the Cordillera are entirely composed of various porphyries, chiefly claystone and greenstone, some intrusive, and others belonging to the porphyritic conglomerate formation, but all quite unlike these submarine lava-streams, we must in all probability look to the plain of Uspallata for the now deeply buried points of eruption.

Comparing our section of the Uspallata range with that of the Cumbre, we see, with the exception of the underlying clay-slate, and perhaps of the intrusive rocks of the axes, a striking dissimilarity in the strata composing them. The great porphyritic conglomerate formation has not extended as far as this range; nor have we here any of the gypseous strata, the magnesian and other limestones, the red sandstones, the siliceous beds with pebbles of quartz, and comparatively little of the conglomerates, all of which form such vast masses over the basal series in the main Cordillera. On the other hand, in the Cordillera, we do not find those endless varieties of indurated tuffs, with their numerous veins and concretionary arrangement, and those grit and mud stones, and singular semi-porcellanic rocks, so abundant in the Uspallata range. The submarine lavas, also, differ considerably; the feldspathic streams of the Cordillera contain much mica, which is absent in those of the Uspallata range: in this latter range we have seen on how grand a scale, basaltic lava has been poured forth, of which there is not a trace in the Cordillera. This dissimilarity is the more striking, considering that these two parallel chains are separated by a plain only between ten and fifteen miles in width; and that the Uspallata lavas, as well as no doubt the alternating tufaceous beds, have proceeded from the west, from points apparently between the two ranges. To imagine that these two piles of strata were contemporaneously deposited in two closely adjoining, very deep, submarine areas, separated from each other by a lofty ridge, where a plain now extends, would be a gratuitous hypothesis. And had they been contemporaneously deposited, without any such dividing ridge, surely some of the gypseous and other sedimentary matter forming such immensely thick masses in the Cordillera, would have extended this short distance eastwards; and surely some of the Uspallata tuffs and basalts also accumulated to so great a thickness, would have extended a little westward. Hence I conclude, that it is far from probable that these two series are not contemporaneous; but that the strata of one of the chains were deposited, and even the chain itself uplifted, before the formation of the other:—which chain, then, is the oldest? Considering that in the Uspallata range the lowest strata on the western flank lie unconformably on the clay- slate, as probably is the case with those on the eastern flank, whereas in the Cordillera all the overlying strata lie conformably on this formation:- - considering that in the Uspallata range some of the beds, both low down and high up in the series, are marked with vegetable impressions, showing the continued existence of neighbouring land;—considering the close general resemblance between the deposits of this range and those of tertiary origin in several parts of the continent;—and lastly, even considering the lesser height and outlying position of the Uspallata range,—I conclude that the strata composing it are in all probability of subsequent origin, and that they were accumulated at a period when a deep sea studded with submarine volcanoes washed the eastern base of the already partially elevated Cordillera.

This conclusion is of much importance, for we have seen that in the Cordillera, during the deposition of the Neocomian strata, the bed of the sea must have subsided many thousand feet: we now learn that at a later period an adjoining area first received a great accumulation of strata, and was upheaved into land on which coniferous trees grew, and that this area then subsided several thousand feet to receive the superincumbent submarine strata, afterwards being broken up, denuded, and elevated in mass to its present height. I am strengthened in this conclusion of there having been two distinct, great periods of subsidence, by reflecting on the thick mass of coarse stratified conglomerate in the valley of Tenuyan, between the Peuquenes and Portillo lines; for the accumulation of this mass seems to me, as previously remarked, almost necessarily to have required a prolonged subsidence; and this subsidence, from the pebbles in the conglomerate having been to a great extent derived from the gypseous or Neocomian strata of the Peuquenes line, we know must have been quite distinct from, and subsequent to, that sinking movement which probably accompanied the deposition of the Peuquenes strata, and which certainly accompanied the deposition of the equivalent beds near the Puente del Inca, in this line of section.

The Uspallata chain corresponds in geographical position, though on a small scale, with the Portillo line; and its clay-slate formation is probably the equivalent of the mica-schist of the Portillo, there metamorphosed by the old white granites and syenites. The coloured beds under the conglomerate in the valley of Tenuyan, of which traces are seen on the crest of the Portillo, and even the conglomerate itself, may perhaps be synchronous with the tufaceous beds and submarine lavas of the Uspallata range; an open sea and volcanic action in the latter case, and a confined channel between two bordering chains of islets in the former case, having been sufficient to account for the mineralogical dissimilarity of the two series. From this correspondence between the Uspallata and Portillo ranges, perhaps in age and certainly in geographical position, one is tempted to consider the one range as the prolongation of the other; but their axes are formed of totally different intrusive rocks; and we have traced the apparent continuation of the red granite of the Portillo in the red porphyries diverging into the main Cordillera. Whether the axis of the Uspallata range was injected before, or as perhaps is more probable, after that of the Portillo line, I will not pretend to decide; but it is well to remember that the highly inclined lava-streams on the eastern flank of the Portillo line, prove that its angular upheavement was not a single and sudden event; and therefore that the anticlinal elevation of the Uspallata range may have been contemporaneous with some of the later angular movements by which the gigantic Portillo range gained its present height above the adjoining plain.

CHAPTER VIII. NORTHERN CHILE. CONCLUSION.

Section from Illapel to Combarbala; gypseous formation with silicified wood. Panuncillo. Coquimbo; mines of Arqueros; section up valley; fossils. Guasco, fossils of. Copiapo, section up valley; Las Amolanas, silicified wood. Conglomerates, nature of former land, fossils, thickness of strata, great subsidence. Valley of Despoblado, fossils, tufaceous deposit, complicated dislocations of. Relations between ancient orifices of eruption and subsequent axes of injection. Iquique, Peru, fossils of, salt-deposits. Metalliferous veins. Summary on the porphyritic conglomerate and gypseous formations. Great subsidence with partial elevations during the cretaceo-oolitic period. On the elevation and structure of the Cordillera. Recapitulation on the tertiary series. Relation between movements of subsidence and volcanic action. Pampean formation. Recent elevatory movements. Long-continued volcanic action in the Cordillera. Conclusion.

VALPARAISO TO COQUIMBO.

I have already described the general nature of the rocks in the low country north of Valparaiso, consisting of granites, syenites, greenstones, and altered feldspathic clay-slate. Near Coquimbo there is much hornblendic rock and various dusky-coloured porphyries. I will describe only one section in this district, namely, from near Illapel in a N.E. line to the mines of Los Hornos, and thence in a north by east direction to Combarbala, at the foot of the main Cordillera.

Near Illapel, after passing for some distance over granite, andesite, and andesitic porphyry, we come to a greenish stratified feldspathic rock, which I believe is altered clay-slate, conformably capped by porphyries and porphyritic conglomerate of great thickness, dipping at an average angle of 20 degrees to N.E. by N. The uppermost beds consist of conglomerates and sandstone only a little metamorphosed, and conformably covered by a gypseous formation of very great thickness, but much denuded. This gypseous formation, where first met with, lies in a broad valley or basin, a little southward of the mines of Los Hornos: the lower half alone contains gypsum, not in great masses as in the Cordillera, but in innumerable thin layers, seldom more than an inch or two in thickness. The gypsum is either opaque or transparent, and is associated with carbonate of lime. The layers alternate with numerous varying ones of a calcareous clay-shale (with strong aluminous odour, adhering to the tongue, easily fusible into a pale green glass), more or less indurated, either earthy and cream-coloured, or greenish and hard. The more indurated varieties have a compact, homogeneous, almost crystalline fracture, and contain granules of crystallised oxide of iron. Some of the varieties almost resemble honestones. There is also a little black, hardly fusible, siliceo- calcareous clay-slate, like some of the varieties alternating with gypsum on the Peuquenes range.

The upper half of this gypseous formation is mainly formed of the same calcareous clay-shale rock, but without any gypsum, and varying extremely in nature: it passes from a soft, coarse, earthy, ferruginous state, including particles of quartz, into compact claystones with crystallised oxide of iron,—into porcellanic layers, alternating with seams of calcareous matter,—and into green porcelain-jasper, excessively hard, but easily fusible. Strata of this nature alternate with much black and brown siliceo-calcareous slate, remarkable from the wonderful number of huge embedded logs of silicified wood. This wood, according to Mr. R. Brown, is (judging from several specimens) all coniferous. Some of the layers of the black siliceous slate contained irregular angular fragments of imperfect pitchstone, which I believe, as in the Uspallata range, has originated in a metamorphic process. There was one bed of a marly tufaceous nature, and of little specific gravity. Veins of agate and calcareous spar are numerous. The whole of this gypseous formation, especially the upper half, has been injected, metamorphosed, and locally contorted by numerous hillocks of intrusive porphyries crowded together in an extraordinary manner. These hillocks consist of purple claystone and of various other porphyries, and of much white feldspathic greenstone passing into andesite; this latter variety included in one case crystals of orthitic and albitic feldspar touching each other, and others of hornblende, chlorite, and epidote. The strata surrounding these intrusive hillocks at the mines of Los Hornos, are intersected by many veins of copper-pyrites, associated with much micaceous iron-ore, and by some of gold: in the neighbourhood of these veins the rocks are blackened and much altered. The gypsum near the intrusive masses is always opaque. One of these hillocks of porphyry was capped by some stratified porphyritic conglomerate, which must have been brought up from below, through the whole immense thickness of the overlying gypseous formation. The lower beds of the gypseous formation resemble the corresponding and probably contemporaneous strata of the main Cordillera; whilst the upper beds in several respects resemble those of the Uspallata chain, and possibly may be contemporaneous with them; for I have endeavoured to show that the Uspallata beds were accumulated subsequently to the gypseous or Neocomian formations of the Cordillera.

This pile of strata dips at an angle of about 20 degrees to N.E. by N., close up to the foot of the Cuesta de Los Hornos, a crooked range of mountains formed of intrusive rocks of the same nature with the above described hillocks. Only in one or two places, on this south-eastern side of the range, I noticed a narrow fringe of the upper gypseous strata brushed up and inclined south-eastward from it. On its north-eastern flank, and likewise on a few of the summits, the stratified porphyritic conglomerate is inclined N.E.: so that, if we disregard the very narrow anticlinal fringe of gypseous strata at its S.E. foot, this range forms a second uniclinal axis of elevation. Proceeding in a north-by-east direction to the village of Combarbala, we come to a third escarpment of the porphyritic conglomerate, dipping eastwards, and forming the outer range of the main Cordillera. The lower beds were here more jaspery than usual, and they included some white cherty strata and red sandstones, alternating with purple claystone porphyry. Higher up in the Cordillera there appeared to be a line of andesitic rocks; and beyond them, a fourth escarpment of the porphyritic conglomerate, again dipping eastwards or inwards. The overlying gypseous strata, if they ever existed here, have been entirely removed.

COPPER MINES OF PANUNCILLO.

From Combarbala to Coquimbo, I traversed the country in a zigzag direction, crossing and recrossing the porphyritic conglomerate and finding in the granitic districts an unusual number of mountain-masses composed of various intrusive, porphyritic rocks, many of them andesitic. One common variety was greenish-black, with large crystals of blackish albite. At Panuncillo a short N.N.W. and S.S.E. ridge, with a nucleus formed of greenstone and of a slate-coloured porphyry including crystals of glassy feldspar, deserves notice, from the very singular nature of the almost vertical strata composing it. These consist chiefly of a finer and coarser granular mixture, not very compact, of white carbonate of lime, of protoxide of iron and of yellowish garnets (ascertained by Professor Miller), each grain being an almost perfect crystal. Some of the varieties consist exclusively of granules of the calcareous spar; and some contain grains of copper ore, and, I believe, of quartz. These strata alternate with a bluish, compact, fusible, feldspathic rock. Much of the above granular mixture has, also, a pseudo-brecciated structure, in which fragments are obscurely arranged in planes parallel to those of the stratification, and are conspicuous on the weathered surfaces. The fragments are angular or rounded, small or large, and consist of bluish or reddish compact feldspathic matter, in which a few acicular crystals of feldspar can sometimes be seen. The fragments often blend at their edges into the surrounding granular mass, and seem due to a kind of concretionary action.

These singular rocks are traversed by many copper veins, and appear to rest conformably on the granular mixture (in parts as fine-grained as a sandstone) of quartz, mica, hornblende, and feldspar; and this on fine- grained, common gneiss; and this on a laminated mass, composed of pinkish ORTHITIC feldspar, including a few specks of hornblende; and lastly, this on granite, which together with andesitic rocks, form the surrounding district.

COQUIMBO: MINING DISTRICT OF ARQUEROS.

At Coquimbo the porphyritic conglomerate formation approaches nearer to the Pacific than in any other part of Chile visited by me, being separated from the coast by a tract only a few miles broad of the usual plutonic rocks, with the addition of a porphyry having a red euritic base. In proceeding to the mines of Arqueros, the strata of porphyritic conglomerate are at first nearly horizontal, an unusual circumstance, and afterwards they dip gently to S.S.E. After having ascended to a considerable height, we come to an undulatory district in which the famous silver mines are situated; my examination was chiefly confined to those of S. Rosa. Most of the rocks in this district are stratified, dipping in various directions, and many of them are of so singular a nature, that at the risk of being tedious I must briefly describe them. The commonest variety is a dull-red, compact, finely brecciated stone, containing much iron and innumerable white crystallised particles of carbonate of lime, and minute extraneous fragments. Another variety is almost equally common near S. Rosa; it has a bright green, scanty basis, including distinct crystals and patches of white carbonate of lime, and grains of red, semi-micaceous oxide of iron; in parts the basis becomes dark green, and assumes an obscure crystalline arrangement, and occasionally in parts it becomes soft and slightly translucent like soapstone. These red and green rocks are often quite distinct, and often pass into each other; the passage being sometimes affected by a fine brecciated structure, particles of the red and green matter being mingled together. Some of the varieties appear gradually to become porphyritic with feldspar; and all of them are easily fusible into pale or dark-coloured beads, strongly attracted by the magnet. I should perhaps have mistaken several of these stratified rocks for submarine lavas, like some of those described at the Puente del Inca, had I not examined, a few leagues eastward of this point, a fine series of analogous but less metamorphosed, sedimentary beds belonging to the gypseous formation, and probably derived from a volcanic source.

This formation is intersected by numerous metalliferous veins, running, though irregularly, N.W. and S.E., and generally at right angles to the many dikes. The veins consist of native silver, of muriate of silver, an amalgam of silver, cobalt, antimony, and arsenic, generally embedded in sulphate of barytes. (See the Report on M. Domeyko's account of those mines, in the "Comptes Rendus" tome 14) I was assured by Mr. Lambert, that native copper without a trace of silver has been found in the same vein with native silver without a trace of copper. At the mines of Aristeas, the silver veins are said to be unproductive as soon as they pass into the green strata, whereas at S. Rosa, only two or three miles distant, the reverse happens; and at the time of my visit, the miners were working through a red stratum, in the hope of the vein becoming productive in the underlying green sedimentary mass. I have a specimen of one of these green rocks, with the usual granules of white calcareous spar and red oxide of iron, abounding with disseminated particles of glittering native and muriate of silver, yet taken at the distance of one yard from any vein,—a circumstance, as I was assured, of very rare occurrence.

SECTION EASTWARD, UP THE VALLEY OF COQUIMBO.

After passing for a few miles over the coast granitic series, we come to the porphyritic conglomerate, with its usual characters, and with some of the beds distinctly displaying their mechanical origin. The strata, where first met with, are, as before stated, only slightly inclined; but near the Hacienda of Pluclaro, we come to an anticlinal axis, with the beds much dislocated and shifted by a great fault, of which not a trace is externally seen in the outline of the hill. I believe that this anticlinal axis can be traced northwards, into the district of Arqueros, where a conspicuous hill called Cerro Blanco, formed of a harsh, cream-coloured euritic rock, including a few crystals of reddish feldspar, and associated with some purplish claystone porphyry, seems to fall on a line of elevation. In descending from the Arqueros district, I crossed on the northern border of the valley, strata inclined eastward from the Pluclaro axis: on the porphyritic conglomerate there rested a mass, some hundred feet thick, of brown argillaceous limestone, in parts crystalline, and in parts almost composed of Hippurites Chilensis, d'Orbigny; above this came a black calcareous shale, and on it a red conglomerate. In the brown limestone, with the Hippurites, there was an impression of a Pecten and a coral, and great numbers of a large Gryphaea, very like, and, according to Professor E. Forbes, probably identical with G. Orientalis, Forbes MS.,—a cretaceous species (probably upper greensand) from Verdachellum, in Southern India. These fossils seem to occupy nearly the same position with those at the Puente del Inca,—namely, at the top of the porphyritic conglomerate, and at the base of the gypseous formation.

A little above the Hacienda of Pluclaro, I made a detour on the northern side of the valley, to examine the superincumbent gypseous strata, which I estimated at 6,000 feet in thickness. The uppermost beds of the porphyritic conglomerate, on which the gypseous strata conformably rest, are variously coloured, with one very singular and beautiful stratum composed of purple pebbles of various kinds of porphyry, embedded in white calcareous spar, including cavities lined with bright-green crystallised epidote. The whole pile of strata belonging to both formations is inclined, apparently from the above-mentioned axis of Pluclaro, at an angle of between 20 and 30 degrees to the east. I will here give a section of the principal beds met with in crossing the entire thickness of the gypseous strata.

Firstly: above the porphyritic conglomerate formation, there is a fine- grained, red, crystalline sandstone.

Secondly: a thick mass of smooth-grained, calcareo-aluminous, shaly rock, often marked with dendritic manganese, and having, where most compact, the external appearance of honestone. It is easily fusible. I shall for the future, for convenience' sake, call this variety pseudo-honestone. Some of the varieties are quite black when freshly broken, but all weather into a yellowish-ash coloured, soft, earthy substance, precisely as is the case with the compact shaly rocks of the Peuquenes range. This stratum is of the same general nature with many of the beds near Los Hornos in the Illapel section. In this second bed, or in the underlying red sandstone (for the surface was partially concealed by detritus), there was a thick mass of gypsum, having the same mineralogical characters with the great beds described in our sections across the Cordillera.

Thirdly: a thick stratum of fine-grained, red, sedimentary matter, easily fusible into a white glass, like the basis of claystone porphyry; but in parts jaspery, in parts brecciated, and including crystalline specks of carbonate of lime. In some of the jaspery layers, and in some of the black siliceous slaty bands, there were irregular seams of imperfect pitchstone, undoubtedly of metamorphic origin, and other seams of brown, crystalline limestone. Here, also, were masses, externally resembling ill-preserved silicified wood.

Fourthly and fifthly: calcareous pseudo-honestone; and a thick stratum concealed by detritus.

Sixthly: a thinly stratified mass of bright green, compact, smooth-grained, calcareo-argillaceous stone, easily fusible, and emitting a strong aluminous odour: the whole has a highly angulo-concretionary structure; and it resembles, to a certain extent, some of the upper tufaceo-infusorial deposits of the Patagonian tertiary formation. It is in its nature allied to our pseudo-honestone, and it includes well characterised layers of that variety; and other layers of a pale green, harder, and brecciated variety; and others of red sedimentary matter, like that of bed Three. Some pebbles of porphyries are embedded in the upper part.

Seventhly: red sedimentary matter or sandstone like that of bed One, several hundred feet in thickness, and including jaspery layers, often having a finely brecciated structure.

Eighthly: white, much indurated, almost crystalline tuff, several hundred feet in thickness, including rounded grains of quartz and particles of green matter like that of bed Six. Parts pass into a very pale green, semi- porcellanic stone.

Ninthly: red or brown coarse conglomerate, three or four hundred feet thick, formed chiefly of pebbles of porphyries, with volcanic particles, in an arenaceous, non-calcareous, fusible basis: the upper two feet are arenaceous without any pebbles.

Tenthly: the last and uppermost stratum here exhibited, is a compact, slate-coloured porphyry, with numerous elongated crystals of glassy feldspar, from one hundred and fifty to two hundred feet in thickness; it lies strictly conformably on the underlying conglomerate, and is undoubtedly a submarine lava.

This great pile of strata has been broken up in several places by intrusive hillocks of purple claystone porphyry, and by dikes of porphyritic greenstone: it is said that a few poor metalliferous veins have been discovered here. From the fusible nature and general appearance of the finer-grained strata, they probably owe their origin (like the allied beds of the Uspallata range, and of the Upper Patagonian tertiary formations), to gentle volcanic eruptions, and to the abrasion of volcanic rocks. Comparing these beds with those in the mining district of Arqueros, we see at both places rocks easily fusible, of the same peculiar bright green and red colours, containing calcareous matter, often having a finely brecciated structure, often passing into each other, and often alternating together: hence I cannot doubt that the only difference between them, lies in the Arqueros beds having been more metamorphosed (in conformity with their more dislocated and injected condition), and consequently in the calcareous matter, oxide of iron and green colouring matter, having been segregated under a more crystalline form.

The strata are inclined, as before stated, from 20 to 30 degrees eastward, towards an irregular north and south chain of andesitic porphyry and of porphyritic greenstone, where they are abruptly cut off. In the valley of Coquimbo, near to the H. of Gualliguaca, similar plutonic rocks are met with, apparently a southern prolongation of the above chain; and eastward of it we have an escarpment of the porphyritic conglomerate, with the strata inclined at a small angle eastward, which makes the third escarpment, including that nearest the coast. Proceeding up the valley we come to another north and south line of granite, andesite, and blackish porphyry, which seem to lie in an irregular trough of the porphyritic conglomerate. Again, on the south side of the R. Claro, there are some irregular granitic hills, which have thrown off the strata of porphyritic conglomerate to the N.W. by W.; but the stratification here has been much disturbed. I did not proceed any farther up the valley, and this point is about two-thirds of the distance between the Pacific and the main Cordillera.

I will describe only one other section, namely, on the north side of the R. Claro, which is interesting from containing fossils: the strata are much dislocated by faults and dikes, and are inclined to the north, towards a mountain of andesite and porphyry, into which they appear to become almost blended. As the beds approach this mountain, their inclination increases up to an angle of 70 degrees, and in the upper part, the rocks become highly metamorphosed. The lowest bed visible in this section, is a purplish hard sandstone. Secondly, a bed two or three hundred feet thick, of a white siliceous sandstone, with a calcareous cement, containing seams of slaty sandstone, and of hard yellowish-brown (dolomitic?) limestone; numerous, well-rounded, little pebbles of quartz are included in the sandstone. Thirdly, a dark coloured limestone with some quartz pebbles, from fifty to sixty feet in thickness, containing numerous silicified shells, presently to be enumerated. Fourthly, very compact, calcareous, jaspery sandstone, passing into (fifthly) a great bed, several hundred feet thick, of conglomerate, composed of pebbles of white, red, and purple porphyries, of sandstone and quartz, cemented by calcareous matter. I observed that some of the finer parts of this conglomerate were much indurated within a foot of a dike eight feet in width, and were rendered of a paler colour with the calcareous matter segregated into white crystallised particles; some parts were stained green from the colouring matter of the dike. Sixthly, a thick mass, obscurely stratified, of a red sedimentary stone or sandstone, full of crystalline calcareous matter, imperfect crystals of oxide of iron, and I believe of feldspar, and therefore closely resembling some of the highly metamorphosed beds at Arqueros: this bed was capped by, and appeared to pass in its upper part into, rocks similarly coloured, containing calcareous matter, and abounding with minute crystals, mostly elongated and glassy, of reddish albite. Seventhly, a conformable stratum of fine reddish porphyry with large crystals of (albitic?) feldspar; probably a submarine lava. Eighthly, another conformable bed of green porphyry, with specks of green earth and cream-coloured crystals of feldspar. I believe that there are other superincumbent crystalline strata and submarine lavas, but I had not time to examine them.

The upper beds in this section probably correspond with parts of the great gypseous formation; and the lower beds of red sandstone conglomerate and fossiliferous limestone no doubt are the equivalents of the Hippurite stratum, seen in descending from Arqueros to Pluclaro, which there lies conformably upon the porphyritic conglomerate formation. The fossils found in the third bed, consist of:—

Pecten Dufreynoyi, d'Orbigny, "Voyage, Part Pal." This species, which occurs here in vast numbers, according to M. D'Orbigny, resembles certain cretaceous forms.

Ostrea hemispherica, d'Orbigny, "Voyage" etc.

Also resembles, according to the same author, cretaceous forms.

Terebratula aenigma, d'Orbigny, "Voyage" etc. (Pl. 22 Figures 10-12.)

Is allied, according to M. d'Orbigny, to T. concinna from the Forest Marble. A series of this species, collected in several localities hereafter to be referred to, has been laid before Professor Forbes; and he informs me that many of the specimens are almost undistinguishable from our oolitic T. tetraedra, and that the varieties amongst them are such as are found in that variable species. Generally speaking, the American specimens of T. aenigma may be distinguished from the British T. tetraedra, by the surface having the ribs sharp and well-defined to the beak, whilst in the British species they become obsolete and smoothed down; but this difference is not constant. Professor Forbes adds, that, possibly, internal characters may exist, which would distinguish the American species from its European allies.

Spirifer linguiferoides, E. Forbes.

Professor Forbes states that this species is very near to S. linguifera of Phillips (a carboniferous limestone fossil), but probably distinct. M. d'Orbigny considers it as perhaps indicating the Jurassic period.

Ammonites, imperfect impression of.

M. Domeyko has sent to France a collection of fossils, which, I presume, from the description given, must have come from the neighbourhood of Arqueros; they consist of:—

Pecten Dufreynoyi, d'Orbigny, "Voyage" Part Pal.

Ostrea hemispherica, d'Orbigny, "Voyage" Part Pal.

Turritella Andii, d'Orbigny, "Voyage" Part Pal. (Pleurotomaria Humboldtii of Von Buch).

Hippurites Chilensis, d'Orbigny, "Voyage" Part Pal.

The specimens of this Hippurite, as well as those I collected in my descent from Arqueros, are very imperfect; but in M. d'Orbigny's opinion they resemble, as does the Turritella Andii, cretaceous (upper greensand) forms.

Nautilus Domeykus, d'Orbigny, "Voyage" Part Pal.

Terebratula aenigma, d'Orbigny, "Voyage" Part Pal.

Terebratula ignaciana, d'Orbigny, "Voyage" Part Pal.

This latter species was found by M. Domeyko in the same block of limestone with the T. aenigma. According to M. d'Orbigny, it comes near to T. ornithocephala from the Lias. A series of this species collected at Guasco, has been examined by Professor E. Forbes, and he states that it is difficult to distinguish between some of the specimens and the T. hastata from the mountain limestone; and that it is equally difficult to draw a line between them and some Marlstone Terebratulae. Without a knowledge of the internal structure, it is impossible at present to decide on their identity with analogous European forms.

The remarks given on the several foregoing shells, show that, in M. d'Orbigny's opinion, the Pecten, Ostrea, Turritella, and Hippurite indicate the cretaceous period; and the Gryphaea appears to Professor Forbes to be identical with a species, associated in Southern India with unquestionably cretaceous forms. On the other hand, the two Terebratulae and the Spirifer point, in the opinion both of M. d'Orbigny and Professor Forbes, to the oolitic series. Hence M. d'Orbigny, not having himself examined this country, has concluded that there are here two distinct formations; but the Spirifer and T. aenigma were certainly included in the same bed with the Pecten and Ostrea, whence I extracted them; and the geologist M. Domeyko sent home the two Terebratulae with the other-named shells, from the same locality, without specifying that they came from different beds. Again, as we shall presently see, in a collection of shells given me from Guasco, the same species, and others presenting analogous differences, are mingled together, and are in the same condition; and lastly, in three places in the valley of Copiapo, I found some of these same species similarly grouped. Hence there cannot be any doubt, highly curious though the fact be, that these several fossils, namely, the Hippurites, Gryphaea, Ostrea, Pecten, Turritella, Nautilus, two Terebratulae, and Spirifer all belong to the same formation, which would appear to form a passage between the oolitic and cretaceous systems of Europe. Although aware how unusual the term must sound, I shall, for convenience' sake, call this formation cretaceo-oolitic. Comparing the sections in this valley of Coquimbo with those in the Cordillera described in the last chapter, and bearing in mind the character of the beds in the intermediate district of Los Hornos, there is certainly a close general mineralogical resemblance between them, both in the underlying porphyritic conglomerate, and in the overlying gypseous formation. Considering this resemblance, and that the fossils from the Puente del Inca at the base of the gypseous formation, and throughout the greater part of its entire thickness on the Peuquenes range, indicate the Neocomian period,—that is, the dawn of the cretaceous system, or, as some have believed, a passage between this latter and the oolitic series—I conclude that probably the gypseous and associated beds in all the sections hitherto described, belong to the same great formation, which I have denominated—cretaceo-oolitic. I may add, before leaving Coquimbo, that M. Gay found in the neighbouring Cordillera, at the height of 14,000 feet above the sea, a fossiliferous formation, including a Trigonia and Pholadomya (D'Orbigny "Voyage" Part Geolog.);—both of which genera occur at the Puente del Inca.

COQUIMBO TO GUASCO.

The rocks near the coast, and some way inland, do not differ from those described northwards of Valparaiso: we have much greenstone, syenite, feldspathic and jaspery slate, and grauwackes having a basis like that of claystone; there are some large tracts of granite, in which the constituent minerals are sometimes arranged in folia, thus composing an imperfect gneiss. There are two large districts of mica-schists, passing into glossy clay-slate, and resembling the great formation in the Chonos Archipelago. In the valley of Guasco, an escarpment of porphyritic conglomerate is first seen high up the valley, about two leagues eastward of the town of Ballenar. I heard of a great gypseous formation in the Cordillera; and a collection of shells made there was given me. These shells are all in the same condition, and appear to have come from the same bed: they consist of:—

Turritella Andii, d'Orbigny, "Voyage" Part Pal.
Pecten Dufreynoyi, d'Orbigny, "Voyage" Part Pal.
Terebatula ignaciana, d'Orbigny, "Voyage" Part Pal.
The relations of these species have been given under the head of Coquimbo.
Terebratula aenigma, d'Orbigny, "Voyage" Part Pal.
This shell M. d'Orbigny does not consider identical with his T. aenigma, but near to T. obsoleta. Professor Forbes thinks that it is certainly a variety of T. aenigma: we shall meet with this variety again at Copiapo.
Spirifer Chilensis, E. Forbes.

Professor Forbes remarks that this fossil resembles several carboniferous limestone Spirifers; and that it is also related to some liassic species, as S. Wolcotii.

If these shells had been examined independently of the other collections, they would probably have been considered, from the characters of the two Terebratulae, and from the Spirifer, as oolitic; but considering that the first species, and according to Professor Forbes, the four first, are identical with those from Coquimbo, the two formations no doubt are the same, and may, as I have said, be provisionally called cretaceo-oolitic.

VALLEY OF COPIAPO.

The journey from Guasco to Copiapo, owing to the utterly desert nature of the country, was necessarily so hurried, that I do not consider my notes worth giving. In the valley of Copiapo some of the sections are very interesting. From the sea to the town of Copiapo, a distance estimated at thirty miles, the mountains are composed of greenstone, granite, andesite, and blackish porphyry, together with some dusky-green feldspathic rocks, which I believe to be altered clay-slate: these mountains are crossed by many brown-coloured dikes, running north and south. Above the town, the main valley runs in a south-east and even more southerly course towards the Cordillera, where it is divided into three great ravines, by the northern one of which, called Jolquera, I penetrated for a short distance. The section, Section 1/3 in Plate 1, gives an eye-sketch of the structure and composition of the mountains on both sides of this valley: a straight east and west line from the town to the Cordillera is perhaps not more than thirty miles, but along the valley the distance is much greater. Wherever the valley trended very southerly, I have endeavoured to contract the section into its true proportion. This valley, I may add, rises much more gently than any other valley which I saw in Chile.

To commence with our section, for a short distance above the town we have hills of the granitic series, together with some of that rock [A], which I suspect to be altered clay-slate, but which Professor G. Rose, judging from specimens collected by Meyen at P. Negro, states is serpentine passing into greenstone. We then come suddenly to the great gypseous formation [B], without having passed over, differently from, in all the sections hitherto described, any of the porphyritic conglomerate. The strata are at first either horizontal or gently inclined westward; then highly inclined in various directions, and contorted by underlying masses of intrusive rocks; and lastly, they have a regular eastward dip, and form a tolerably well pronounced north and south line of hills. This formation consists of thin strata, with innumerable alternations, of black, calcareous slate-rock, of calcareo-aluminous stones like those at Coquimbo, which I have called pseudo-honestones of green jaspery layers, and of pale-purplish, calcareous, soft rotten-stone, including seams and veins of gypsum. These strata are conformably overlaid by a great thickness of thinly stratified, compact limestone with included crystals of carbonate of lime. At a place called Tierra Amarilla, at the foot of a mountain thus composed there is a broad vein, or perhaps stratum, of a beautiful and curious crystallised mixture, composed, according to Professor G. Rose, of sulphate of iron under two forms, and of the sulphates of copper and alumina (Meyen's "Reise" etc. Th. 1, s. 394.): the section is so obscure that I could not make out whether this vein or stratum occurred in the gypseous formation, or more probably in some underlying masses [A], which I believe are altered clay-slate.

SECOND AXIS OF ELEVATION.

After the gypseous masses [B], we come to a line of hills of unstratified porphyry [C], which on their eastern side blend into strata of great thickness of porphyritic conglomerate, dipping eastward. This latter formation, however, here has not been nearly so much metamorphosed as in most parts of Central Chile; it is composed of beds of true purple claystone porphyry, repeatedly alternating with thick beds of purplish-red conglomerate with the well-rounded, large pebbles of various porphyries, not blended together.

THIRD AXIS OF ELEVATION.

Near the ravine of Los Hornitos, there is a well-marked line of elevation, extending for many miles in a N.N.E. and S.S.W. direction, with the strata dipping in most parts (as in the second axis) only in one direction, namely, eastward at an average angle of between 30 and 40 degrees. Close to the mouth of the valley, however, there is, as represented in the section, a steep and high mountain [D], composed of various green and brown intrusive porphyries enveloped with strata, apparently belonging to the upper parts of the porphyritic conglomerate, and dipping both eastward and westward. I will describe the section seen on the eastern side of this mountain [D], beginning at the base with the lowest bed visible in the porphyritic conglomerate, and proceeding upwards through the gypseous formation. Bed 1 consists of reddish and brownish porphyry varying in character, and in many parts highly amygdaloidal with carbonate of lime, and with bright green and brown bole. Its upper surface is throughout clearly defined, but the lower surface is in most parts indistinct, and towards the summit of the mountain [D] quite blended into the intrusive porphyries. Bed 2, a pale lilac, hard but not heavy stone, slightly laminated, including small extraneous fragments, and imperfect as well as some perfect and glassy crystals of feldspar; from one hundred and fifty to two hundred feet in thickness. When examining it in situ, I thought it was certainly a true porphyry, but my specimens now lead me to suspect that it possibly may be a metamorphosed tuff. From its colour it could be traced for a long distance, overlying in one part, quite conformably to the porphyry of bed 1, and in another not distant part, a very thick mass of conglomerate, composed of pebbles of a porphyry chiefly like that of bed 1: this fact shows how the nature of the bottom formerly varied in short horizontal distances. Bed 3, white, much indurated tuff, containing minute pebbles, broken crystals, and scales of mica, varies much in thickness. This bed is remarkable from containing many globular and pear-shaped, externally rusty balls, from the size of an apple to a man's head, of very tough, slate-coloured porphyry, with imperfect crystals of feldspar: in shape these balls do not resemble pebbles, AND I BELIEVE THAT THEY ARE SUBAQUEOUS VOLCANIC BOMBS; they differ from SUBAERIAL bombs only in not being vesicular. Bed 4; a dull purplish-red, hard conglomerate, with crystallised particles and veins of carbonate of lime, from three hundred to four hundred feet in thickness. The pebbles are of claystone porphyries of many varieties; they are tolerably well rounded, and vary in size from a large apple to a man's head. This bed includes three layers of coarse, black, calcareous, somewhat slaty rock: the upper part passes into a compact red sandstone.

In a formation so highly variable in mineralogical nature, any division not founded on fossil remains, must be extremely arbitrary: nevertheless, the beds below the last conglomerate may, in accordance with all the sections hitherto described, be considered as belonging to the porphyritic conglomerate, and those above it to the gypseous formation, marked [E] in the section. The part of the valley in which the following beds are seen is near Potrero Seco. Bed 5, compact, fine-grained, pale greenish-grey, non- calcareous, indurated mudstone, easily fusible into a pale green and white glass. Bed 6, purplish, coarse-grained, hard sandstone, with broken crystals of feldspar and crystallised particles of carbonate of lime; it possesses a slightly nodular structure. Bed 7, blackish-grey, much indurated, calcareous mudstone, with extraneous particles of unequal size; the whole being in parts finely brecciated. In this mass there is a stratum, twenty feet in thickness, of impure gypsum. Bed 8, a greenish mudstone, with several layers of gypsum. Bed 9, a highly indurated, easily fusible, white tuff, thickly mottled with ferruginous matter, and including some white semi-porcellanic layers, which are interlaced with ferruginous veins. This stone closely resembles some of the commonest varieties in the Uspallata chain. Bed 10, a thick bed of rather bright green, indurated mudstone or tuff, with a concretionary nodular structure so strongly developed that the whole mass consists of balls. I will not attempt to estimate the thickness of the strata in the gypseous formation hitherto described, but it must certainly be very many hundred feet. Bed 11 is at least 800 feet in thickness: it consists of thin layers of whitish, greenish, or more commonly brown, fine-grained, indurated tuffs, which crumble into angular fragments: some of the layers are semi-porcellanic, many of them highly ferruginous, and some are almost composed of carbonate of lime and iron with drusy cavities lined with quartzf-crystals. Bed 12, dull purplish or greenish or dark-grey, very compact and much indurated mudstone: estimated at 1,500 feet in thickness: in some parts this rock assumes the character of an imperfect coarse clay-slate; but viewed under a lens, the basis always has a mottled appearance, with the edges of the minute component particles blending together. Parts are calcareous, and there are numerous veins of highly crystalline carbonate of lime charged with iron. The mass has a nodular structure, and is divided by only a few planes of stratification: there are, however, two layers, each about eighteen inches thick, of a dark brown, finer-grained stone, having a conchoidal, semi-porcellanic fracture, which can be followed with the eye for some miles across the country.

I believe this last great bed is covered by other nearly similar alternations; but the section is here obscured by a tilt from the next porphyritic chain, presently to be described. I have given this section in detail, as being illustrative of the general character of the mountains in this neighbourhood; but it must not be supposed that any one stratum long preserves the same character. At a distance of between only two and three miles the green mudstones and white indurated tuffs are to a great extent replaced by red sandstone and black calcareous shaly rocks, alternating together. The white indurated tuff, bed 11, here contains little or no gypsum, whereas on the northern and opposite side of the valley, it is of much greater thickness and abounds with layers of gypsum, some of them alternating with thin seams of crystalline carbonate of lime. The uppermost, dark-coloured, hard mudstone, bed 12, is in this neighbourhood the most constant stratum. The whole series differs to a considerable extent, especially in its upper part, from that met with at [BB], in the lower part of the valley; nevertheless, I do not doubt that they are equivalents.

FOURTH AXIS OF ELEVATION (VALLEY OF COPIAPO).

This axis is formed of a chain of mountains [F], of which the central masses (near La Punta) consist of andesite containing green hornblende and coppery mica, and the outer masses of greenish and black porphyries, together with some fine lilac-coloured claystone porphyry; all these porphyries being injected and broken up by small hummocks of andesite. The central great mass of this latter rock, is covered on the eastern side by a black, fine-grained, highly micaceous slate, which, together with the succeeding mountains of porphyry, are traversed by numerous white dikes, branching from the andesite, and some of them extending in straight lines, to a distance of at least two miles. The mountains of porphyry eastward of the micaceous schist soon, but gradually, assume (as observed in so many other cases) a stratified structure, and can then be recognised as a part of the porphyritic conglomerate formation. These strata [G] are inclined at a high angle to the S.E., and form a mass from fifteen hundred to two thousand feet in thickness. The gypseous masses to the west already described, dip directly towards this axis, with the strata only in a few places (one of which is represented in the section) thrown from it: hence this fourth axis is mainly uniclinal towards the S.E., and just like our third axis, only locally anticlinal.

The above strata of porphyritic conglomerate [G] with their south-eastward dip, come abruptly up against beds of the gypseous formation [H], which are gently, but irregularly, inclined westward: so that there is here a synclinal axis and great fault. Further up the valley, here running nearly north and south, the gypseous formation is prolonged for some distance; but the stratification is unintelligible, the whole being broken up by faults, dikes, and metalliferous veins. The strata consist chiefly of red calcareous sandstones, with numerous veins in the place of layers, of gypsum; the sandstone is associated with some black calcareous slate-rock, and with green pseudo-honestones, passing into porcelain-jasper. Still further up the valley, near Las Amolanas [I], the gypseous strata become more regular, dipping at an angle of between 30 and 40 degrees to W.S.W., and conformably overlying, near the mouth of the ravine of Jolquera, strata [K] of porphyritic conglomerate. The whole series has been tilted by a partially concealed axis [L], of granite, andesite, and a granitic mixture of white feldspar, quartz, and oxide of iron.

FIFTH AXIS OF ELEVATION (VALLEY OF COPIAPO, NEAR LOS AMOLANAS).

I will describe in some detail the beds [I] seen here, which, as just stated, dip to W.S.W., at an angle of from 30 to 40 degrees. I had not time to examine the underlying porphyritic conglomerate, of which the lowest beds, as seen at the mouth of the Jolquera, are highly compact, with crystals of red oxide of iron; and I am not prepared to say whether they are chiefly of volcanic or metamorphic origin. On these beds there rests a coarse purplish conglomerate, very little metamorphosed, composed of pebbles of porphyry, but remarkable from containing one pebble of granite;- -of which fact no instance has occurred in the sections hitherto described. Above this conglomerate, there is a black siliceous claystone, and above it numerous alternations of dark-purplish and green porphyries, which may be considered as the uppermost limit of the porphyritic conglomerate formation.

Above these porphyries comes a coarse, arenaceous conglomerate, the lower half white and the upper half of a pink colour, composed chiefly of pebbles of various porphyries, but with some of red sandstone and jaspery rocks. In some of the more arenaceous parts of the conglomerate, there was an oblique or current lamination; a circumstance which I did not elsewhere observe. Above this conglomerate, there is a vast thickness of thinly stratified, pale-yellowish, siliceous sandstone, passing into a granular quartz-rock, used for grindstones (hence the name of the place Las Amolanas), and certainly belonging to the gypseous formation, as does probably the immediately underlying conglomerate. In this yellowish sandstone there are layers of white and pale-red siliceous conglomerate; other layers with small, well-rounded pebbles of white quartz, like the bed at the R. Claro at Coquimbo; others of a greenish, fine-grained, less siliceous stone, somewhat resembling the pseudo-honestones lower down the valley; and lastly, others of a black calcareous shale-rock. In one of the layers of conglomerate, there was embedded a fragment of mica-slate, of which this is the first instance; hence perhaps, it is from a formation of mica-slate, that the numerous small pebbles of quartz, both here and at Coquimbo, have been derived. Not only does the siliceous sandstone include layers of the black, thinly stratified, not fissile, calcareous shale-rock, but in one place the whole mass, especially the upper part, was, in a marvellously short horizontal distance, after frequent alternations, replaced by it. When this occurred, a mountain-mass, several thousand feet in thickness was thus composed; the black calcareous shale-rock, however, always included some layers of the pale-yellowish siliceous sandstone, of the red conglomerate, and of the greenish jaspery and pseudo-honestone varieties. It likewise included three or four widely separated layers of a brown limestone, abounding with shells immediately to be described. This pile of strata was in parts traversed by many veins of gypsum. The calcareous shale-rock, though when freshly broken quite black, weathers into an ash- colour: in which respect and in general appearance, it perfectly resembles those great fossiliferous beds of the Peuquenes range, alternating with gypsum and red sandstone, described in the last chapter.

The shells out of the layers of brown limestone, included in the black calcareous shale-rock, which latter, as just stated, replaces the white siliceous sandstone, consist of:—

Pecten Dufreynoyi, d'Orbigny, "Voyage" Part Pal.
Turritella Andii, d'Orbigny, "Voyage" Part Pal.
Astarte Darwinii, E. Forbes.
Gryphaea Darwinii, E. Forbes.
An intermediate form between G. gigantea and G. incurva.
Gryphaea nov. spec.?, E. Forbes.
Perna Americana, E. Forbes.
Avicula, nov. spec.
Considered by Mr. G.B. Sowerby as the A. echinata, by M. d'Orbigny as certainly a new and distinct species, having a Jurassic aspect. The specimen has been unfortunately lost.
Terebratula aenigma, d'Orbigny, (var. of do. E. Forbes.)
This is the same variety, with that from Guasco, considered by M. D'Orbigny to be a distinct species from his T. aenigma, and related to T. obsoleta.
Plagiostoma and Ammonites, fragments of.
The lower layers of the limestone contained thousands of the Gryphaea; and the upper

ones as many of the Turritella, with the Gryphaea (nov. species) and Serpulae adhering to them; in all the layers, the Terebratula and fragments of the Pecten were included. It was evident, from the manner in which species were grouped together, that they had lived where now embedded. Before making any further remarks, I may state, that higher up this same valley we shall again meet with a similar association of shells; and in the great Despoblado Valley, which branches off near the town from that of Copiapo, the Pecten Dufreynoyi, some Gryphites (I believe G. Darwinii), and the TRUE Terebratula aenigma of d'Orbigny were found together in an equivalent formation, as will be hereafter seen. A specimen also, I may add, of the true T. aenigma, was given me from the neighbourhood of the famous silver mines of Chanuncillo, a little south of the valley of the Copiapo, and these mines, from their position, I have no doubt, lie within the great gypseous formation: the rocks close to one of the silver veins, judging from fragments shown me, resemble those singular metamorphosed deposits from the mining district of Arqueros near Coquimbo.

I will reiterate the evidence on the association of these several shells in the several localities.

COQUIMBO.
In the same bed, Rio Claro:
Pecten Dufreynoyi.
Ostrea hemispherica.
Terebratula aenigma.
Spirifer linguiferoides.
Same bed, near Arqueros:
Hippurites Chilensis.
Gryphaea orientalis.
Collected by M. Domeyko from the same locality, apparently near Arqueros:
Terebratula aenigma and Terebratula ignaciana, in same block of limestone:
Pecten Dufreynoyi.
Ostrea hemispherica.
Hippurites Chilensis.
Turritella Andii.
Nautilus Domeykus.
GUASCO.
In a collection from the Cordillera, given me: the specimens all in the
same condition:
Pecten Dufreynoyi.
Turritella Andii.
Terebratula ignaciana.
Terebratula aenigma, var.
Spirifer Chilensis.
COPIAPO.
Mingled together in alternating beds in the main valley of Copiapo near Las Amolanas, and likewise higher up the valley:
Pecten Dufreynoyi.
Turritella Andii.
Terebratula aenigma, var. as at Guasco.
Astarte Darwinii.
Gryphaea Darwinii.
Gryphaea nov. species?
Perna Americana.
Avicula, nov. species.
Main valley of Copiapo, apparently same formation with that of Amolanas:
Terebratula aenigma (true).
In the same bed, high up the great lateral valley of the Despoblado, in the

ravine of Maricongo:
Terebratula aenigma (true).
Pecten Dufreynoyi.
Gryphaea Darwinii?

Considering this table, I think it is impossible to doubt that all these fossils belong to the same formation. If, however, the species from Las Amolanas, in the Valley of Copiapo, had, as in the case of those from Guasco, been separately examined, they would probably have been ranked as oolitic; for, although no Spirifers were found here, all the other species, with the exception of the Pecten, Turritella, and Astarte, have a more ancient aspect than cretaceous forms. On the other hand, taking into account the evidence derived from the cretaceous character of these three shells, and of the Hippurites, Gryphaea orientalis, and Ostrea, from Coquimbo, we are driven back to the provisional name already used of cretaceo-oolitic. From geological evidence, I believe this formation to be the equivalent of the Neocomian beds of the Cordillera of Central Chile.

To return to our section near Las Amolanas:—Above the yellow siliceous sandstone, or the equivalent calcareous slate-rock, with its bands of fossil-shells, according as the one or other prevails, there is a pile of strata, which cannot be less than from two to three thousand feet in thickness, in main part composed of a coarse, bright red conglomerate, with many intercalated beds of red sandstone, and some of green and other coloured porcelain-jaspery layers. The included pebbles are well-rounded, varying from the size of an egg to that of a cricket-ball, with a few larger; and they consist chiefly of porphyries. The basis of the conglomerate, as well as some of the alternating thin beds, are formed of a red, rather harsh, easily fusible sandstone, with crystalline calcareous particles. This whole great pile is remarkable from the thousands of huge, embedded, silicified trunks of trees, one of which was eight feet long, and another eighteen feet in circumference: how marvellous it is, that every vessel in so thick a mass of wood should have been converted into silex! I brought home many specimens, and all of them, according to Mr. R. Brown, present a coniferous structure.

Above this great conglomerate, we have from two to three hundred feet in thickness of red sandstone; and above this, a stratum of black calcareous slate-rock, like that which alternates with and replaces the underlying yellowish-white, siliceous sandstone. Close to the junction between this upper black slate-rock and the upper red sandstone, I found the Gryphaea Darwinii, the Turritella Andii, and vast numbers of a bivalve, too imperfect to be recognised. Hence we see that, as far as the evidence of these two shells serves—and the Turritella is an eminently characteristic species—the whole thickness of this vast pile of strata belongs to the same age. Again, above the last-mentioned upper red sandstone, there were several alternations of the black, calcareous slate-rock; but I was unable to ascend to them. All these uppermost strata, like the lower ones, vary extremely in character in short horizontal distances. The gypseous formation, as here seen, has a coarser, more mechanical texture, and contains much more siliceous matter than the corresponding beds lower down the valley. Its total thickness, together with the upper beds of the porphyritic conglomerate, I estimated at least at 8,000 feet; and only a small portion of the porphyritic conglomerate, which on the eastern flank of the fourth axis of elevation appeared to be from fifteen hundred to two thousand feet thick, is here included. As corroborative of the great thickness of the gypseous formation, I may mention that in the Despoblado Valley (which branches from the main valley a little above the town of Copiapo) I found a corresponding pile of red and white sandstones, and of dark, calcareous, semi-jaspery mudstones, rising from a nearly level surface and thrown into an absolutely vertical position; so that, by pacing, I ascertained their thickness to be nearly two thousand seven hundred feet; taking this as a standard of comparison, I estimated the thickness of the strata ABOVE the porphyritic conglomerate at 7,000 feet.

The fossils before enumerated, from the limestone-layers in the whitish siliceous sandstone, are now covered, on the least computation, by strata from 5,000 to 6,000 feet in thickness. Professor E. Forbes thinks that these shells probably lived at a depth of from about 30 to 40 fathoms, that is from 180 to 240 feet; anyhow, it is impossible that they could have lived at the depth of from 5,000 to 6,000 feet. Hence in this case, as in that of the Puente del Inca, we may safely conclude that the bottom of the sea on which the shells lived, subsided, so as to receive the superincumbent submarine strata: and this subsidence must have taken place during the existence of these shells; for, as I have shown, some of them occur high up as well as low down in the series. That the bottom of the sea subsided, is in harmony with the presence of the layers of coarse, well- rounded pebbles included throughout this whole pile of strata, as well as of the great upper mass of conglomerate from 2,000 to 3,000 feet thick; for coarse gravel could hardly have been formed or spread out at the profound depths indicated by the thickness of the strata. The subsidence, also, must have been slow to have allowed of this often-recurrent spreading out of the pebbles. Moreover, we shall presently see that the surfaces of some of the streams of porphyritic lava beneath the gypseous formation, are so highly amygdaloidal that it is scarcely possible to believe that they flowed under the vast pressure of a deep ocean. The conclusion of a great subsidence during the existence of these cretaceo-oolitic fossils, may, I believe, be extended to the district of Coquimbo, although owing to the fossiliferous beds there not being directly covered by the upper gypseous strata, which in the section north of the valley are about 6,000 feet in thickness, I did not there insist on this conclusion.

The pebbles in the above conglomerates, both in the upper and lower beds, are all well rounded, and, though chiefly composed of various porphyries, there are some of red sandstone and of a jaspery stone, both like the rocks intercalated in layers in this same gypseous formation; there was one pebble of mica-slate and some of quartz, together with many particles of quartz. In these respects there is a wide difference between the gypseous conglomerates and those of the porphyritic-conglomerate formation, in which latter, angular and rounded fragments, almost exclusively composed of porphyries, are mingled together, and which, as already often remarked, probably were ejected from craters deep under the sea. From these facts I conclude, that during the formation of the conglomerates, land existed in the neighbourhood, on the shores of which the innumerable pebbles were rounded and thence dispersed, and on which the coniferous forests flourished—for it is improbable that so many thousand logs of wood should have drifted from any great distance. This land, probably islands, must have been mainly formed of porphyries, with some mica-slate, whence the quartz was derived, and with some red sandstone and jaspery rocks. This latter fact is important, as it shows that in this district, even previously to the deposition of the lower gypseous or cretaceo-oolitic beds, strata of an analogous nature had elsewhere, no doubt in the more central ranges of the Cordillera, been elevated; thus recalling to our minds the relations of the Cumbre and Uspallata chains. Having already referred to the great lateral valley of the Despoblado, I may mention that above the 2,700 feet of red and white sandstone and dark mudstone, there is a vast mass of coarse, hard, red conglomerate, some thousand feet in thickness, which contains much silicified wood, and evidently corresponds with the great upper conglomerate at Las Amolanas: here, however, the conglomerate consists almost exclusively of pebbles of granite, and of disintegrated crystals of reddish feldspar and quartz firmly recemented together. In this case, we may conclude that the land whence the pebbles were derived, and on which the now silicified trees once flourished, was formed of granite.

The mountains near Las Amolanas, composed of the cretaceo-oolitic strata, are interlaced with dikes like a spider's web, to an extent which I have never seen equalled, except in the denuded interior of a volcanic crater: north and south lines, however, predominate. These dikes are composed of green, white, and blackish rocks, all porphyritic with feldspar, and often with large crystals of hornblende. The white varieties approach closely in character to andesite, which composes as we have seen, the injected axes of so many of the lines of elevation. Some of the green varieties are finely laminated, parallel to the walls of the dikes.

SIXTH AXIS OF ELEVATION (VALLEY OF COPIAPO).

This axis consists of a broad mountainous mass [O] of andesite, composed of albite, brown mica, and chlorite, passing into andesitic granite, with quartz: on its western side it has thrown off, at a considerable angle, a thick mass of stratified porphyries, including much epidote [NN], and remarkable only from being divided into very thin beds, as highly amygdaloidal on their surfaces as subaerial lava-streams are often vesicular. This porphyritic formation is conformably covered, as seen some way up the ravine of Jolquera, by a mere remnant of the lower part of the cretaceo-oolitic formation [MM], which in one part encases, as represented in the coloured section, the foot of the andesitic axis [L], of the already described fifth line, and in another part entirely conceals it: in this latter case, the gypseous or cretaceo-oolitic strata falsely appeared to dip under the porphyritic conglomerate of the fifth axis. The lowest bed of the gypseous formation, as seen here [M], is of yellowish siliceous sandstone, precisely like that of Amolanas, interlaced in parts with veins of gypsum, and including layers of the black, calcareous, non-fissile slate-rock: the Turritella Andii, Pecten Dufreynoyi, Terebratula aenigma, var., and some Gryphites were embedded in these layers. The sandstone varies in thickness from only twenty to eighty feet; and this variation is caused by the inequalities in the upper surface of an underlying stream of purple claystone porphyry. Hence the above fossils here lie at the very base of the gypseous or cretaceo-oolitic formation, and hence they were probably once covered up by strata about seven thousand feet in thickness: it is, however, possible, though from the nature of all the other sections in this district not probable, that the porphyritic claystone lava may in this case have invaded a higher level in the series. Above the sandstone there is a considerable mass of much indurated, purplish-black, calcareous claystone, allied in nature to the often-mentioned black calcareous slate- rock. Eastward of the broad andesitic axis of this sixth line, and penetrated by many dikes from it, there is a great formation [P] of mica-schist, with its usual variations, and passing in one part into a ferruginous quartz-rock. The folia are curved and highly inclined, generally dipping eastward. It is probable that this mica-schist is an old formation, connected with the granitic rocks and metamorphic schists near the coast; and that the one fragment of mica-slate, and the pebbles of quartz low down in the gypseous formation at Las Amolanas, have been derived from it. The mica-schist is succeeded by stratified porphyritic conglomerate [Q] of great thickness, dipping eastward with a high inclination: I have included this latter mountain-mass in the same anticlinal axis with the porphyritic streams [NN]; but I am far from sure that the two masses may not have been independently upheaved.

SEVENTH AXIS OF ELEVATION.

Proceeding up the ravine, we come to another mass [R] of andesite; and beyond this, we again have a very thick, stratified porphyritic formation [S], dipping at a small angle eastward, and forming the basal part of the main Cordillera. I did not ascend the ravine any higher; but here, near Castano, I examined several sections, of which I will not give the details, only observing, that the porphyritic beds, or submarine lavas, preponderate greatly in bulk over the alternating sedimentary layers, which have been but little metamorphosed: these latter consist of fine-grained red tuffs and of whitish volcanic grit-stones, together with much of a singular, compact rock, having an almost crystalline basis, finely brecciated with red and green fragments, and occasionally including a few large pebbles. The porphyritic lavas are highly amygdaloidal, both on their upper and lower surfaces; they consist chiefly of claystone porphyry, but with one common variety, like some of the streams at the Puente del Inca, having a grey mottled basis, abounding with crystals of red hydrous oxide of iron, green ones apparently of epidote, and a few glassy ones of feldspar. This pile of strata differs considerably from the basal strata of the Cordillera in Central Chile, and may possibly belong to the upper and gypseous series: I saw, however, in the bed of the valley, one fragment of porphyritic breccia-conglomerate, exactly like those great masses met with in the more southern parts of Chile.

Finally, I must observe, that though I have described between the town of Copiapo and the western flank of the main Cordillera seven or eight axes of elevation, extending nearly north and south, it must not be supposed that they all run continuously for great distances. As was stated to be the case in our sections across the Cordillera of Central Chile, so here most of the lines of elevation, with the exception of the first, third, and fifth, are very short. The stratification is everywhere disturbed and intricate; nowhere have I seen more numerous faults and dikes. The whole district, from the sea to the Cordillera, is more or less metalliferous; and I heard of gold, silver, copper, lead, mercury, and iron veins. The metamorphic action, even in the lower strata, has certainly been far less here than in Central Chile.

VALLEY OF THE DESPOBLADO.

This great barren valley, which has already been alluded to, enters the main valley of Copiapo a little above the town: it runs at first northerly, then N.E., and more easterly into the Cordillera; I followed its dreary course to the foot of the first main ridge. I will not give a detailed section, because it would be essentially similar to that already given, and because the stratification is exceedingly complicated. After leaving the plutonic hills near the town, I met first, as in the main valley, with the gypseous formation, having the same diversified character as before, and soon afterwards with masses of porphyritic conglomerate, about one thousand feet in thickness. In the lower part of this formation there were very thick beds composed of fragments of claystone porphyries, both angular and rounded, with the smaller ones partially blended together and the basis rendered porphyritic; these beds separated distinct streams, from sixty to eighty feet in thickness, of claystone lavas. Near Paipote, also, there was much true porphyritic breccia-conglomerate: nevertheless, few of these masses were metamorphosed to the same degree with the corresponding formation in Central Chile. I did not meet in this valley with any true andesite, but only with imperfect andesitic porphyry, including large crystals of hornblende: numerous as have been the varieties of intrusive porphyries already mentioned, there were here mountains composed of a new kind, having a compact, smooth, cream-coloured basis, including only a few crystals of feldspar, and mottled with dendritic spots of oxide of iron. There were also some mountains of a porphyry with a brick-red basis, containing irregular, often lens-shaped, patches of compact feldspar, and crystals of feldspar, which latter to my surprise I find to be orthite.

At the foot of the first ridge of the main Cordillera, in the ravine of Maricongo, and at an elevation which, from the extreme coldness and appearance of the vegetation, I estimated at about ten thousand feet, I found beds of white sandstone and of limestone including the Pecten Dufreynoyi, Terebratula aenigma, and some Gryphites. This ridge throws the water on the one hand into the Pacific, and on the other, as I was informed, into a great gravel-covered, basin-like plain, including a salt-lake, and without any drainage-exit. In crossing the Cordillera by this Pass, it is said that three principal ridges must be traversed, instead of two, or only one as in Central Chile.

The crest of this first main ridge and the surrounding mountains, with the exception of a few lofty pinnacles, are capped by a great thickness of a horizontally stratified, tufaceous deposit. The lowest bed is of a pale purple colour, hard, fine-grained, and full of broken crystals of feldspar and scales of mica. The middle bed is coarser, and less hard, and hence weathers into very sharp pinnacles; it includes very small fragments of granite, and innumerable ones of all sizes of grey vesicular trachyte, some of which were distinctly rounded. The uppermost bed is about two hundred feet in thickness, of a darker colour and apparently hard: but I had not time to ascend to it. These three horizontal beds may be seen for the distance of many leagues, especially westward or in the direction of the Pacific, capping the summits of the mountains, and standing on the opposite sides of the immense valleys at exactly corresponding heights. If united they would form a plain, inclined very slightly towards the Pacific; the beds become thinner in this direction, and the tuff (judging from one point to which I ascended, some way down the valley) finer-grained and of less specific gravity, though still compact and sonorous under the hammer. The gently inclined, almost horizontal stratification, the presence of some rounded pebbles, and the compactness of the lowest bed, though rendering it probable, would not have convinced me that this mass had been of subaqueous origin, for it is known that volcanic ashes falling on land and moistened by rain often become hard and stratified; but beds thus originating, and owing their consolidation to atmospheric moisture, would have covered almost equally every neighbouring summit, high and low, and would not have left those above a certain exact level absolutely bare; this circumstance seems to me to prove that the volcanic ejections were arrested at their present, widely extended, equable level, and there consolidated by some other means than simple atmospheric moisture; and this no doubt must have been a sheet of water. A lake at this great height, and without a barrier on any one side, is out of the question; consequently we must conclude that the tufaceous matter was anciently deposited beneath the sea. It was certainly deposited before the excavation of the valleys, or at least before their final enlargement (I have endeavoured to show in my "Journal" etc. (2nd edition), that this arid valley was left by the retreating sea, as the land slowly rose, in the state in which we now see it.); and I may add, that Mr. Lambert, a gentleman well acquainted with this country, informs me, that in ascending the ravine of Santandres (which branches off from the Despoblado) he met with streams of lava and much erupted matter capping all the hills of granite and porphyry, with the exception of some projecting points; he also remarked that the valleys had been excavated subsequently to these eruptions.

This volcanic formation, which I am informed by Mr. Lambert extends far northward, is of interest, as typifying what has taken place on a grander scale on the corresponding western side of the Cordillera of Peru. Under another point of view, however, it possesses a far higher interest, as confirming that conclusion drawn from the structure of the fringes of stratified shingle which are prolonged from the plains at the foot of the Cordillera far up the valleys,—namely, that this great range has been elevated in mass to a height of between eight and nine thousand feet (I may here mention that on the south side of the main valley of Copiapo, near Potrero Seco, the mountains are capped by a thick mass of horizontally stratified shingle, at a height which I estimated at between fifteen hundred and two thousand feet above the bed of the valley. This shingle, I believe, forms the edge of a wide plain, which stretches southwards between two mountain ranges.); and now, judging from this tufaceous deposit, we may conclude that the horizontal elevation has been in the district of Copiapo about ten thousand feet.

(FIGURE 24.)

In the valley of the Despoblado, the stratification, as before remarked has been much disturbed, and in some points to a greater degree than I have anywhere else seen. I will give two cases: a very thick mass of thinly stratified red sandstone, including beds of conglomerate, has been crushed together (as represented in Figure 24) into a yoke or urn-formed trough, so that the strata on both sides have been folded inwards: on the right hand the properly underlying porphyritic claystone conglomerate is seen overlying the sandstone, but it soon becomes vertical, and then is inclined towards the trough, so that the beds radiate like the spokes of a wheel: on the left hand, the inverted porphyritic conglomerate also assumes a dip towards the trough, not gradually, as on the right hand, but by means of a vertical fault and synclinal break; and a little still further on towards the left, there is a second great oblique fault (both shown by the arrow- lines), with the strata dipping to a directly opposite point; these mountains are intersected by infinitely numerous dikes, some of which can be seen to rise from hummocks of greenstone, and can be traced for thousands of feet. In the second case, two low ridges trend together and unite at the head of a little wedge-shaped valley: throughout the right- hand ridge, the strata dip at 45 degrees to the east; in the left-hand ridge, we have the very same strata and at first with exactly the same dip; but in following this ridge up the valley, the strata are seen very regularly to become more and more inclined until they stand vertical, they then gradually fall over (the basset edges forming symmetrical serpentine lines along the crest), till at the very head of the valley they are reversed at an angle of 45 degrees: so that at this point the beds have been turned through an angle of 135 degrees; and here there is a kind of anticlinal axis, with the strata on both sides dipping to opposite points at an angle of 45 degrees, but those on the left hand upside down.

ON THE ERUPTIVE SOURCES OF THE PORPHYRITIC CLAYSTONE AND GREENSTONE LAVAS.

In Central Chile, from the extreme metamorphic action, it is in most parts difficult to distinguish between the streams of porphyritic lava and the porphyritic breccia-conglomerate, but here, at Copiapo, they are generally perfectly distinct, and in the Despoblado, I saw for the first time, two great strata of purple claystone porphyry, after having been for a considerable space closely united together, one above the other, become separated by a mass of fragmentary matter, and then both thin out;—the lower one more rapidly than the upper and greater stream. Considering the number and thickness of the streams of porphyritic lava, and the great thickness of the beds of breccia-conglomerate, there can be little doubt that the sources of eruption must originally have been numerous: nevertheless, it is now most difficult even to conjecture the precise point of any one of the ancient submarine craters. I have repeatedly observed mountains of porphyries, more or less distinctly stratified towards their summits or on their flanks, without a trace of stratification in their central and basal parts: in most cases, I believe this is simply due either to the obliterating effects of metamorphic action, or to such parts having been mainly formed of intrusive porphyries, or to both causes conjoined; in some instances, however, it appeared to me very probable that the great central unstratified masses of porphyry were the now partially denuded nuclei of the old submarine volcanoes, and that the stratified parts marked the points whence the streams flowed. In one case alone, and it was in this Valley of the Despoblado, I was able actually to trace a thick stratum of purplish porphyry, which for a space of some miles conformably overlay the usual alternating beds of breccia-conglomerates and claystone lavas, until it became united with, and blended into, a mountainous mass of various unstratified porphyries.

The difficulty of tracing the streams of porphyries to their ancient and doubtless numerous eruptive sources, may be partly explained by the very general disturbance which the Cordillera in most parts has suffered; but I strongly suspect that there is a more specific cause, namely, THAT THE ORIGINAL POINTS OF ERUPTION TEND TO BECOME THE POINTS OF INJECTION. This in itself does not seem improbable; for where the earth's crust has once yielded, it would be liable to yield again, though the liquified intrusive matter might not be any longer enabled to reach the submarine surface and flow as lava. I have been led to this conclusion, from having so frequently observed that, where part of an unstratified mountain-mass resembled in mineralogical character the adjoining streams or strata, there were several other kinds of intrusive porphyries and andesitic rocks injected into the same point. As these intrusive mountain-masses form most of the axes-lines in the Cordillera, whether anticlinal, uniclinal, or synclinal, and as the main valleys have generally been hollowed out along these lines, the intrusive masses have generally suffered much denudation. Hence they are apt to stand in some degree isolated, and to be situated at the points where the valleys abruptly bend, or where the main tributaries enter. On this view of there being a tendency in the old points of eruption to become the points of subsequent injection and disturbance, and consequently of denudation, it ceases to be surprising that the streams of lava in the porphyritic claystone conglomerate formation, and in other analogous cases, should most rarely be traceable to their actual sources.

IQUIQUE, SOUTHERN PERU.

Differently from what we have seen throughout Chile, the coast here is formed not by the granitic series, but by an escarpment of the porphyritic conglomerate formation, between two and three thousand feet in height. (The lowest point, where the road crosses the coast-escarpment, is 1,900 feet by the barometer above the level of the sea.) I had time only for a very short examination; the chief part of the escarpment appears to be composed of various reddish and purple, sometimes laminated, porphyries, resembling those of Chile; and I saw some of the porphyritic breccia-conglomerate; the stratification appeared but little inclined. The uppermost part, judging from the rocks near the famous silver mine of Huantajaya, consists of laminated, impure, argillaceous, purplish-grey limestone, associated, I believe, with some purple sandstone. (Mr. Bollaert has described "Geological Proceedings" volume 2, a singular mass of stratified detritus, gravel, and sand, eighty-one yards in thickness, overlying the limestone, and abounding with loose masses of silver ore. The miners believe that they can attribute these masses to their proper veins.) In the limestone shells are found: the three following species were given me:—

Lucina Americana, E. Forbes.
Terebratula inca, E. Forbes.
Terebratula aenigma, D'Orbigny.

This latter species we have seen associated with the fossils of which lists have been given in this chapter, in two places in the valley of Coquimbo, and in the ravine of Maricongo at Copiapo. Considering this fact, and the superposition of these beds on the porphyritic conglomerate formation; and, as we shall immediately see, from their containing much gypsum, and from their otherwise close general resemblance in mineralogical nature with the strata described in the valley of Copiapo, I have little doubt that these fossiliferous beds of Iquique belong to the great cretaceo-oolitic formation of Northern Chile. Iquique is situated seven degrees latitude north of Copiapo; and I may here mention, that an Ammonites, nov. species, and an Astarte, nov. species, were given me from the Cerro Pasco, about ten degrees of latitude north of Iquique, and M. D'Orbigny thinks that they probably indicate a Neocomian formation. Again, fifteen degrees of latitude northward, in Colombia, there is a grand fossiliferous deposit, now well known from the labours of Von Buch, Lea, d'Orbigny, and Forbes, which belongs to the earlier stages of the cretaceous system. Hence, bearing in mind the character of the few fossils from Tierra del Fuego, there is some evidence that a great portion of the stratified deposits of the whole vast range of the South American Cordillera belongs to about the same geological epoch.

Proceeding from the coast escarpment inwards, I crossed, in a space of about thirty miles, an elevated undulatory district, with the beds dipping in various directions. The rocks are of many kinds,—white laminated, sometimes siliceous sandstone,—purple and red sandstone, sometimes so highly calcareous as to have a crystalline fracture,—argillaceous limestone,—black calcareous slate-rock, like that so often described at Copiapo and other places,—thinly laminated, fine-grained, greenish, indurated, sedimentary, fusible rocks, approaching in character to the so- called pseudo-honestone of Chile, including thin contemporaneous veins of gypsum,—and lastly, much calcareous, laminated porcelain jasper, of a green colour, with red spots, and of extremely easy fusibility: I noticed one conformable stratum of a freckled-brown, feldspathic lava. I may here mention that I heard of great beds of gypsum in the Cordillera. The only novel point in this formation, is the presence of innumerable thin layers of rock-salt, alternating with the laminated and hard, but sometimes earthy, yellowish, or bright red and ferruginous sandstones. The thickest layer of salt was only two inches, and it thinned out at both ends. On one of these saliferous masses I noticed a stratum about twelve feet thick, of dark-brown, hard brecciated, easily fusible rock, containing grains of quartz and of black oxide of iron, together with numerous imperfect fragments of shells. The problem of the origin of salt is so obscure, that every fact, even geographical position, is worth recording. (It is well known that stratified salt is found in several places on the shores of Peru. The island of San Lorenzo, off Lima, is composed of a pile of thin strata, about eight hundred feet in thickness, composed of yellowish and purplish, hard siliceous, or earthy sandstones, alternating with thin layers of shale, which in places passes into a greenish, semi-porcellanic, fusible rock. There are some thin beds of reddish mudstone, and soft ferruginous rotten-stones, with layers of gypsum. In nearly all these varieties, especially in the softer sandstones, there are numerous thin seams of rock-salt: I was informed that one layer has been found two inches in thickness. The manner in which the minutest fissures of the dislocated beds have been penetrated by the salt, apparently by subsequent infiltration, is very curious. On the south side of the island, layers of coal and of impure limestone have been discovered. Hence we here have salt, gypsum, and coal associated together. The strata include veins of quartz, carbonate of lime, and iron pyrites; they have been dislocated by an injected mass of greenish-brown feldspathic trap. Not only is salt abundant on the extreme western limits of the district between the Cordillera and the Pacific, but, according to Helms, it is found in the outlying low hills on the eastern flank of the Cordillera. These facts appear to me opposed to the theory, that rock-salt is due to the sinking of water, charged with salt, in mediterranean spaces of the ocean. The general character of the geology of these countries would rather lead to the opinion, that its origin is in some way connected with volcanic heat at the bottom of the sea: see on this subject Sir R. Murchison "Anniversary Address to the Geological Society" 1843) With the exception of these saliferous beds, most of the rocks as already remarked, present a striking general resemblance with the upper parts of the gypseous or cretaceo-oolitic formation of Chile.

METALLIFEROUS VEINS.

I have only a few remarks to make on this subject: in nine mining districts, some of them of considerable extent, which I visited in CENTRAL Chile, I found the PRINCIPAL veins running from between [N. and N.W.] to [S. and S.E.] (These mining districts are Yaquil near Nancagua, where the direction of the chief veins, to which only in all cases I refer, is north and south; in the Uspallata range, the prevailing line is N.N.W. and S.S.E.; in the C. de Prado, it is N.N.W. and S.S.E.; near Illapel, it is N. by W. and S. by E.; at Los Hornos the direction varies from between [N. and N.W.] to [S. and S.E.]; at the C. de los Hornos (further northward), it is N.N.W. and S.S.E.; at Panuncillo, it is N.N.W. and S.S.E.; and, lastly, at Arqueros, the direction is N.W. and S.E.): in some other places, however, their courses appeared quite irregular, as is said to be generally the case in the whole valley of Copiapo: at Tambillos, south of Coquimbo, I saw one large copper vein extending east and west. It is worthy of notice, that the foliation of the gneiss and mica-slate, where such rocks occur, certainly tend to run like the metalliferous veins, though often irregularly, in a direction a little westward of north. At Yaquil, I observed that the principal auriferous veins ran nearly parallel to the grain or imperfect cleavage of the surrounding GRANITIC rocks. With respect to the distribution of the different metals, copper, gold, and iron are generally associated together, and are most frequently found (but with many exceptions, as we shall presently see) in the rocks of the lower series, between the Cordillera and the Pacific, namely, in granite, syenite, altered feldspathic clay-slate, gneiss, and as near Guasco mica-schist. The copper-ores consist of sulphurets, oxides, and carbonates, sometimes with laminae of native metal: I was assured that in some cases (as at Panuncillo S.E. of Coquimbo), the upper part of the same vein contains oxides, and the lower part sulphurets of copper. (The same fact has been observed by Mr. Taylor in Cuba: "London Philosophical Journal" volume 11.) Gold occurs in its native form; it is believed that, in many cases, the upper part of the vein is the most productive part: this fact probably is connected with the abundance of this metal in the stratified detritus of Chile, which must have been chiefly derived from the degradation of the upper portions of the rocks. These superficial beds of well-rounded gravel and sand, containing gold, appeared to me to have been formed under the sea close to the beach, during the slow elevation of the land: Schmidtmeyer remarks that in Chile gold is sought for in shelving banks at the height of some feet on the sides of the streams, and not in their beds, as would have been the case had this metal been deposited by common alluvial action. ("Travels in Chile") Very frequently the copper-ores, including some gold, are associated with abundant micaceous specular iron. Gold is often found in iron-pyrites: at two gold mines at Yaquil (near Nancagua), I was informed by the proprietor that in one the gold was always associated with copper-pyrites, and in the other with iron-pyrites: in this latter case, it is said that if the vein ceases to contain iron-pyrites, it is yet worth while to continue the search, but if the iron-pyrites, when it reappears, is not auriferous, it is better at once to give up working the vein. Although I believe copper and gold are most frequently found in the lower granitic and metamorphic schistose series, yet these metals occur both in the porphyritic conglomerate formation (as on the flanks of the Bell of Quillota and at Jajuel), and in the superincumbent strata. At Jajuel I was informed that the copper-ore, with some gold, is found only in the greenstones and altered feldspathic clay-slate, which alternate with the purple porphyritic conglomerate. Several gold veins and some of copper- ore are worked in several parts of the Uspallata range, both in the metamorphosed strata, which have been shown to have been of probably subsequent origin to the Neocomian or gypseous formation of the main Cordillera, and in the intrusive andesitic rocks of that range. At Los Hornos (N.E. of Illapel), likewise, there are numerous veins of copper- pyrites and of gold, both in the strata of the gypseous formation and in the injected hills of andesite and various porphyries.

Silver, in the form of a chloride, sulphuret, or an amalgam, or in its native state, and associated with lead and other metals, and at Arqueros with pure native copper, occurs chiefly in the upper great gypseous or cretaceo-oolitic formation which forms probably the richest mass in Chile. We may instance the mining districts of Arqueros near Coquimbo, and of nearly the whole valley of Copiapo, and of Iquique (where the principal veins run N.E. by E. and S.W. by W.), in Peru. Hence comes Molina's remark, that silver is born in the cold and solitary deserts of the Upper Cordillera. There are, however, exceptions to this rule: at Paral (S.E. of Coquimbo) silver is found in the porphyritic conglomerate formation; as I suspect is likewise the case at S. Pedro de Nolasko in the Peuquenes Pass. Rich argentiferous lead is found in the clay-slate of the Uspallata range; and I saw an old silver-mine in a hill of syenite at the foot of the Bell of Quillota: I was also assured that silver has been found in the andesitic and porphyritic region between the town of Copiapo and the Pacific. I have stated in a previous part of this chapter, that in two neighbouring mines at Arqueros the veins in one were productive when they traversed the singular green sedimentary beds, and unproductive when crossing the reddish beds; whereas at the other mine exactly the reverse takes place; I have also described the singular and rare case of numerous particles of native silver and of the chloride being disseminated in the green rock at the distance of a yard from the vein. Mercury occurs with silver both at Arqueros and at Copiapo: at the base of C. de los Hornos (S.E. of Coquimbo, a different place from Los Hornos, before mentioned) I saw in a syenitic rock numerous quartzose veins, containing a little cinnabar in nests: there were here other parallel veins of copper and of a ferrugino-auriferous ore. I believe tin has never been found in Chile.

From information given me by Mr. Nixon of Yaquil (At the Durazno mine, the gold is associated with copper-pyrites, and the veins contain large prisms of plumbago. Crystallised carbonate of lime is one of the commonest minerals in the matrix of the Chilean veins.), and by others, it appears that in Chile those veins are generally most permanently productive, which, consisting of various minerals (sometimes differing but slightly from the surrounding rocks), include parallel strings RICH in metals; such a vein is called a veta real. More commonly the mines are worked only where one, two, or more thin veins or strings running in a different direction, intersect a POOR "veta real:" it is unanimously believed that at such points of intersection (cruceros), the quantity of metal is much greater than that contained in other parts of the intersecting veins. In some cruceros or points of intersection, the metals extend even beyond the walls of the main, broad, stony vein. It is said that the greater the angle of intersection, the greater the produce; and that nearly parallel strings attract each other; in the Uspallata range, I observed that numerous thin auri-ferruginous veins repeatedly ran into knots, and then branched out again. I have already described the remarkable manner in which rocks of the Uspallata range are indurated and blackened (as if by a blast of gunpowder) to a considerable distance from the metallic veins.

Finally, I may observe, that the presence of metallic veins seems obviously connected with the presence of intrusive rocks, and with the degree of metamorphic action which the different districts of Chile have undergone. (Sir R. Murchison and his fellow travellers have given some striking facts on this subject in their account of the Ural Mountains ("Geological Proceedings" volume 3) Such metamorphosed areas are generally accompanied by numerous dikes and injected masses of andesite and various porphyries: I have in several places traced the metalliferous veins from the intrusive masses into the encasing strata. Knowing that the porphyritic conglomerate formation consists of alternate streams of submarine lavas and of the debris of anciently erupted rocks, and that the strata of the upper gypseous formation sometimes include submarine lavas, and are composed of tuffs, mudstones, and mineral substances, probably due to volcanic exhalations,—the richness of these strata is highly remarkable when compared with the erupted beds, often of submarine origin, but NOT METAMORPHOSED, which compose the numerous islands in the Pacific, Indian, and Atlantic Oceans; for in these islands metals are entirely absent, and their nature even unknown to the aborigines.

SUMMARY OF THE GEOLOGICAL HISTORY OF THE CHILEAN CORDILLERA, AND OF THE SOUTHERN PARTS OF SOUTH AMERICA.

We have seen that the shores of the Pacific, for a space of 1,200 miles from Tres Montes to Copiapo, and I believe for a very much greater distance, are composed, with the exception of the tertiary basins, of metamorphic schists, plutonic rocks, and more or less altered clay-slate. On the floor of the ocean thus constituted, vast streams of various purplish claystone and greenstone porphyries were poured forth, together with great alternating piles of angular and rounded fragments of similar rocks ejected from the submarine craters. From the compactness of the streams and fragments, it is probable that, with the exception of some districts in Northern Chile, the eruptions took place in profoundly deep water. The orifices of eruption appear to have been studded over a breadth, with some outliers, of from fifty to one hundred miles: and closely enough together, both north and south, and east and west, for the ejected matter to form a continuous mass, which in Central Chile is more than a mile in thickness. I traced this mould-like mass, for only 450 miles; but judging from what I saw at Iquique, from specimens, and from published accounts, it appears to have a manifold greater length. In the basal parts of the series, and especially towards the flanks of the range, mud, since converted into a feldspathic slaty rock, and sometimes into greenstone, was occasionally deposited between the beds of erupted matter: with this exception the uniformity of the porphyritic rocks is very remarkable.

At the period when the claystone and greenstone porphyries nearly or quite ceased being erupted, that great pile of strata which, from often abounding with gypsum, I have generally called the gypseous formation was deposited, and feldspathic lavas, together with other singular volcanic rocks, were occasionally poured forth: I am far from pretending that any distinct line of demarcation can be drawn between this formation and the underlying porphyries and porphyritic conglomerate, but in a mass of such great thickness, and between beds of such widely different mineralogical nature, some division was necessary. At about the commencement of the gypseous period, the bottom of the sea here seems first to have been peopled by shells, not many in kind, but abounding in individuals. At the P. del Inca the fossils are embedded near the base of the formation; in the Peuquenes range, at different levels, halfway up, and even higher in the series; hence, in these sections, the whole pile of strata belongs to the same period: the same remark is applicable to the beds at Copiapo, which attain a thickness of between seven and eight thousand feet. The fossil shells in the Cordillera of Central Chile, in the opinion of all the palaeontologists who have examined them, belong to the earlier stages of the cretaceous system; whilst in Northern Chile there is a most singular mixture of cretaceous and oolitic forms: from the geological relations, however, of these two districts, I cannot but think that they all belong to nearly the same epoch, which I have provisionally called cretaceo-oolitic.

The strata in this formation, composed of black calcareous shaly-rocks of red and white, and sometimes siliceous sandstone, of coarse conglomerates, limestones, tuffs, dark mudstones, and those singular fine-grained rocks which I have called pseudo-honestones, vast beds of gypsum, and many other jaspery and scarcely describable varieties, vary and replace each other in short horizontal distances, to an extent, I believe, unequalled even in any tertiary basin. Most of these substances are easily fusible, and have apparently been derived either from volcanoes still in quiet action, or from the attrition of volcanic products. If we picture to ourselves the bottom of the sea, rendered uneven in an extreme degree, with numerous craters, some few occasionally in eruption, but the greater number in the state of solfataras, discharging calcareous, siliceous, ferruginous matters, and gypsum or sulphuric acid to an amount surpassing, perhaps, even the existing sulphureous volcanoes of Java (Von Buch's "Description Physique des Iles Canaries"), we shall probably understand the circumstances under which this singular pile of varying strata was accumulated. The shells appear to have lived at the quiescent periods when only limestone or calcareo-argillaceous matter was depositing. From Dr. Gillies' account, this gypseous or cretaceo-oolitic formation extends as far south as the Pass of Planchon, and I followed it northward at intervals for 500 miles: judging from the character of the beds with the Terebratula aenigma, at Iquique, it extends from four to five hundred miles further: and perhaps even for ten degrees of latitude north of Iquique to the Cerro Pasco, not far from Lima: again, we know that a cretaceous formation, abounding with fossils, is largely developed north of the equator, in Colombia: in Tierra del Fuego, at about this same period, a wide district of clay-slate was deposited, which in its mineralogical characters and external features, might be compared to the Silurian regions of North Wales. The gypseous formation, like that of the porphyritic breccia-conglomerate on which it rests, is of inconsiderable breadth; though of greater breadth in Northern than in Central Chile.

As the fossil shells in this formation are covered, in the Peuquenes ridge, by a great thickness of strata; at the Puente del Inca, by at least five thousand feet; at Coquimbo, though the superposition there is less plainly seen, by about six thousand feet; and at Copiapo, certainly by five or six thousand, and probably by seven thousand feet (the same species there recurring in the upper and lower parts of the series), we may feel confident that the bottom of the sea subsided during this cretaceo-oolitic period, so as to allow of the accumulation of the superincumbent submarine strata. This conclusion is confirmed by, or perhaps rather explains, the presence of the many beds at many levels of coarse conglomerate, the well- rounded pebbles in which we cannot believe were transported in very deep water. Even the underlying porphyries at Copiapo. with their highly amygdaloidal surfaces, do not appear to have flowed under great pressure. The great sinking movement thus plainly indicated, must have extended in a north and south line for at least four hundred miles, and probably was co- extensive with the gypseous formation.

The beds of conglomerate just referred to, and the extraordinarily numerous silicified trunks of fir-trees at Los Hornos, perhaps at Coquimbo and at two distant points in the valley of Copiapo, indicate that land existed at this period in the neighbourhood. This land, or islands, in the northern part of the district of Copiapo, must have been almost exclusively composed, judging from the nature of the pebbles of granite: in the southern parts of Copiapo, it must have been mainly formed of claystone porphyries, with some mica-schist, and with much sandstone and jaspery rocks exactly like the rocks in the gypseous formation, and no doubt belonging to its basal series. In several other places also, during the accumulation of the gypseous formation, its basal parts and the underlying porphyritic conglomerate must likewise have been already partially upheaved and exposed to wear and tear; near the Puente del Inca and at Coquimbo, there must have existed masses of mica-schist or some such rock, whence were derived the many small pebbles of opaque quartz. It follows from these facts, that in some parts of the Cordillera the upper beds of the gypseous formation must lie unconformably on the lower beds; and the whole gypseous formation, in parts, unconformably on the porphyritic conglomerate; although I saw no such cases, yet in many places the gypseous formation is entirely absent; and this, although no doubt generally caused by quite subsequent denudation, may in others be due to the underlying porphyritic conglomerate having been locally upheaved before the deposition of the gypseous strata, and thus having become the source of the pebbles of porphyry embedded in them. In the porphyritic conglomerate formation, in its lower and middle parts, there is very rarely any evidence, with the exception of the small quartz pebbles at Jajuel near Aconcagua, and of the single pebble of granite at Copiapó, of the existence of neighbouring land: in the upper parts, however, and especially in the district of Copiapó, the number of thoroughly well-rounded pebbles of compact porphyries make me believe, that, as during the prolonged accumulation of the gypseous formation the lower beds had already been locally upheaved and exposed to wear and tear, so it was with the porphyritic conglomerate. Hence in following thus far the geological history of the Cordillera, it may be inferred that the bed of a deep and open, or nearly open, ocean was filled up by porphyritic eruptions, aided probably by some general and some local elevations, to that comparatively shallow level at which the cretaceo- oolitic shells first lived. At this period, the submarine craters yielded at intervals a prodigious supply of gypsum and other mineral exhalations, and occasionally, in certain places poured forth lavas, chiefly of a feldspathic nature: at this period, islands clothed with fir-trees and composed of porphyries, primary rocks, and the lower gypseous strata had already been locally upheaved, and exposed to the action of the waves;—the general movement, however, at this time having been over a very wide area, one of slow subsidence, prolonged till the bed of the sea sank several thousand feet.

In Central Chile, after the deposition of a great thickness of the gypseous strata, and after their upheaval, by which the Cumbre and adjoining ranges were formed, a vast pile of tufaceous matter and submarine lava was accumulated, where the Uspallata chain now stands; also after the deposition and upheaval of the equivalent gypseous strata of the Peuquenes range, the great thick mass of conglomerate in the valley of Tenuyan was accumulated: during the deposition of the Uspallata strata, we know absolutely, from the buried vertical trees, that there was a subsidence of some thousand feet; and we may infer from the nature of the conglomerate in the valley of Tenuyan, that a similar and perhaps contemporaneous movement there took place. We have, then, evidence of a second great period of subsidence; and, as in the case of the subsidence which accompanied the accumulation of the cretaceo-oolitic strata, so this latter subsidence appears to have been complicated by alternate or local elevatory movement— for the vertical trees, buried in the midst of the Uspallata strata, must have grown on dry land, formed by the upheaval of the lower submarine beds. Presently I shall have to recapitulate the facts, showing that at a still later period, namely, at nearly the commencement of the old tertiary deposits of Patagonia and of Chile, the continent stood at nearly its present level, and then, for the third time, slowly subsided to the amount of several hundred feet, and was afterwards slowly re-uplifted to its present level.

The highest peaks of the Cordillera appear to consist of active or more commonly dormant volcanoes,—such as Tupungato, Maypu, and Aconcagua, which latter stands 23,000 feet above the level of the sea, and many others. The next highest peaks are formed of the gypseous and porphyritic strata, thrown into vertical or highly inclined positions. Besides the elevation thus gained by angular displacements, I infer, without any hesitation— from the stratified gravel-fringes which gently slope up the valleys of the Cordillera from the gravel-capped plains at their base, which latter are connected with the plains, still covered with recent shells on the coast— that this great range has been upheaved in mass by a slow movement, to an amount of at least 8,000 feet. In the Despoblado Valley, north of Copiapó, the horizontal elevation, judging from the compact, stratified tufaceous deposit, capping the distant mountains at corresponding heights, was about ten thousand feet. It is very possible, or rather probable, that this elevation in mass may not have been strictly horizontal, but more energetic under the Cordillera, than towards the coast on either side; nevertheless, movements of this kind may be conveniently distinguished from those by which strata have been abruptly broken and upturned. When viewing the Cordillera, before having read Mr. Hopkins's profound "Researches on Physical Geology," the conviction was impressed on me, that the angular dislocations, however violent, were quite subordinate in importance to the great upward movement in mass, and that they had been caused by the edges of the wide fissures, which necessarily resulted from the tension of the elevated area, having yielded to the inward rush of fluidified rock, and having thus been upturned.

The ridges formed by the angularly upheaved strata are seldom of great length: in the central parts of the Cordillera they are generally parallel to each other, and run in north and south lines; but towards the flanks they often extend more or less obliquely. The angular displacement has been much more violent in the central than in the exterior MAIN lines; but it has likewise been violent in some of the MINOR lines on the extreme flanks. The violence has been very unequal on the same short lines; the crust having apparently tended to yield on certain points along the lines of fissures. These points, I have endeavoured to show, were probably first foci of eruption, and afterwards of injected masses of porphyry and andesite. (Sir R. Murchison and his companions state "Geological Proceedings" volume 3, that no true granite appears in the higher Ural Mountains; but that syenitic greenstone—a rock closely analogous to our andesite—is far the most abundant of the intrusive masses.) The close similarity of the andesitic granites and porphyries, throughout Chile, Tierra del Fuego, and even in Peru, is very remarkable. The prevalence of feldspar cleaving like albite, is common not only to the andesites, but (as I infer from the high authority of Professor G. Rose, as well as from my own measurements) to the various claystone and greenstone porphyries, and to the trachytic lavas of the Cordillera. The andesitic rocks have in most cases been the last injected ones, and they probably form a continuous dome under this great range: they stand in intimate relationship with the modern lavas; and they seem to have been the immediate agent in metamorphosing the porphyritic conglomerate formation, and often likewise the gypseous strata, to the extraordinary extent to which they have suffered.

With respect to the age at which the several parallel ridges composing the Cordillera were upthrown, I have little evidence. Many of them may have been contemporaneously elevated and injected in the same manner as in volcanic archipelagoes lavas are contemporaneously ejected on the parallel lines of fissure. ("Volcanic Islands" etc.) But the pebbles apparently derived from the wear and tear of the porphyritic conglomerate formation, which are occasionally present in the upper parts of this same formation, and are often present in the gypseous formation, together with the pebbles from the basal parts of the latter formation in its upper strata, render it almost certain that portions, we may infer ridges, of these two formations were successively upheaved. In the case of the gigantic Portillo range, we may feel almost certain that a preexisting granitic line was upraised (not by a single blow, as shown by the highly inclined basaltic streams in the valley on its eastern flank) at a period long subsequent to the upheavement of the parallel Peuquenes range. (I have endeavoured to show in my "Journal" 2nd edition, that the singular fact of the river, which drains the valley between these two ranges, passing through the Portillo and higher line, is explained by its slow and subsequent elevation. There are many analogous cases in the drainage of rivers: see "Edinburgh New Philosophical Journal" volume 28) Again, subsequently to the upheavement of the Cumbre chain, that of Uspallata was formed and elevated; and afterwards, I may add, in the plain of Uspallata, beds of sand and gravel were violently upthrown. The manner in which the various kinds of porphyries and andesites have been injected one into the other, and in which the infinitely numerous dikes of various composition intersect each other, plainly show that the stratified crust has been stretched and yielded many times over the same points. With respect to the age of the axes of elevation between the Pacific and the Cordillera, I know little: but there are some lines which must—namely, those running north and south in Chiloe, those eight or nine east and west, parallel, far-extended, most symmetrical uniclinal lines at P. Rumena, and the short N.W.-S.E. and N.E.- S.W. lines at Concepcion—have been upheaved long after the formation of the Cordillera. Even during the earthquake of 1835, when the linear north and south islet of St. Mary was uplifted several feet above the surrounding area, we perhaps see one feeble step in the formation of a subordinate mountain-axis. In some cases, moreover, for instance, near the baths of Cauquenes, I was forcibly struck with the small size of the breaches cut through the exterior mountain-ranges, compared with the size of the same valleys higher up where entering the Cordillera; and this circumstance appeared to me scarcely explicable, except on the idea of the exterior lines having been subsequently upthrown, and therefore having been exposed to a less amount of denudation. From the manner in which the fringes of gravel are prolonged in unbroken slopes up the valleys of the Cordillera, I infer that most of the greater dislocations took place during the earlier parts of the great elevation in mass: I have, however, elsewhere given a case, and M. de Tschudi has given another, of a ridge thrown up in Peru across the bed of a river, and consequently after the final elevation of the country above the level of the sea. ("Reise in Peru" Band 2 s.8: Author's "Journal" 2nd edition)

Ascending to the older tertiary formations, I will not again recapitulate the remarks already given at the end of the Fifth Chapter,—on their great extent, especially along the shores of the Atlantic—on their antiquity, perhaps corresponding with that of the eocene deposits of Europe,—on the almost entire dissimilarity, though the formations are apparently contemporaneous, of the fossils from the eastern and western coasts, as is likewise the case, even in a still more marked degree, with the shells now living in these opposite though approximate seas,—on the climate of this period not having been more tropical than what might have been expected from the latitudes of the places under which the deposits occur; a circumstance rendered well worthy of notice, from the contrast with what is known to have been the case during the older tertiary periods of Europe, and likewise from the fact of the southern hemisphere having suffered at a much later period, apparently at the same time with the northern hemisphere, a colder or more equable temperature, as shown by the zones formerly affected by ice-action. Nor will I recapitulate the proofs of the bottom of the sea, both on the eastern and western coast, having subsided seven or eight hundred feet during this tertiary period; the movement having apparently been co-extensive, or nearly co-extensive, with the deposits of this age. Nor will I again give the facts and reasoning on which the proposition was founded, that when the bed of the sea is either stationary or rising, circumstances are far less favourable than when its level is sinking, to the accumulation of conchiferous deposits of sufficient thickness, extension, and hardness to resist, when upheaved, the ordinary vast amount of denudation. We have seen that the highly remarkable fact of the absence of any EXTENSIVE formations containing recent shells, either on the eastern or western coasts of the continent,—though these coasts now abound with living mollusca,—though they are, and apparently have always been, as favourable for the deposition of sediment as they were when the tertiary formations were copiously deposited,—and though they have been upheaved to an amount quite sufficient to bring up strata from the depths the most fertile for animal life—can be explained in accordance with the above proposition. As a deduction, it was also attempted to be shown, first, that the want of close sequence in the fossils of successive formations, and of successive stages in the same formation, would follow from the improbability of the same area continuing slowly to subside from one whole period to another, or even during a single entire period; and secondly, that certain epochs having been favourable at distant points, in the same quarter of the world for the synchronous accumulation of fossiliferous strata, would follow from movements of subsidence having apparently, like those of elevation, contemporaneously affected very large areas.

There is another point which deserves some notice, namely, the analogy between the upper parts of the Patagonian tertiary formation, as well as of the upper possibly contemporaneous beds at Chiloe and Concepcion, with the great gypseous formation of Cordillera; for in both formations, the rocks, in their fusible nature, in their containing gypsum, and in many other characters, show a connection, either intimate or remote, with volcanic action; and as the strata in both were accumulated during subsidence, it appears at first natural to connect this sinking movement with a state of high activity in the neighbouring volcanoes. During the cretaceo-oolitic period this certainly appears to have been the case at the Puente del Inca, judging from the number of intercalated lava-streams in the lower 3,000 feet of strata; but generally, the volcanic orifices seem at this time to have existed as submarine solfataras, and were certainly quiescent compared with their state during the accumulation of the porphyritic conglomerate formation. During the deposition of the tertiary strata we know that at S. Cruz, deluges of basaltic lava were poured forth; but as these lie in the upper part of the series, it is possible that the subsidence may at that time have ceased: at Chiloe, I was unable to ascertain to what part of the series the pile of lavas belonged. The Uspallata tuffs and great streams of submarine lavas, were probably intermediate in age between the cretaceo- oolitic and older tertiary formations, and we know from the buried trees that there was a great subsidence during their accumulation; but even in this case, the subsidence may not have been strictly contemporaneous with the great volcanic eruptions, for we must believe in at least one intercalated period of elevation, during which the ground was upraised on which the now buried trees grew. I have been led to make these remarks, and to throw some doubt on the strict contemporaneousness of high volcanic activity and movements of subsidence, from the conviction impressed on my mind by the study of coral formations, that these two actions do not generally go on synchronously;—on the contrary, that in volcanic districts, subsidence ceases as soon as the orifices burst forth into renewed action, and only recommences when they again have become dormant. ("The Structure and Distribution of Coral Reefs.")

At a later period, the Pampean mud, of estuary origin, was deposited over a wide area,— in one district conformably on the underlying old tertiary strata, and in another district unconformably on them, after their upheaval and denudation. During and before the accumulation, however, of these old tertiary strata, and, therefore, at a very remote period, sediment, strikingly resembling that of the Pampas, was deposited; showing during how long a time in this case the same agencies were at work in the same area. The deposition of the Pampean estuary mud was accompanied, at least in the southern parts of the Pampas, by an elevatory movement, so that the M. Hermoso beds probably were accumulated after the upheaval of those round the S. Ventana; and those at P. Alta after the upheaval of the M. Hermoso strata; but there is some reason to suspect that one period of subsidence intervened, during which mud was deposited over the coarse sand of the Barrancas de S. Gregorio, and on the higher parts of Banda Oriental. The mammiferous animals characteristic of this formation, many of which differ as much from the present inhabitants of South America, as do the eocene mammals of Europe from the present ones of that quarter of the globe, certainly co-existed at B. Blanca with twenty species of mollusca, one balanus, and two corals, all now living in the adjoining sea: this is likewise the case in Patagonia with the Macrauchenia, which co-existed with eight shells, still the commonest kinds on that coast. I will not repeat what I have elsewhere said, on the place of habitation, food, wide range, and extinction of the numerous gigantic mammifers, which at this late period inhabited the two Americas.

The nature and grouping of the shells embedded in the old tertiary formations of Patagonia and Chile show us, that the continent at that period must have stood only a few fathoms below its present level, and that afterwards it subsided over a wide area, seven or eight hundred feet. The manner in which it has since been rebrought up to its actual level, was described in detail in the First and Second Chapters. It was there shown that recent shells are found on the shores of the Atlantic, from Tierra del Fuego northward for a space of at least 1,180 nautical miles, and at the height of about 100 feet in La Plata, and of 400 feet in Patagonia. The elevatory movements on this side of the continent have been slow; and the coast of Patagonia, up to the height in one part of 950 feet and in another of 1,200 feet, is modelled into eight great, step-like, gravel-capped plains, extending for hundreds of miles with the same heights; this fact shows that the periods of denudation (which, judging from the amount of matter removed, must have been long continued) and of elevation were synchronous over surprisingly great lengths of coasts. On the shores of the Pacific, upraised shells of recent species, generally, though not always, in the same proportional numbers as in the adjoining sea, have actually been found over a north and south space of 2,075 miles, and there is reason to believe that they occur over a space of 2,480 miles. The elevation on this western side of the continent has not been equable; at Valparaiso, within the period during which upraised shells have remained undecayed on the surface, it has been 1,300 feet, whilst at Coquimbo, 200 miles northward, it has been within this same period only 252 feet. At Lima, the land has been uplifted at least 80 feet since Indian man inhabited that district; but the level within historical times apparently has subsided. At Coquimbo, in a height of 364 feet, the elevation has been interrupted by five periods of comparative rest. At several places the land has been lately, or still is, rising both insensibly and by sudden starts of a few feet during earthquake-shocks; this shows that these two kinds of upward movement are intimately connected together. For a space of 775 miles, upraised recent shells are found on the two opposite sides of the continent; and in the southern half of this space, it may be safely inferred from the slope of the land up to the Cordillera, and from the shells found in the central part of Tierra del Fuego, and high up the River Santa Cruz, that the entire breadth of the continent has been uplifted. From the general occurrence on both coasts of successive lines of escarpments, of sand-dunes and marks of erosion, we must conclude that the elevatory movement has been normally interrupted by periods, when the land either was stationary, or when it rose at so slow a rate as not to resist the average denuding power of the waves, or when it subsided. In the case of the present high sea-cliffs of Patagonia and in other analogous instances, we have seen that the difficulty in understanding how strata can be removed at those depths under the sea, at which the currents and oscillations of the water are depositing a smooth surface of mud, sand, and sifted pebbles, leads to the suspicion that the formation or denudation of such cliffs has been accompanied by a sinking movement.

In South America, everything has taken place on a grand scale, and all geological phenomena are still in active operation. We know how violent at the present day the earthquakes are, we have seen how great an area is now rising, and the plains of tertiary origin are of vast dimensions; an almost straight line can be drawn from Tierra del Fuego for 1,600 miles northward, and probably for a much greater distance, which shall intersect no formation older than the Patagonian deposits; so equable has been the upheaval of the beds, that throughout this long line, not a fault in the stratification or abrupt dislocation was anywhere observable. Looking to the basal, metamorphic, and plutonic rocks of the continent, the areas formed of them are likewise vast; and their planes of cleavage and foliation strike over surprisingly great spaces in uniform directions. The Cordillera, with its pinnacles here and there rising upwards of twenty thousand feet above the level of the sea, ranges in an unbroken line from Tierra del Fuego, apparently to the Arctic circle. This grand range has suffered both the most violent dislocations, and slow, though grand, upward and downward movements in mass; I know not whether the spectacle of its immense valleys, with mountain-masses of once liquified and intrusive rocks now bared and intersected, or whether the view of those plains, composed of shingle and sediment hence derived, which stretch to the borders of the Atlantic Ocean, is best adapted to excite our astonishment at the amount of wear and tear which these mountains have undergone.

The Cordillera from Tierra del Fuego to Mexico, is penetrated by volcanic orifices, and those now in action are connected in great trains. The intimate relation between their recent eruptions and the slow elevation of the continent in mass, appears to me highly important, for no explanation of the one phenomenon can be considered as satisfactory which is not applicable to the other. (On the Connection of certain Volcanic Phenomena in South America: "Geological Transactions" volume 5) The permanence of the volcanic action on this chain of mountains is, also, a striking fact; first, we have the deluges of submarine lavas alternating with the porphyritic conglomerate strata, then occasionally feldspathic streams and abundant mineral exhalations during the gypseous or cretaceo- oolitic period: then the eruptions of the Uspallata range, and at an ancient but unknown period, when the sea came up to the eastern foot of the Cordillera, streams of basaltic lava at the foot of the Portillo range; then the old tertiary eruptions; and lastly, there are here and there amongst the mountains, much worn and apparently very ancient volcanic formations without any craters; there are, also, craters quite extinct, and others in the condition of solfataras, and others occasionally or habitually in fierce action. Hence it would appear that the Cordillera has been, probably with some quiescent periods, a source of volcanic matter from an epoch anterior to our cretaceo-oolitic formation to the present day; and now the earthquakes, daily recurrent on some part of the western coast, give little hope that the subterranean energy is expended.

Recurring to the evidence by which it was shown that some at least of the parallel ridges, which together compose the Cordillera, were successively and slowly upthrown at widely different periods; and that the whole range certainly once, and almost certainly twice, subsided some thousand feet, and being then brought up by a slow movement in mass, again, during the old tertiary formations, subsided several hundred feet, and again was brought up to its present level by a slow and often interrupted movement; we see how opposed is this complicated history of changes slowly effected, to the views of those geologists who believe that this great mountain-chain was formed in late times by a single blow. I have endeavoured elsewhere to show, that the excessively disturbed condition of the strata in the Cordillera, so far from indicating single periods of extreme violence, presents insuperable difficulties, except on the admission that the masses of once liquified rocks of the axes were repeatedly injected with intervals sufficiently long for their successive cooling and consolidation. ("Geological Transactions" volume 5) Finally, if we look to the analogies drawn from the changes now in progress in the earth's crust, whether to the manner in which volcanic matter is erupted, or to the manner in which the land is historically known to have risen and sunk: or again, if we look to the vast amount of denudation which every part of the Cordillera has obviously suffered, the changes through which it has been brought into its present condition, will appear neither to have been too slowly effected, nor to have been too complicated.

NOTE.

As, both in France and England, translations of a passage in Professor Ehrenberg's Memoir, often referred to in the Fourth Chapter of this volume, have appeared, implying that Professor Ehrenberg believes, from the character of the infusoria, that the Pampean formation was deposited by a sea-debacle rushing over the land, I may state, on the authority of a letter to me, that these translations are incorrect. The following is the passage in question:—

"Durch Beachtung der mikroscopischen Formen hat sich nun feststellen lassen, das die Mastodonten-Lager am La Plata und die Knochen-Lager am Monte Hermoso, who wie die der Riesen-Gurtelthiere in den Dunenhugeln bei Bahia Blanca, beides in Patagonien, unveranderte brakische Susswasserbildungen sind, die einst wohl sammtlich zum obersten Fluthgebiethe des Meeres im tieferen Festlande gehorten."—"Monatsberichten der konigl. Akad. etc." zu Berlin vom April 1845.

INDEX.

Abich, on a new variety of feldspar.

Abrolhos islands.

Absence of recent formations on the S. American coasts.

Aguerros on elevation of Imperial.

Albite, constituent mineral in andesite. —in rocks of Tierra del Fuego. —in porphyries. —crystals of, with orthite.

Alison, Mr., on elevation of Valparaiso.

Alumina, sulphate of.

Ammonites from Concepcion.

Amolanas, Las.

Amygdaloid, curious varieties of.

Amygdaloids of the Uspallata range. —of Copiapo.

Andesite of Chile. —in the valley of Maypu. —of the Cumbre pass. —of the Uspallata range. —of Los Hornos. —of Copiapo.

Anhydrite, concretions of.

Araucaria, silicified wood of.

Arica, elevation of.
Arqueros, mines of.
Ascension, gypsum deposited on. —laminated volcanic rocks of.
Augite in fragments, in gneiss. —with albite, in lava.
Austin, Mr. R.A.C., on bent cleavage lamina.
Austin, Captain, on sea-bottom.
Australia, foliated rocks of.
Azara labiata, beds of, at San Pedro.
Baculites vagina.
Bahia Blanca, elevation of. —formations near. —character of living shells of.
Bahia (Brazil), elevation near. —crystalline rocks of.
Ballard, M., on the precipitation of sulphate of soda.
Banda Oriental, tertiary formations of. —crystalline rocks of.
Barnacles above sea-level. —adhering to upraised shells.
Basalt of S. Cruz. —streams of, in the Portillo range. —in the Uspallata range.
Basin chains of Chile.
Beagle Channel.
Beaumont, Elie de, on inclination of lava-streams. —on viscid quartz-rocks.
Beech-tree, leaves of fossil.
Beechey, Captain, on sea-bottom.
Belcher, Lieutenant, on elevated shells from Concepcion.
Bella Vista, plain of.
Benza, Dr., on decomposed granite.
Bettington, Mr., on quadrupeds transported by rivers.
Blake, Mr., on the decay of elevated shells near Iquique. —on nitrate of soda.
Bole.
Bollaert, Mr., on mines of Iquique.
Bones, silicified. —fossil, fresh condition of.
Bottom of sea off Patagonia.
Bougainville, on elevation of the Falkland islands.
Boulder formation of S. Cruz. —of Falkland islands. —anterior to certain extinct quadrupeds. —of Tierra del Fuego.
Boulders in the Cordillera. —transported by earthquake-waves. —in fine-grained tertiary deposits.
Brande, Mr., on a mineral spring.
Bravais, M., on elevation of Scandinavia.
Brazil, elevation of. —crystalline rocks of.
Broderip, Mr., on elevated shells from Concepcion.
Brown, Mr. R., on silicified wood of Uspallata range.
Brown, on silicified wood.
Bucalema, elevated shells near.
Buch, Von, on cleavage. —on cretaceous fossils of the Cordillera. —on the sulphureous volcanoes of Java.
Buenos Ayres.
Burchell, Mr., on elevated shells of Brazil.
Byron, on elevated shells.
Cachapual, boulders in valley of.
Caldcleugh, Mr., on elevation of Coquimbo. —on rocks of the Portillo range.
Callao, elevation near. —old town of.
Cape of Good Hope, metamorphic rocks of.
Carcharias megalodon.
Carpenter, Dr., on microscopic organisms.
Castro (Chiloe), beds near.

Cauquenes Baths, boulders near.
—pebbles in porphyry near.
—volcanic formation near.
—stratification near.
Caves above sea-level.
Cervus pumilus, fossil-horns of.
Chevalier, M., on elevation near Lima.
Chile, structure of country between the Cordillera and the Pacific. —tertiary formations of. —crystalline rocks in. —central, geology of. —northern, geology of.
Chiloe, gravel on coast. —elevation of. —tertiary formation of. —crystalline rocks of.
Chlorite-schist, near M. Video.
Chonos archipelago, tertiary formations of. —crystalline rocks of.
Chupat, Rio, scoriae transported by.
Claro, Rio, fossiliferous beds of.
Clay-shale of Los Hornos.
Clay-slate, formation of, Tierra del Fuego. —of Concepcion. —feldspathic, of Chile. —of the Uspallata range. —black siliceous, band of, in porphyritic formations of Chile.
Claystone porphyry, formation of, in Chile. —origin of. —eruptive sources of.
Cleavage, definition of.
—at Bahia.
—Rio de Janeiro.
—Maldonado.
—Monte Video.
—S. Guitru-gueyu.
—Falkland I.
—Tierra del Fuego.
—Chonos I.
—Chiloe.
—Concepcion.
—Chile.
—discussion on.
Cleavage-laminae superficially bent.
Cliffs, formation of.
Climate, late changes in. —of Chile during tertiary period.
Coal of Concepcion.
—S. Lorenzo.
Coast-denudation of St. Helena.
Cobija, elevation of.
Colombia, cretaceous formation of.
Colonia del Sacramento, elevation of.
—Pampean formation near Colorado, Rio, gravel of.
—sand-dunes of.
—Pampean formation near.
Combarbala.
Concepcion, elevation of. —deposits of. —crystalline rocks of.
Conchalee, gravel-terraces of.
Concretions of gypsum, at Iquique. —in sandstone at S. Cruz. —in tufaceous tuff of Chiloe. —in gneiss. —in claystone-porphyry at Port Desire. —in gneiss at Valparaiso. —in metamorphic rocks. —of anhydrite. —relations of, to veins.
Conglomerate claystone of Chile. —of Tenuyan. —of the Cumbre Pass. —of Rio Claro. —of Copiapo.
Cook, Captain, on form of sea-bottom.
Copiapo, elevation of. —tertiary formations of. —secondary formations of.

Copper, sulphate of. —native, at Arqueros. —mines of, at Panuncillo. —veins, distribution of.

Coquimbo, elevation and terraces of. —tertiary formations of. —secondary formations of.

Corallines living on pebbles.

Cordillera, valleys bordered by gravel fringes. —basal strata of. —fossils of. —elevation of. —gypseous formations of. —claystone-porphyries of. —andesitic rocks of. —volcanoes of.

Coste, M., on elevation of Lemus.

Coy inlet, tertiary formation of.

Crassatella Lyellii.

Cruickshanks, Mr., on elevation near Lima.

Crystals of feldspar, gradual formation of, at Port Desire.

Cumbre, Pass of, in Cordillera.

Cuming, Mr., on habits of the Mesodesma. —on range of living shells on west coast.

Dana, Mr., on foliated rocks. —on amygdaloids.

Darwin, Mount.

D'Aubuisson, on concretions. —on foliated rocks. Decay, gradual, of upraised shells.

Decomposition of granite rocks.

De la Beche, Sir H., his theoretical researches in geology. —on the action of salt on calcareous rocks. —on bent cleavage-laminae.

Denudation on coast of Patagonia. —great powers of. —of the Portillo range.

Deposits, saline.

Despoblado, valley of.

Detritus, nature of, in Cordillera.

Devonshire, bent cleavage in.

Dikes, in gneiss of Brazil. —near Rio de Janeiro. —pseudo, at Port Desire. —in Tierra del Fuego. —in Chonos archipelago, containing quartz. —near Concepcion, with quartz. —granitic-porphyritic, at Valparaiso. —rarely vesicular in Cordillera. —absent in the central ridges of the Portillo pass. —of the Portillo range, with grains of quartz. —intersecting each other often. —numerous at Copiapo.

Domeyko, M., on the silver mines of Coquimbo. on the fossils of Coquimbo.

D'Orbigny, M. A., on upraised shells of Monte Video. —on elevated shells at St. Pedro. —on elevated shells near B. Ayres. —on elevation of S. Blas. —on the sudden elevation of La Plata. —on elevated shells near Cobija. —on elevated shells near Arica. —on the climate of Peru. —on salt deposits of Cobija. —on crystals of gypsum in salt-lakes. —on absence of gypsum in the Pampean formation. —on fossil remains from Bahia Blanca. —on fossil remains from the banks of the Parana. —on the geology of St. Fe. —on the age of Pampean formation. —on the Mastodon Andium. —on the geology of the Rio Negro. —on the character of the Patagonian fossils. —on fossils from Concepcion. — —from Coquimbo. — —from Payta. —on fossil tertiary shells of Chile. —on cretaceous fossils of Tierra del Fuego. — —from the Cordillera of Chile.

Earth, marine origin of.

Earthenware, fossil.

Earthquake, effect of, at S. Maria. —elevation during, at Lemus. —of 1822, at Valparaiso. —effects of, in shattering surface. —fissures made by. —probable effects on cleavage.

Earthquakes in Pampas.

Earthquake-waves, power of, in throwing up shells. —effects of, near Lima. —power of, in transporting boulders.

Edmonston, Mr., on depths at which shells live at Valparaiso.

Ehrenberg, Professor, on infusoria in the Pampean formation. —on infusoria in the Patagonian formation.

Elevation of La Plata. —Brazil. —Bahia Blanca. —San Blas. —Patagonia. —Tierra del Fuego. —Falkland islands. —Pampas. —Chonos archipelago. —Chiloe. —Chile. — Valparaiso. —Coquimbo. —Guasco. —Iquique. —Cobija. —Lima. —sudden, at S. Maria. — —at Lemus. —insensible, at Chiloe. — —at Valparaiso. — —at Coquimbo. —axes of, at Chiloe. — —at P. Rumena. —at Concepcion. —unfavourable for the accumulation of permanent deposits. —lines of, parallel to cleavage and foliation. —lines of, oblique to foliation. —areas of, causing lines of elevation and cleavage. —lines of, in the Cordillera. —slow, in the Portillo range. —two periods of, in Cordillera of Central Chile. —of the Uspallata range. —two periods of, in Cumbre Pass. —horizontal, in the Cordillera of Copiapo. —axes of, coincident with volcanic orifices. —of the Cordillera, summary on.

Elliott, Captain, on human remains.
Ensenada, elevated shells of.
Entre Rios, geology of.
Equus curvidens.
Epidote in Tierra del Fuego. —in gneiss. —frequent in Chile. —in the Uspallata range. —in porphyry of Coquimbo.
Erman, M., on andesite.
Escarpments, recent, of Patagonia.
Extinction of fossil mammifers.
Falkland islands, elevation of. —pebbles on coast. —geology of.
Falkner, on saline incrustations.
Faults, great, in Cordillera.
Feldspar, earthy, metamorphosis of, at Port Desire. —albitic. —crystals of, with albite. —orthitic, in conglomerate of Tenuyan. —in granite of Portillo range. —in porphyries in the Cumbre Pass.
Feuillee on sea-level at Coquimbo.
Fissures, relations of, to concretions. —upfilled, at Port Desire. —in clay-slate.
Fitton, Dr., on the geology of Tierra del Fuego.
Fitzroy, Captain, on the elevation of the Falkland islands. —on the elevation of Concepcion.
Foliation, definition of.
—of rocks at Bahia.
—Rio de Janeiro.
—Maldonado.
—Monte Video.
—S. Guitru-gueyu.
—Falkland I.
—Tierra del Fuego.
—Chonos archipelago.
—Chiloe.
—Concepcion.
—Chile.
—discussion on.

Forbes, Professor E., on cretaceous fossils of Concepcion. —on cretaceous fossils and subsidence in Cumbre Pass. —on fossils from Guasco. — —from Coquimbo. — —from Copiapo. —on depths at which shells live.

Formation, Pampean. — —area of. — —estuary origin. —tertiary of Entre Rios. —of Banda Oriental. —volcanic, in Banda Oriental. —of Patagonia. —summary on. —tertiary of Tierra del Fuego. — —of the Chonos archipelago. — —of Chiloe. — —of Chile. — —of Concepcion. — —of Navidad. — —of Coquimbo. — —of Peru. — —subsidence during. —volcanic, of Tres Montes. — —of Chiloe. — —old, near Maldonado. — —with laminar structure. — —ancient, in Tierra del Fuego. —recent, absent on S. American coast. —metamorphic, of claystone-porphyry of Patagonia. —foliation of. —plutonic, with laminar structure. —palaeozoic, of the Falkland I. —claystone, at Concepcion. —Jurassic, of Cordillera. —Neocomian, of the Portillo Pass. —volcanic, of Cumbre Pass. —gypseous, of Los Hornos. — —of Coquimbo. — —of Guasco. — —of Copiapo. — —of Iquique. — cretaceo-oolitic, of Coquimbo. — —of Guasco. — —of Copiapo. — —of Iquique.

Fossils, Neocomian, of Portillo Pass. — —of Cumbre Pass. —secondary, of Coquimbo. — —of Guasco. — —of Copiapo. — —of Iquique. —palaeozoic, from the Falklands.

Fragments of hornblende-rock in gneiss. —of gneiss in gneiss.

Freyer, Lieutenant, on elevated shells of Arica.

Frezier on sea-level at Coquimbo.

Galapagos archipelago, pseudo-dikes of.

Gallegos, Port, tertiary formation of.

Garnets in gneiss. —in mica-slate. —at Panuncillo.

Gardichaud, M., on granites of Brazil.

Gay, M., on elevated shells. —on boulders in the Cordillera. —on fossils from Cordillera of Coquimbo.

Gill, Mr., on brickwork transported by an earthquake-wave.

Gillies, Dr., on heights in the Cordillera. —on extension of the Portillo range.

Glen Roy, parallel roads of. —sloping terraces of.

Gneiss, near Bahia. —of Rio de Janeiro. —decomposition of.

Gold, distribution of.

Gorodona, formations near. Granite, axis of oblique, to foliation. —andesitic. —of Portillo range. —veins of, quartzose. —pebble of, in porphyritic conglomerate. —conglomerate.

Grauwacke of Uspallata range.

Gravel at bottom of sea. —formation of, in Patagonia. —means of transportation of. —strata of, inclined.

Gravel-terraces in Cordillera.

Greenough, Mr., on quartz veins.

Greenstone, resulting from metamorphose hornblende-rock. —of Tierra del Fuego. —on the summit of the Campana of Quillota. —porphyry. —relation of, to clay-slate.

Gryphaea orientalis.

Guasco, elevation of. —secondary formation of.

Guitru-gueyu, Sierra.

Guyana, gneissic rocks of.

Gypsum, nodules of, in gravel at Rio Negro. —deposited from sea-water. —deposits of, at Iquique. —crystals of, in salt lakes. —in Pampean formation. —in tertiary formation of Patagonia. —great formation of, in the Portillo Pass. — —in the Cumbre Pass. — —near Los Hornos. — —at Coquimbo. — —at Copiapo. — —near Iquique. —of San Lorenzo.

Hall, Captain, on terraces at Coquimbo.

Hamilton, Mr., on elevation near Tacna.

Harlan, Dr., on human remains.

Hayes, Mr. A., on nitrate of soda.

Henslow, Professor, on concretions.

Herbert, Captain, on valleys in the Himalaya.

Herradura Bay, elevated shells of. —tertiary formations of.

Himalaya, valleys in.

Hippurites Chilensis.

Hitchcock, Professor, on dikes.

Honestones, pseudo, of Coquimbo. —of Copiapo.

Hooker, Dr. J.D., on fossil beech-leaves.

Hopkins, Mr., on axes of elevation oblique to foliation. —on origin of lines of elevation.

Hornblende-rock, fragments of, in gneiss.

Hornblende-schist, near M. Video.

Hornos, Los, section near.
Hornstone, dike of.
Horse, fossil tooth of.
Huafo island. —subsidence at.
Huantajaya, mines of.
Humboldt, on saline incrustations. —on foliations of gneiss. —on concretions in gneiss.
Icebergs, action on cleavage.
Illapel, section near.
Imperial, beds of shells near.
Incrustations, saline.
Infusoria in Pampean formation. —in Patagonian formation.
Iodine, salts of.
Iquique, elevation of. —saliferous deposits of. —cretaceo-oolitic formation of.
Iron, oxide of, in lavas. —in sedimentary beds. —tendency in, to produce hollow concretions. —sulphate of.
Isabelle, M., on volcanic rocks of Banda Oriental.
Joints in clay-slate.
Jukes, Mr., on cleavage in Newfoundland.
Kamtschatka, andesite of.
Kane, Dr., on the production of carbonate of soda.
King George's sound, calcareous beds of.
Lakes, origin of. —fresh-water, near salt lakes.
Lava, basaltic, of S. Cruz. —claystone-porphyry, at Chiloe. — —ancient submarine. — basaltic, of the Portillo range. —feldspathic, of the Cumbre Pass. —submarine, of the Uspallata range. —basaltic, of the Uspallata range. —submarine, of Coquimbo. —of Copiapo.
Lemus island.
Lemuy islet.
Lignite of Chiloe. —of Concepcion.
Lima, elevation of.
Lime, muriate of.
Limestone of Cumbre Pass. —of Coquimbo. —of Copiapo.
Lund and Clausen on remains of caves in Brazil.
Lund, M., on granites of Brazil.
Lyell, M., on upraised shells retaining their colours. —on terraces at Coquimbo. —on elevation near Lima. —on fossil horse's tooth. —on the boulder-formation being anterior to the extinction of North American mammifers. —on quadrupeds washed down by floods. — on age of American fossil mammifers. —on changes of climate. —on denudation. —on foliation.
MacCulloch, Dr., on concretions. —on beds of marble.
Maclaren, Mr., letter to, on coral-formations.
Macrauchenia Patachonica.
Madeira, subsidence of.
Magellan, Strait, elevation near, of.
Magnesia, sulphate of, in veins.
Malcolmson, Dr., on trees carried out to sea.
Maldonado, elevation of. —Pampean formation of. —crystalline rocks of.
Mammalia, fossil, of Bahia Blanca. — —near St. Fe. — —of Banda Oriental. — —of St. Julian. — —at Port Gallegos. —washed down by floods. —number of remains of, and range of, in Pampas.
Man, skeletons of (Brazil). —remains of, near Lima. —Indian, antiquity of.
Marble, beds of.
Maricongo, ravine of.

Marsden, on elevation of Sumatra.
Mastodon Andium, remains of. —range of.
Maypu, Rio, mouth of, with upraised shells. —gravel fringes of. —debouchement from the Cordillera.
Megalonyx, range of.
Megatherium, range of.
Miers, Mr., on elevated shells. —on the height of the Uspallata plain.
Minas, Las.
Mocha Island, elevation of. —tertiary form of. —subsidence at.
Molina, on a great flood.
Monte Hermoso, elevation of. —fossils of.
Monte Video, elevation of. —Pampean formation of. —crystalline rocks of.
Morris and Sharpe, Messrs., on the palaeozoic fossils of the Falklands.
Mud, Pampean. —long deposited on the same area.
Murchison, Sir R., on cleavage. —on waves transporting gravel. —on origin of salt formations. —on the relations of metalliferous veins and intrusive rocks. —on the absence of granite in the Ural.
Nautilus d'Orbignyanus.
Navidad, tertiary formations of, subsidence of.
Negro, Rio, pumice of pebbles of.
—gravel of.
—salt lakes of.
—tertiary strata of.
North America, fossil remains of.
North Wales, sloping terraces absent in. —bent cleavage of.
Neuvo Gulf, plains of. —tertiary formation of.
Owen, Professor, on fossil mammiferous remains.
Palmer, Mr., on transportation of gravel.
Pampas, elevation of. —earthquakes of. —formation of. —localities in which fossil mammifers have been found.
Panuncillo, mines of.
Parana, Rio, on saline incrustations. —Pampean formations near. —on the S. Tandil.
Parish, Sir W., on elevated shells near Buenos Ayres. —on earthquakes in the Pampas. —on fresh-water near salt lakes. —on origin of Pampean formation.
Patagonia, elevation and plains of. —denudation of. —gravel-formation of. —sea-cliffs of. —subsidence during tertiary period. —crystalline rocks of.
Payta, tertiary formations of.
Pebbles of pumice. —decrease in size on the coast of Patagonia. —means of transportation. —encrusted with living corallines. —distribution of, at the eastern foot of Cordillera. —dispersal of, in the Pampas. —zoned with colour.
Pentland, Mr., on heights in the Cordillera. —on fossils of the Cordillera.
Pernambuco.
Peru, tertiary formations of.
Peuquenes, Pass of, in the Cordillera. —ridge of.
Pholas, elevated shells of.
Pitchstone of Chiloe. —of Port Desire. —near Cauquenes. —layers of, in the Uspallata range. —of Los Hornos. —of Coquimbo.
Plains of Patagonia. —of Chiloe. —of Chile. —of Uspallata. —on eastern foot of Cordillera. —of Iquique.

Plata, La, elevation of. —tertiary formation of. —crystalline rocks of.
Playfair, Professor, on the transportation of gravel.
Pluclaro, axis of.

Pondicherry, fossils of.

Porcelain rocks of Port Desire. —of the Uspallata range.

Porphyry, pebbles of, strewed over Patagonia.

Porphyry, claystone, of Chiloe, — —of Patagonia. — —of Chile. —greenstone, of Chile. —doubly columnar. —claystone, rare, on the eastern side of the Portillo Pass. —brick-red and orthitic, of Cumbre Pass. —intrusive, repeatedly injected. —claystone of the Uspallata range. — —of Copiapo. — —eruptive sources of.

Port Desire, elevation and plains of. —tertiary formation of. —porphyries of.

Portillo Pass in the Cordillera.

Portillo chain. —compared with that of the Uspallata.

Prefil or sea-wall of Valparaiso.

Puente del Inca, section of.

Pumice, pebbles of. —conglomerate of R. Negro. —hills of, in the Cordillera.

Punta Alta, elevation of. —beds of.

Quartz-rock of the S. Ventana. —C. Blanco. —Falkland islands. —Portillo range. —viscidity of. —veins of, near Monte Video. — —in dike of greenstone. —grains of, in mica slate. — —in dikes. —veins of, relations to cleavage.

Quillota, Campana of.

Quintero, elevation of.

Quiriquina, elevation of. —deposits of.

Rancagua, plain of.

Rapel, R. elevation near.

Reeks, Mr. T., his analysis of decomposed shells. —his analysis of salts.

Remains, human.

Rio de Janeiro, elevation near. —crystalline rocks of.

Rivers, small power of transporting pebbles. —small power of, in forming valleys. —drainage of, in the Cordillera.

Roads, parallel, of Glen Roy.

Rocks, volcanic, of Banda Oriental.

—Tres Montes.

—Chiloe.

—Tierra del Fuego.

—with laminar structure.

Rodents, fossil, remains of.

Rogers, Professor, address to Association of American Geologists.

Rose, Professor G., on sulphate of iron at Copiapo.

S. Blas, elevation of.

S. Cruz, elevation and plains of. —valley of. —nature of gravel in valley of. —boulder formation of. —tertiary formation of. —subsidence at.

S. Fe Bajada, formations of.

S. George's bay, plains of.

S. Helena island, sea-cliffs, and subsidence of.

S. Josef, elevation of. —tertiary formation of.

S. Juan, elevation near.

S. Julian, elevation and plains of. —salt lake of. —earthy deposit with mammiferous remains. —tertiary formations of. —subsidence at.

S. Lorenzo, elevation of. —old salt formation of.

S. Mary, island of, elevation of.

S. Pedro, elevation of.

Salado, R., elevated shells of.

—Pampean formation of.

Salines.

Salt, with upraised shell. —lakes of. —purity of, in salt lakes. —deliquescent, necessary for the preservation of meat. —ancient formation of, at Iquique. — —at S. Lorenzo. —strata of, origin of.

Salts, superficial deposits of.

Sand-dunes of the Uruguay. —of the Pampas. —near Bahia Blanca. —of the Colorado. —of S. Cruz. —of Arica.

Sarmiento, Mount.

Schmidtmeyer on auriferous detritus.

Schomburghk, Sir R., on sea-bottom. —on the rocks of Guyana.

Scotland, sloping terraces of.

Sea, nature of bottom of, off Patagonia. —power of, in forming valleys.

Sea cliffs, formation of.

Seale, Mr., model of St. Helena.

Sebastian Bay, tertiary formation of.

Sedgwick, Professor, on cleavage.

Serpentine of Copiapo.

Serpulae, on upraised rocks.

Shale-rock, of the Portillo Pass. —of Copiapo.

Shells, upraised state of, in Patagonia. —elevated, too small for human food. —transported far inland, for food. —upraised, proportional numbers varying. — —gradual decay of. — —absent on high plains of Chile. — —near Bahia Blanca. —preserved in concretions. —living and fossil range of, on west coast. —living, different on the east and west coast.

Shingle of Patagonia.

Siau, M., on sea-bottom.

Silver mines of Arqueros. —of Chanuncillo. —of Iquique. —distribution of.

Slip, great, at S. Cruz.

Smith, Mr., of Jordan Hill, on upraised shells retaining their colours. —on Madeira. —on elevated seaweed. —on inclined gravel beds.

Soda, nitrate of. —sulphate of, near Bahia Blanca. —carbonate of.

Soundings off Patagonia. —in Tierra del Fuego.

Spirifers.

Spix and Martius on Brazil.

Sprengel on the production of carbonate of soda.

Springs, mineral, in the Cumbre Pass.

Stratification of sandstone in metamorphic rocks. —of clay-slate in Tierra del Fuego. —of the Cordillera of Central Chile. —little disturbed in Cumbre Pass. —disturbance of, near Copiapo.

Streams of lava at S. Cruz, inclination of. —in the Portillo range.

String of cotton with fossil-shells.

Struthiolaria ornata.

Studer, M., on metamorphic rocks.

Subsidence during formation of sea-cliffs. —near Lima. —probable, during Pampean formation. —necessary for the accumulation of permanent deposits. —during the tertiary formations of Chile and Patagonia. —probable during the Neocomian formation of the Portillo Pass. —probable during the formation of conglomerate of Tenuyan. —during the Neocomian formation of the Cumbre Pass. —of the Uspallata range. —great, at Copiapo. — —during the formation of the Cordillera.

Sulphur, volcanic exhalations of.

Sumatra, promontories of.

Summary on the recent elevatory movements. —on the Pampean formation. —on the tertiary formations of Patagonia and Chile. —on the Chilean Cordillera. —on the cretaceo-oolitic formation. —on the subsidences of the Cordillera. —on the elevation of the Cordillera.

Tacna, elevation of.
Tampico, elevated shells near.
Tandil, crystalline rocks of.
Tapalguen, Pampean formation of. —crystalline rocks of.
Taylor, Mr., on copper veins of Cuba.
Temperature of Chile during the tertiary period.
Tension, lines of, origin of, axes of elevation and of cleavage.
Tenuy Point, singular section of.
Tenuyan, valley of.
Terraces of the valley of S. Cruz. —of equable heights throughout Patagonia. —of Patagonia, formation of. —of Chiloe. —at Conchalee. —of Coquimbo. —not horizontal at Coquimbo. —of Guasco. —of S. Lorenzo. —of gravel within the Cordillera.
Theories on the origin of the Pampean formation.
Tierra Amarilla.
Tierra del Fuego, form of sea-bottom. —tertiary formations of. —clay-slate formation of. —cretaceous formation of. —crystalline rocks of. —cleavage of clay-slate.
Tosca rock.
Trachyte of Chiloe. —of Port Desire. —in the Cordillera.
Traditions of promontories having been islands. —on changes of level near Lima.
Trees buried in plain of Iquique. —silicified, vertical, of the Uspallata range.
Tres Montes, elevation of. —volcanic rocks of.
Trigonocelia insolita.
Tristan Arroyo, elevated shells of.
Tschudi, Mr., on subsidence near Lima.
Tuff, calcareous, at Coquimbo. —on basin-plain near St. Jago. —structure of, in Pampas. —origin of, in Pampas. —pumiceous, of R. Negro. —Nuevo Gulf. —Port Desire. —S. Cruz. —Patagonia, summary on Chiloe. —formation of, in Portillo chain. —great deposit of, at Copiapo.
Tuffs, volcanic, metamorphic, of Uspallata. —of Coquimbo.
Ulloa, on rain in Peru. —on elevation near Lima.
Uruguay, Rio, elevation of country near.
Uspallata, plain of. —pass of. —range of. —concluding remarks on.
Valdivia, tertiary beds of. —mica-slate of.
Valley of S. Cruz, structure of.
—Coquimbo.
—Guasco, structure of.
—Copiapo, structure of.
—S. Cruz, tertiary formations of.
—Coquimbo, geology of.
—Guasco, secondary formations of.
—Copiapo, secondary formations of.
—Despoblado.
Valleys in the Cordillera bordered by gravel fringes. —formation of. —in the Cordillera.
Valparaiso, elevation of. —gneiss of.
Vein of quartz near Monte Video. —in mica-slate. —relations of, to cleavage. —in a trap dike. —of granite, quartzose. —remarkable, in gneiss, near Valparaiso.
Veins, relations of, to concretions. —metalliferous, of the Uspallata range. —metalliferous, discussion on.
Venezuela, gneissic rocks of.
Ventana, Sierra, Pampean formation near. —quartz-rock of.
Villa Vincencio Pass.
Volcan, Rio, mouth of. —fossils of.
Volcanoes of the Cordillera. —absent, except near bodies of water. —ancient

submarine, in Cordillera. —action of, in relation to changes of level. —long action of, in the Cordillera.

Wafer on elevated shells.

Waves caused by earthquakes, power of, in transporting boulders. —power of, in throwing up shells.

Weaver, Mr., on elevated shells.

White, Martin., on sea-bottom.

Wood, silicified, of Entre Rios.

—S. Cruz.

—Chiloe.

—Uspallata range.

—Los Hornos.

—Copiapo.

Yeso, Rio, and plain of.

Ypun Island, tertiary formation of.

Zeagonite.

End of Project Gutenberg Etext South American Geology, by Charles Darwin